*Geographical Voices*

SPACE, PLACE, AND SOCIETY

John Rennie Short, *Series Editor*

# Geographical
# VOICES

## Fourteen Autobiographical Essays

Edited by PETER GOULD
and FORREST R. PITTS

SYRACUSE UNIVERSITY PRESS

First Edition 2002
02   03   04   05   06   07        6   5   4   3   2   1

**Border Lord.** Words and music by Kris Kristofferson, Donnie Fritts, Stephen Bruton, and Terry Paul © 1972 (renewed 2000) RESACA MUSIC PUBLISHING CO. All rights controlled and administered by EMI BLACKWOOD MUSIC INC. All rights reserved. International copyright secured. Used by permission.

The paper used in this publication meets the minimum requirements of American National Standard for Information Sciences—Permanence of Paper for Printed Library Materials, ANSI Z39.48–1984.∞™

**Library of Congress Cataloging-in-Publication Data**

Geographical voices : fourteen autobiographical essays / edited by Peter Gould and Forrest R. Pitts.—1st ed.
p. cm.—(Space, place, and society)
ISBN 0-8156-2940-0 (pbk. : alk. paper)
1. Geographers—Biography. 2.Geography—Philosophy. I. Gould, Peter, 1932–  . II. Pitts, Forrest Ralph, 1924–  . III. Series.
G67 .G48 2002
910'.92'2—dc21
2002001713

# Contents

# Illustrations

# Acknowledgments

The editors would like to acknowledge the kindness of the American Council of Learned Societies for permission to reproduce the Haskins Lectures of Professors Donald Meinig and Yi-Fu Tuan.

We also thank Rosemary Hibbler, Ann and Michael Hearty, and Chris Wagner for valuable help in preparing the original manuscript.

# Introduction

The idea for *Geographical Voices* traces back to the stunning autobiographical essay "In Search for the Sources of Concepts" by the Swedish geographer Torsten Hägerstrand, contained in *The Practice of Geography*, edited by Anne Buttimer.[1] It certainly inspired our own personal and professional recollections,[2] and raised the intriguing notion of how insightful it would be to collect such autobiographical recollections from many prominent geographers of the past. Geographers such as Halford MacKinder, Ellen Semple, Carl Sauer, Paul Vidal de la Blache, and August Lösch appear to us as immediate candidates, but many other geographical lives are equally informative and fascinating. Yet today such prominent scholars only confirm our human finitude, for we are all creatures moving toward death, and the opportunities to invite such geographers to write their own most personal statements will never come again.

It was in this spirit that we set out to capture a few "voices" and immediately faced an *embarras de richesses*. Even confining the set of people to those who entered geography around the late 1950s and early 1960s still left us with a host of difficult choices. Citations played a part in our choices, but only a part, for contributions to the field appear in other ways. In the end, we decided we wanted to capture the wide range of different influences that led people of very different backgrounds to the wonderful field of geographic in-

---

1. T. Hägerstrand, "In Search for the Sources of Concepts," in *The Practice of Geography*, edited by A. Buttimer, 238–56 (London and New York: Longman, 1983).

2. P. Gould, "Introduction," in *Becoming a Geographer*, 1–36 (Syracuse, N.Y.: Syracuse Univ. Press, 1999); and F. Pitts, "Sliding Sideways into Geography," this volume.

quiry. For those familiar with Torsten Hägerstrand's space-time diagrams, in which people trace out graphically three-dimensional daily paths, we saw in our imaginations "life lines" starting at different places and different times, generating quite different personal experiences, yet all merging in the project we call geography.

Putting together *Geographical Voices* has been a most rewarding experience, perhaps the first of a continuing series—for none of us is getting any younger! Young entrants to the field become middle-aged, and the middle-aged become the older generation in their turn. But from our experience in editing this book, we would like to make four observations, and we would be delighted to receive from readers any further comments and observations.

First, there is an absence of women's voices here, something we regret. Not many doctoral degrees were awarded to women in the late 1950s and early 1960s. Some scholarly voices, such as Jacqueline Beaujeu-Garnier's in France[3] and Gerd Enequist's in Sweden,[4] have already been recorded; others were quickly stilled by high-level administrative duties in the academic world. We invited several women geographers to participate, but unfortunately they indicated that they faced a press of other research and teaching responsibilities and were unable to participate despite generous and elastic deadlines. Any sequel to this volume obviously has the opportunity to reach down further into a period when many more women were entering the field.

Second, we have been obliged to confine our choices to those geographers of the English-speaking world and those who have published extensively in English. We feel this constraint acutely, particularly because of personal access to work written in French, Japanese, and Korean. There is inevitably a thwarting "linguistic gulf" that few, if any, geographers can overcome. Nearly twenty years ago one of us pointed to this grievous barrier to greater geographic awareness and understanding,[5] but nothing has been done to overcome it. We

3. J. Beaujeu-Garnier, "Autobiographical Essay," in *The Practice of Geography*, edited by A. Buttimer, 141–52 (London and New York: Longman, 1983).

4. G. Enequist, "Landscape and Life: A Personal Story," in *Geographers of Norden: Reflections on Career Experiences*, edited by T. Hägerstrand and A. Buttimer, 71–82 (Lund, Sweden: Lund Univ. Press, 1988).

5. P. Gould, "International Geography: Strengthening the Fabric," in *The Geographer at Work*, 331–38 (London: Routledge, 1985).

feel it is a project worthy of the best efforts of the International Geographical Union. A prominent European geographer has already pointed regretfully to many worthy exclusions in the collection, and it is our hope that these omissions can be repaired in a subsequent volume, perhaps published jointly by an American and a European press.

Third, we have noticed a vivid contrast in the narratives of British and American geographers, a contrast based on a small sample to be sure, but one that struck us nevertheless. We are reluctant to disclose what it is, but when you close this book at the end, see if something hits you about these personal and professional stories as it did us.

Fourth, and this feature has come out of correspondence as well as the essays themselves, we have discovered that at *least* four geographers prominent for their contributions in the 1960s—William Garrison, John Borchert, Edward Taaffe, and William Warntz—served as meteorological officers in the U.S. Air Force during the Second World War. Their daily tasks were saturated with a concern for dynamic maps, for they were responsible for forecasting what the next maps in the sequence would look like, often with precious little data to work with compared to the satellite images we take for granted today. Emerging from such "spatiotemporal" concern, in which time was an integral part of and a coordinate with space, they entered graduate schools still intellectually stunned by the atemporal chorographic tradition that declared that time, change, and process were none of geography's business. All of these men became highly influential teachers—three went on to become presidents of the Association of American Geographers, and the fourth displayed an extraordinary imagination by drawing a direct analogy between dynamic weather and "income" fronts.[6] Time and change are taken for granted now that the authoritative yoke of chorographic inquiry has been discarded, but how were the influences of these men felt through their own research and through the students they helped? It is an intriguing footnote to the history of the discipline, and we invite readers to let us know if there are any further examples, particularly from Britain and the Commonwealth.[7]

6. W. Warntz, *Macrogeography and Income Fronts* (Philadelphia: Regional Science Research Institute, 1965).

7. W. Balchin, "United Kingdom Geographers in the Second World War," *Geographical Journal* 153, part 2 (July 1987): 159–80.

Finally, we would like to thank all the authors who contributed to this book for their patience and generosity. From extraordinarily diverse backgrounds, including far-ranging experiences in applying geographic principles and knowledge to the problems society faces, they have painted collectively a rich and colorful picture of one era of geography. It is our hope that subsequent generations will find joy, hope, and inspiration from these narratives, many of which describe difficult circumstances that the authors overcame to become—geographers!

*Geographical Voices*

# Clara Voce Cognito

## Brian J. L. Berry

Courtesy of Brian J. L. Berry

*Brian J. L. Berry (b. 1934) obtained his B.S. in economics from University College, London, in 1955 and his Ph.D. from the University of Washington in 1958. He was a member of the faculty at Chicago (1958–76) and Harvard (1976–81) and a dean at Carnegie-Mellon (1981–86) before becoming a Founder's Professor at the University of Texas. In the 1960s, his urban and regional research played a crucial part in the "quantitative revolution" and made him the world's most-cited geographer for more than a quarter of a century. He has also chaired more than ninety doctoral dissertations. In 1976, Harvard awarded him an honorary degree, and two years later he became the youngest president ever of the Association of American Geographers. A member of the National Academy of Sciences and a fellow of the American Academy of Arts and Letters, he has received the Medal of Honor from the Association of American Geographers and the Victoria Medal from the Royal Geographical Society. His book* America's Utopian Experiments—*one of 450 books, articles, and professional publications to his name—won the Rockefeller Prize in 1992. In 1999, he was elected the first geographer to serve on the council of the National Academy of Sciences. With a concern to blend both theory and practice, he has fre-*

Editors' note. Professor Berry's recollections of the beginnings of geography's quantitative revolution have been published elsewhere: Brian J. L. Berry, "Geography's Quantitative Revolution: Initial Conditions, 1954–1960—A Personal Memoir," *Urban Geography* 14 (1993): 434–41. He promises a fuller autobiography detailing his life as a geographer in the United States.

1

*quently been involved in urban and regional development planning, and has given his expertise in many countries, including India, Indonesia, and Australia. He is the Lloyd Viel Berkner Regental Professor of Political Economy at the University of Texas (Dallas).*

What led me to become a geographer? I have no easy answer, so indulge me while I think out loud.

It's a long way from February 17, 1934, the day after my birth in Sedgley, Staffordshire. The factory manager had waited until after my delivery to tell my father that the business was closing, a victim of the Great Depression. Soon thereafter we were back in Beckingham, a Nottinghamshire village three miles west of the Trent Valley river port of Gainsborough in Lincolnshire. Both of my parents had been born in Gainsborough, and Granddad Berry had built a house in Beckingham a few years earlier.

Granddad was born in West Yorkshire. He brought his new bride to Gainsborough from Huddersfield in 1908 to become foreman of the Northern Manufacturing Company, a subsidiary of Roses' Engineering Works. In the late nineteenth century, William Rose had invented a machine for twist wrapping half-ounce packets of tobacco, an invention that started the packaging industry. In 1898, he also began to produce automobiles—Rose Nationals— and needed proper gearing for both. Northern Manufacturing Company was established as the gear-cutting specialist. Granddad had apprenticed with the John Brown Engineering Works in Huddersfield, breaking several centuries of family involvement with the woolen industry, and brought the requisite skills to Roses'. He was foreman his entire working life. The managerial positions were all held by Rose family members.

The West Yorkshire Berrys were descended from the family that held Bury in Lancashire, first from Saxon overlords, then from the Norman de Montbegons. Earlier spellings of the name were Byri, Buri, and Berri. The de Montbegons married their natural daughters into their knights' families, including that of the Byris. In the thirteenth century, the de Lascys acquired the de Montbegon barony. The West Yorkshire branch of the Berrys originated when a younger son was granted the rights to a more distant part of the Lancashire family's holdings. After Almondbury (now part of greater Huddersfield) was given market rights by Henry de Lascy in 1294, the West Yorkshire Berrys became clothiers—weavers of woolen cloth. When mechanization destroyed the livelihood of the hand-loom weavers in the nineteenth century, Great-great-granddad Berry ended his weaving career and became a grocer. Great-granddad Berry was a cloth dyer in one of the new mills.

Like his father, my dad left school at fourteen and apprenticed for seven years at Roses' Engineering Works. There, he learned to build, maintain, and repair the company's packaging machinery. By the time he was eighteen, he was traveling the United Kingdom for the firm, troubleshooting their installations. Where the installations were big enough, Roses' provided one of their journeymen as a resident engineer, which is why my parents moved to Staffordshire after their marriage. It is also why, within a year and a half, we had moved to London. Dad had been given charge of the twist-wrapping and packaging machinery at another of Roses' installations, Walters' Palm Toffee, located in West Acton. This was another family business managed by the sons of the founder. The Walters were refugees from the Russian pale who spent their money to make their children as "U" as was possible for immigrants of Jewish ancestry.

What about this "U" and "non-U" stuff? It has everything to do with the English class system. Before his marriage in 1932, Dad was a champion bicycle racer, the protégé of England's professional champion Lal White. Until well into the twentieth century, amateur sports were "U": public school and country estate, with aristocratic accent; sherry and cucumber sandwiches at garden parties; Rugby Union, tennis, and the Olympic Games. Professional sports were "non-U": factory town and common school; ale and pork pies at brass band concerts; earthy regional accents, Rugby League, and "football" (soccer). Dad's bailiwick was north country and definitely non-U.

Neither Granddad nor Dad had any respect for the lesser Rose or Walters middle-management family members, and it was undoubtedly from them that I acquired an abiding disregard for those with unwarranted pretensions obtained by inheritance rather than by effort. Mother was a different story, always sensitive to class and the appropriateness of everyday behavior. She, too, left school when she was fourteen, the youngest surviving daughter of a family of eleven, eight of whom lived to adulthood—four girls, three boys, and Uncle Henry, a sweet man who had been a golden-voiced (but sadly abused) choir boy at Lincoln Cathedral, a lifetime nurse who managed operating rooms in London during the blitz, and a retiree who relished elderhood as an aging queen. Although originating in North Yorkshire, the Lobleys had for centuries been coopers who worked out of the village of Harpswell, a few miles east of Gainsborough. Mother's great-great-grandfather moved into town in the early nineteenth century, becoming a cabinetmaker. His sons took different paths. One ran a local hostelry, and his daughters married the town's principal auctioneers and fishmongers. Another had a son who migrated to Australia, serving in the Tasmanian Mounted Police. My great-grandfather Lobley continued in the joinery business, but Granddad Lobley turned to Gainsborough's other large engineering firm, Marshall's, apprenticed as an iron molder, but married early.

For a time, the young newlyweds lived on a smallholding outside town. Later, they moved back to Gainsborough, and Grandma Lobley ran a boarding house to make ends meet. She was a Langton, descended from a family that had been the high sheriffs of Lincolnshire. Her branch was that of a third son who inherited property to the north of Gainsborough from his mother. Something happened to the gentleman farmer's family in the middle of the nineteenth century, however. Her grandmother Sally had five out-of-wedlock children, including Great-grandfather Moses—the family joked that she found him in the bulrushes. The father was supposedly an already-married North Lincolnshire gentleman whose "reputation" was protected by Sarah in return for support of the children. Moses was a blacksmith and a pig farmer. His wife, Great-grandmother Jane, became a Gainsborough justice of the peace. Of their fifteen children, one became a Mormon and migrated with husband and children to Salt Lake City, but Grandma Lobley, marrying young, also produced many children and always had difficulty making ends meet. When Granddad Lobley worked, he had to be met at the factory gate on pay night to head him off before he reached the local pub. I remember him sitting in the bar at the Yarborough during World War II, a line of half-pints in front of him, the nightly yield from playing dominoes, something that he did very well.

When mother left school at fourteen, she went to live in the household of Polish immigrant Abe Levine, looking after his children. Abe ran a haberdashery business from stalls in the markets at Gainsborough, Lincoln, Market Rasen, and Doncaster—the merchants followed a regular weekly market-day cycle—and she later staffed a stall, a welcome relief from being the Levines' shiksa. Later, she moved back home after she was able to get a job as a cashier in the local cooperative store, turning over her weekly pay to Grandma Lobley. Very bright, she was overwhelmed by the English class system, fearful about social station and the gaffe that would disclose all: "What will 'they' think?" Mindful of her parents' descent to the margins of poverty, she was especially sensitive about doing things that might lead others to make life more difficult. Granddad Lobley had lost his job in the molding shop at Marshall's in a dispute with his manager. Because he pushed the manager into the pit of molding sand as he left, he was a marked man and never had a steady job thereafter. Mother moved up a step on the social ladder when she married a mechanical engineer, but my father says that she always felt inadequate, her insecurities heightened, no doubt, by the vagaries of employment during the Great Depression and by her own mother's resentment that her daughter had achieved more comfortable living conditions and had to be depended on for occasional financial support. Mother's insecurities were translated to me as an expectation that I should always perform well so that others would not think poorly of me or of the family: it was necessary that I be better to ensure that I be treated with respect. Yet she was always fearful when I did succeed. Was I "getting above my station" and likely to suffer a fall as a consequence? This undertone of insecurity led Mother to brood too often about imagined hurts, and in me it probably led to a combination of shyness bred by feelings of ineptness in many social situations and aggressiveness when "put down." I have been told that I do not suffer fools gladly.

It is of West Acton that I have my first cogent memories. We lived in a block of flats overlooking a railway line, and I have vivid recollections of people crowding the balconies looking down at a cloth-swathed train passing slowly beneath. It was early 1936, and King George V was being carried to his burial. In one of the ground-floor flats, I played with retired Yangtze (Chang) skipper Barney Walsh and his wife. They made boats for me in their living room, and we braved rapids and fought river pirates as he showed me his photographs of Asia and told me of his adventures. The flats were located near Western Avenue, where I went walking with my uncle Tom Lobley when he came to visit. He made me sit outside on the steps of the "vinegar shop" while he went in for a quick pint. It was to Western Avenue that I went to look for

Granddad and Grandma Berry and my aunt Mabel when they came from Beckingham to visit us in their new black Morris 8 (license plate EVO 270—why should I remember that after sixty years?). Grandma Berry had suffered a stroke when she was relatively young, and Mabel stayed home to help care for the household. I remember starting kindergarten at Wales Farm Road School in Acton. We bought a puppy at London's Petticoat Lane Market, but we had him only for a fortnight because a harridan of a next-door neighbor complained about his barking. My only image of that neighbor is of her screaming at me. She went ballistic when workmen disassembling a woodpile nearby found a rifle and threw it to me. I dragged it home—it was too heavy to carry—and heard her start her performance as I rounded the corner.

We spent summer holidays on the coast of Yorkshire, usually at Bridlington, where Granddad Berry rented a cottage each year for the family. We were standing on the cliffs at Flamborough Head in the summer of 1939 when the Home Fleet steamed by on its way to its wartime base at Scapa Flow. "Where are they going, Granddad?" When we got back to the cottage, he took out his road atlas and showed me. We had taken photographs of the fleet, but when we took them to be developed, the rolls of film were returned blank.

When hostilities broke out between England and Germany in 1939, plans were made to evacuate children from London. Uncle Tom was visiting, and because he was on his way back to Gainsborough, my parents packed me off with him to live with my grandparents. For the next two years, I attended the village school in Beckingham and walked the countryside with Auntie Mabel learning about crops and trees, hedgerows and wildlife, and tussling with her dog. Massive shire horses tilled the fields, and there were great puffing steam-driven threshing machines. A monster steamroller maintained the roads. The blacksmith's shop was always an attraction, with its bellows and forge, hammers, anvil, sweat, and the smell of burning hoof. Early in the morning, we would go out into the fields to gather great dinner-plate-size mushrooms for breakfast, and when the blackberries were ripe, we picked them to be made into pies and jam. When the wheat was harvested, the sheaves of grain were stacked in stooks to dry before the threshers arrived. The farmers worked in circles, from the outside in, and stood outside the final island of grain with shotguns, waiting for the rabbits to make a panicked dash from their refuge as the last cuts began. Harvest time was the time for game-and-kidney pies.

The first winter in Beckingham was extremely cold, and I recall the great snow drifts and going to walk across the river Trent, which had frozen. The spring brought floods that reached three miles west from Gainsborough to the railway line along the edge of Beckingham.

I heard stories about the family from my ninety-year-old great-grandmother Berry, who had lived with my grandparents since Great-granddad died. I wanted to know more about where the family members had gone because they were all over the world. Great-grandma told of her brother who had deserted the British army, changed his name to Thomas Sheldon, moved to South Africa, and opened a business called Ostrich and Ivory Manufactures. There were examples of his wares around the house. Great-grandma was a ramrod-straight lady with beautiful snow-white hair. I learned to tie shoelaces one evening after I had undone hers. She simply sat and made me retie them. It took an hour, but she would not compromise. Later, she gave me a threepenny bit for tying them when she put on her "Sunday best." She lived life by a strict regimen and expected others to do the same. It was only very much later, as I worked on a genealogy of the family, that I learned that her first daughter was born several years before she married Great-granddad Berry—the child of the owner of the woolen mill where she had worked.

Grandma Berry had wonderful stories, too. After her father Albert Dyson had died young, her mother had remarried and had two additional children. Grandma's stepfather Thomas Donovan had been a member of Kitchener's Irish Guards when they marched south through the Sudan fighting the "fuzzy-wuzzies" in the attempt to relieve General Gordon at Khartoum. He was at Omdurman when the Mahdi's forces were defeated. After Great-grandma Donovan died, Thomas migrated to upstate New York and retired to a small fruit farm. Grandma's half-brother Walter Donovan ran away from home in 1914, lying about his age in order to join the Royal Artillery. He fought in France and later in Dublin. After the war, he served in Bangalore, India, but followed his father to America once his period of service was over. He was a big man who played Irish football and managed a grocery store in the Bronx until he lost the job because of his under-the-counter activity during Prohibition. For a while, he drove trucks to and from Canada, and then vanished from view until after World War II. It was Walter who met me when I disembarked from the *Queen Mary* in New York in early September 1955.

He lived in the middle of the hard-drinking Irish community on Staten Island, and he evidently commanded great "respect" on the waterfront in lower Manhattan. As we walked by, merchants brought him gifts of tomatoes, fresh bread, and bagels. He bought me my first-ever hamburger on the street outside Madison Square Garden, garnished with a great slice of white onion.

Of all the books in my grandparents' home, the ones I enjoyed most were the automobile club's guidebook that told me all about places in England, with its maps and pictures and the descriptions that I soon learned to read. Another favorite was a large, two-volume, gold-trimmed, leather-bound set that told the stories of brave Britons during the Indian Mutiny and the Afghan Wars, at Rourke's Drift, Ladysmith, and in World War I, with wonderful illustrations of people and places and lots of maps. I was fascinated and spent many a dark night in bed going to sleep imagining my own adventures in exotic places. Somehow, reading the maps and relating pictures to places came naturally. Later, during World War II, it became routine for me to try to find out where Tobruk, Benghazi, Anzio, Stalingrad, or the Marshall Islands were located when such places were mentioned on the nightly BBC news broadcast. It was a family ritual to sit in front of the fire and listen to the nine o'clock news before Great-grandma went to bed.

The tranquility of Beckingham was soon disturbed. My countryside walks with Aunt Mabel were limited to weekends after she went into war work at Roses'. The Germans dropped sea mines by parachute along the railway line at the edge of the village, resulting in thunderous explosions that gouged deep holes that soon filled up with water, but yielded buckets of shrapnel to enterprising six-year-olds and yards of parachute silk to the village ladies. A West Sussex regiment was in training for service overseas, and the village was lined with troops and their lorries and bren-gun carriers. Each day the units clumped by, off on long route marches, complete with all their gear on their backs. Aunt Mabel fell for one of the soldiers, Herbert "Buck" Jones. He went off to war, seconded to the Royal Electrical and Mechanical Engineers, and served with the Eighth Army at El Alamein, across North Africa, and in Sicily in the first British wave at Anzio, through Rome, and finally in Austria. He was wounded in Italy. Mabel and Buck married after the war, in 1947, and went to live in Sevenoaks, Kent. Buck was a rogue, but more on that later. Their move to Kent was to be of extraordinary importance to my career choice later on.

My parents returned to Gainsborough in 1941 and produced a new sister, Susan (who now lives in New Zealand with her daughters and their Maori and Rarotongan boyfriends and husbands, while another daughter lives in Australia with her Anatolian spouse). I rejoined their household at the end of the summer of 1941, in time to start a new school year at Ropery Road Elementary School, the school that my father and aunt had attended. The headmistress, Miss Dodds, had been their teacher. Dad had been recalled by Roses', which was manufacturing a new type of rear-gun turret for the Royal Air Force (RAF) Lancaster heavy bombers. Gainsborough was located amidst a network of RAF bases, and Dad's responsibility was to oversee the installation and maintenance of the turrets on the night bomber force. I saw relatively little of him for the next four years, but his home guard uniform, sten gun, ammunition, and grenades were in the wardrobe. Whenever possible, the family did gather for Sunday dinner at Beckingham. There were times when Dad took me to one of the aerodromes early in the morning to greet survivors returning from the previous night's raids. He would put me in a gun turret of a returning plane as it was towed across the airfield, amidst the smell of gunpowder, fuel, sweat, and excrement. Where had they been? Berlin, Hamburg, Essen. . . ? We had breakfast in the mess with the tired, hollow-eyed pilots and gunners. Later I learned that Dad helped prepare the planes for the raid that destroyed the Ruhr Valley dams (remember *The Dam Busters?*). I recall looking over one of the dam sites in 1954 with the German geographer Carl Troll as he talked about the RAF's use of torpedoes to destroy the dams. He was taken aback when I, an upstart English undergraduate from University College, London, on a field trip, dared to correct the great professor by explaining to him how bombs that bounced along the surface of the water had been delivered by low-flying Lancasters and had sunk inside the dams, blowing them downstream—and was even more surprised when I told him how I knew. I had seen the planes training, low over the Trent Valley marshes, but it was only after the war that Dad told me why they had trained that way.

Our house on Campbell Street was part of a three-story terrace built in the decade preceding World War I for the skilled journeymen who worked in the town's engineering firms. Narrow yards fronted on the street, and each house had an attached wash house with boilers to heat water, a coal shed, and a flush toilet—all entered from the backyard. The flush toilets had been added when the town ran a sewer line between the backs of the terraces. It was a freezing trip on a wet winter night, flashlight in hand. When my friend Ken Davey and I received chemistry sets for Christmas, we decided to make gunpowder. We bought sulfur, carbon black, and saltpeter from three different chemists' shops and mixed a batch, put it in the coal shed, lit the fuse, and blew the door off with a great bang! The result: no more chemistry.

Campbell Street was four blocks north and two blocks west of the town center. As you walked home from the market square, it was one block to the Old Hall, the best-preserved Tudor manor house in England, where Catherine Parr had lived before becoming Henry VIII's last wife. One more block north and one west was Roses' Engineering Works. Two blocks farther north and one west and you were on Campbell Street, which ran parallel to and two blocks east of the river Trent. Two blocks due north of Campbell Street was Ropery Road Elementary School. A few blocks east was an escarpment, the bluffs on the eastern edge of the Trent's floodplain. Just a few blocks north, at Castle Hills, the bluffs wrapped westward toward the river, forming the northern edge of the town.

Life on Campbell Street had its daily rhythms. Early in the morning, the milkman would arrive with his horse and cart and deliver bottles to the front doorstep, and the paperboys would put the newspapers through the mail slots in the doors. Later, after the older children had left for school, the women would be out, scrubbing their front steps with holystones until they were a brilliant white. The younger children spent the day playing on the sidewalks and in the street. The street sweeper would pass by, brushing the debris ahead of him and shoveling it into his pushcart. The women would take off for their daily trip to the bakers, the greengrocers, the grocers, the fishmongers, and the butchers—some on bicycle, some walking, and some with their babies in prams—returning later with the makings of that night's meal. Afternoons, Mr. Smart would be on the street with his pair of massive horses, pulling a cartload of coal. Dressed in a leather hat with flap to protect his neck and a studded leather cape that protected his back, he would carry sacked hundredweights on his back to the coal sheds in the rear yards. The housewives quickly scooped up his horses' droppings for use on the flower or vegetable gardens. The town still produced its own gas to illuminate the streetlights, and after the early blackouts were eased, the lamplighter would bicycle by in late afternoon, long pole in hand, raising the flames. On many nights, the teenagers who already had jobs would be off to the cinema or to one of the town's social clubs. Only Sundays were different. A few went to church. Families would gather for Sunday dinner and spend afternoons walking in the country or simply relaxing.

Ropery Road School was crowded and uncomfortable. We arrived each morning and had our gas masks checked—we carried these over-the-shoulder boxes. At lunch, we had to take a cod liver oil pill, drink a glass of reconstituted orange juice, and eat two ounces of cheese before anything else was served. Miss Dodds patrolled the room with cane in hand to make sure that we ate it all. "Nitty Nora" was always there to check heads for lice. We must have been a scruffy, smelly bunch. None of the houses had hot water, and few of us bathed more than once a week (when the boilers were fired up to do the laundry), other than a "stand up wash." Between the river and Campbell Street were two blocks of subsidized housing for folks who had not yet made it to the first rungs of the social ladder, *diddicois*—the part-Romany who had stopped wandering and their hangers-on. The *diddicoi* children were always unkempt—particular targets for Nitty Nora—and epidemics ran riot through the school, not simply the usual childhood measles, scarlet fever, mumps, and chicken pox, but nastier things such as impetigo and "jaundice" (hepatitis A). I caught them all.

The streets were lined with army equipment, and empty bedrooms were used to billet the tommies whose regiments were training in the town or the war workers transferred in from elsewhere. The billeting officers would walk along the street, find out how many lived in each house, and make chalk marks by the front door indicating the number to be assigned. We had three factory workers in our back bedrooms. The town was a target for German bombs because Roses' was building gun turrets and Marshall's midget submarines. The night that the town center was destroyed by a heavy raid that dropped its bombs midway between Roses' and Marshall's, the hotels were full of one regiment's noncommissioned officers, and forty were killed. The thump of the bombs falling so close knocked us from bed. I walked over early the next morning to see the bodies being extracted from the smoking piles of debris. A year later a German plane was hit by a night fighter and screamed over our rooftop in flames, crashing just a couple of blocks away. I was in bed and vividly remember looking out of my window at the plane passing overhead and seeing the crash and flash of the explosion. By then we had stopped using the bomb shelters, and several of my schoolmates were killed.

Yet, in spite of the war, Gainsborough was a magical place for a growing boy. My friends were all several years older, so I had to be on my toes to keep up. There was great excitement in town when the first Americans were stationed nearby at Upton and at Sturgate. They came into town, big men from Texas and Kansas, Indiana and New York, walking with a swagger, in separate groups of blacks and whites—"overpaid, oversexed, and over here." The black Americans were the first of their race many of the townspeople had seen, excepting Hollywood depictions at the cinema, and there was fear mixed with curiosity about them. The *diddicoi* girls—early bloomers—quickly overcame what fears they may have had, however, and switched from the tommies to the new, more lucrative trade. They soon sported silk stockings and packs of American cigarettes. There were frequent brawls among the troops. The tommies, in particular, resented the access that the better-paid black Americans had to the girls. Our tangle of snotty-nosed boys would sometimes harass the black soldiers and sometimes try to learn about reputed differences in physiology, but for a pack of chewing gum or sixpence for fish and chips we would serve as lookouts to ensure the privacy of the girls' Trent bank trysting places. We soon observed that there was more to life than the birds and the bees, and the *diddicois* taught us a colorful four-letter language and provided us with the makings of the town's best water bombs.

As part of our schoolboy contribution to the war effort, we joined the St. John's Ambulance Brigade. They ignored their age requirements for me and let me take their course "First Aid for the Injured." So qualified and proudly wearing my uniform, I got to accompany a senior member of the brigade to the cinema twice a week, once to each of the town's two cinemas, free of charge. One night the senior did not show up. During the performance, the street outside was strafed by a German plane. There was glass and blood everywhere, and I bandaged cuts until the town's emergency service arrived. The course had served its purpose: the uniform and the over-the-shoulder first-aid kit persuaded reluctant townsfolk that a ten-year-old child could help.

On weekends, our group of boys explored the town and its environs. Figure 1 reproduces part of Adam Stark's wonderful 1827 map of the

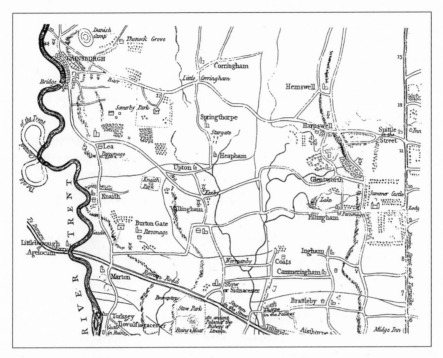

1. A portion of "A Map of Part of Lincolnshire, published by A. Stark, Gainsburgh, 1827." Adam Stark had privately published *History and Antiquities of Gainsburgh* in 1816, revised and republished as *History of Gainsborough* in 1843.

region. Gainsborough is the St. Ogg's of George Eliot's *Mill on the Floss*. It was a town where in 1860 Eliot could write that as the twice-daily tidal bore, the aegir, swept upstream from the sea, "black ships, laden with fresh-scented fir planks, with rounded sacks of oil-bearing seed, or with the dark glitter of coal, are borne along to the town . . . which shows its aged, fluted red roofs and the broad gables of its wharves between the low wooded hills and the river bank."

The railroad took away part of the town's river-port function, and the engineering trades provided new employment, but the wharves and the medieval streets still were there in the 1940s; barges plied the river and docked at Farley's wharves, and the air was filled with the smells from the riverside maltsters. The Old Hall was locked and barred, but we found ways to explore it, and we played along the river as the aegir swept upstream, despite our mothers' admonitions not to play on the Trent bank. The willows along the riverbank supplied us with our arrows, and the yews in the churchyard with our bows.

On the bluffs northeast of town, we played on the Castle Hills, the remnants of the strong fortress that King Sweyn of Denmark and his son Canute had built at the limit of navigation for seagoing vessels in A.D. 1014. Sweyn had died here and been buried in the great tumulus beside the fortress before being returned to the burial place of the Danish kings at Roskilde. With our bows and arrows, we manned the Danes' ramparts and fought off many an imagined German invasion. Was this place also, I later thought as I learned Anglo-Saxon history at Queen Elizabeth's Grammar School, where King Alfred had married Eahlswith, daughter of Ethelred Mucil, ealdorman of the Ganae, in 868? On these bluffs, there was a feeling of connection to the mythical figures in English history.

Next to Castle Hills was Thonock, the country estate of Sir Hickman Bacon, first baronet of England. The Bacon family had relocated from the Old Hall to their new home at Thonock in the eighteenth century. We sneaked to Thonock to peek into the front parlor at the Hall, reputed to hold Sir Hickman's coffin, but instead saw only a great medieval chest. Sir Hickman drove to the public library each week in a massive 1920s roadster, open to the elements, in tweed suit, Sherlock Holmes overcoat, and deerstalker, with a full bushy gray beard and a gold-tipped walking stick. Once he was in the library, we would sneak onto the car's running boards to see the interior of his wonderful machine.

Somehow we all acquired bicycles—I had my dad's racing bike—and we used them to explore the surrounding countryside. It was wartime, and there were very few vehicles on the roads, so a pack of bicycling boys could ride with impunity. Lincoln was a great attraction, with its cathedral and great walled fortress perched atop the highest point for miles. On a clear day, you could see more than thirty miles from the top of the cathedral's towers. Running straight north from Lincoln along the top of a limestone escarpment, the Lincoln Edge, was a Roman road, Ermine Street. Beneath the Edge was a succession of spring-line villages—Aisthorpe, Fillingham, Harpswell, Hemswell—and we visited them all. The head of the Rose clan, Alfred, lived at Fillingham's Summer Castle, which, like Thonock, had been built in the eighteenth century as the gentry moved to landscaped country estates. After the war, when "Alfie" received an OBE (Order of the British Empire) from King George VI for Roses' wartime service, Granddad said it was for Other Buggers' Efforts. It was Granddad who had solved a gearing problem in a new naval gun in 1942. The designers' specifications were inadequate, and the gears would not mesh properly. Granddad realized that the problem could be solved by taking pi ($\pi$) from three to seven significant digits: the difficulty lay in round-off error.

Westward beyond Beckingham, past the "trusty" Italian prisoners digging drainage ditches around the osier plantations in the Beckingham marshes, we rode to the mainline railway station at Retford to spend days train spotting. It was always exciting to see the streamlined *Flying Scotsman* come racing through.

As we bicycled the Lincolnshire and Nottinghamshire countryside, we taught ourselves to read maps and to judge times and distances. It was easier to ride uphill when the contour lines were farther apart and great fun to freewheel down when they were close, so we learned to choose our routes accordingly. The scars on my elbows and knees are proof that sometimes the lines were too close, but nonetheless *scarp* and *dip* became part of our language, and we learned just how far boys on bicycles could go in a day and be back home before dark.

A group of us also took English watercolor-painting lessons from eccentric artist Karl Wood (his forte was English windmills—he painted more than thirteen hundred of them). It was in his second-floor studio on the town's market square that we learned that some men like boys, how to read the warning signs, and how to take protective action. Not all boys were as observant. After our family returned to London, we heard that Karl suffered Oscar Wilde's fate and spent his declining years in prison. Time cures all. On a July 1998 visit to England, my wife and I learned that Karl had been "rediscovered"—a major exhibit of his work had just opened in London.

When the potatoes had to be harvested, the farmers' lorries came to the end of Campbell Street at dawn and picked up a motley crew of local children and *diddicoi* women, whose conversations taught us many things that our parents didn't. In the fields, we had twenty-pace "stints." The potatoes were grown in long, hilled rows, and a tractor would come along the row with a spinner and scatter the tubers across the adjacent yard and a half. We had to pick them and transfer them to a row of hampers before the tractor came back. Another tractor pulled a trailer into which the hampers were dumped. The crop was stored in straw in long earth-covered hillocks until needed. The days ended only because we had to be delivered back to town by dusk. I repeated this two or three weeks of hard labor three years in a row. The third year I got a nasty infection that abscessed and required hospitalization and surgery. There was a single ward for males with twenty or thirty beds for everyone from age five to ninety. Overhead, as usual, was the drone of the Lancasters forming their squadrons before heading to Germany. Next to me, a boy dying of a brain tumor screamed "Stop them planes" for forty-eight hours before he died. Later, I had nightmares of the operating room, the gauze pad held over my nose and mouth, the chloroform, and the restraints to prevent me from pulling the sutures, as well as of the boy's death and the deaths of three old men.

It was not long thereafter that I sat in class listening to the BBC broadcast on the morning of D-Day. We suspected that the invasion was imminent when the Lancasters switched their raids to the coast of France. The teacher was crying. Her husband had been stationed on the south coast. Within a few months, the Nazis responded with their "secret weapons." Some were directed at northern industrial cities. We stood on the hills at Gringley, just west of Beckingham, watching the "buzz bombs" (V-1 rockets) passing overhead on their way to Sheffield. When the *pop-pop-pop* of their engines stopped, we sought cover.

Nineteen forty-five brought changes. I passed the "eleven plus" examination and transferred to Queen Elizabeth's Grammar School, the upper level of secondary education. Under the new scholarship system, my parents would not have to pay tuition, which made the transfer possible. Otherwise, had the family stayed in Gainsborough, I would have spent three more years at Ropery Road School, leaving for the factory at fourteen to begin a seven-year apprenticeship, just like my dad.

At Queen Elizabeth's, we started with classes in English, mathematics, English history, geography, Latin, French, biology, art, woodworking, mechanical drawing, and divinity. The geography teacher, Mr. Billington, let us know that one of the school's former pupils, Sir Halford Mackinder, had become England's most distinguished geographer, and regaled us with stories of

Mackinder's life and work. He had grown up only three blocks south of Campbell Street in a house by the river, just east of the town center. It now bears a plaque in his memory. I don't think that Mackinder's name resonated for me until seven years later, however, when I became a student at University College, London. In 1945, I was more interested in Gainsborough's history. My home room and history teacher, Mr. Edge, went to great pains to make us understand the relationship between Canute's fortress and the local landscape of the time: river, marshes, woods, escarpments, arable land on the terraces and the flinty dip slope, and the daily rhythms of the aegir. Only later did I learn that the Mackinders came to Gainsborough from the ancestral village of my maternal grandmother Langton. The Langtons had held Langton-by-Spilsby at least from the twelfth century, and in the village graveyard the Mackinders' graves are next to those of the Langtons. I was very pleased when my friend Gordon Clark became the Mackinder Professor of Geography at Oxford University a few years ago; it just felt right.

I spent two years at Queen Elizabeth's, bicycling back and forth from Beckingham during the second year because, with the war over, my parents had returned to London, and I went to live with my grandparents again. There were organized school sports, and I discovered that I could run and jump—valuable skills later on in 1947, when I was back with my parents in London, transferring to Acton County Grammar School. There was a terrible housing shortage, and we lived in several converted rooms on the ground floor of the Walters' Palm Toffee office block, which was attached to the front of the factory. I rode my bicycle two miles to school each morning, home for dinner—we ate breakfast, dinner, tea, and supper—back to school again, and then home following after-school activities. As a latecomer, I had to fight to establish a position in the schoolboy pecking order. In the initial unhappy days, I tried to persuade my parents to let me enlist on the Royal Navy's full-rigged training ship *Arethusa,* but my cousin Frank Waterhouse, serving in the navy, persuaded them otherwise. A peaceful accommodation with my new classmates came only because I could run faster than they, initially for self-preservation but later on in the sports field. I had to catch up on the two years of German that I had missed (I was assigned to a student teacher from Köln for one-on-one instruction and caught up in three months), and I also learned to use my speed on the soccer field.

I soon developed a new rhythm of studying hard and playing hard. Food was still strictly rationed, and I spent many an hour in queues at the greengrocers to buy ten pounds of potatoes, sometimes twice on a Saturday afternoon at different shops. Dad much preferred gardening or reading novels in the evenings, so I became my mother's companion at the whist drives and ballroom dances that she loved, learning to play cards and to dance. I stopped at the library once a week to check out four or five mystery novels for Dad—I read them, too—plus an array of the travel and adventure books that were not part of the school's required reading. Money was tight, and as soon as I was old enough, I spent holidays working in the toffee factory, sometimes manning the toffee cauldrons, pouring the scalding mix on the cooling slabs behind the twist-wrap machines, sometimes packing sweets, and sometimes wheeling half-ton pallets of boxed toffees into the warehouse and loading the lorries. Saturday mornings were devoted to scrubbing floors. The workers I came to know were a mix of Cockneys from Hammersmith and Shepherd's Bush, housewives from the nearby West Acton railway workers' estate, Jewish refugees planning to make their way to Israel, and immigrants from the Caribbean, largely from Jamaica and Trinidad. A Cockney girl introduced me to the Hammersmith Palais, a Jewish warehouse coworker to seder, and the Jamaicans to steel bands, calypso, and the limbo.

The money I earned in the factory helped buy my clothes and pay for my sports activities. I played the usual games for the school—soccer and cricket—and ran and jumped well enough to become my age-group Middlesex County champion in the 100 and 220 yards and in the long jump (remember that English counties are roughly the equivalent of American states), and to anchor the school's winning relay teams. Mother was scandalized when I spent some of my factory earnings for a pair of hand-made running spikes and an expensive imported American jock strap. The sports took me all over London, and I was spotted by one of the Polytechnic Harriers' coaches and invited to train with their club. Olympians Macdonald Bailey and Arthur Wint were Harriers and heroes to a teenage puppy dog. I ran wind sprints behind them and tried to match Bailey out of the starting blocks. Occasional visitor Harold Abrahams (*Chariots of Fire*) was a god and even gave words of encouragement. I began to run for Polytechnic teams, which took me farther afield across England and to major national competitions at White City and Wembley Stadium.

I did well enough in the classroom to receive annual prizes for overall academic performance and for English and geography. The geography master, H. W. S. Urch, also my housemaster, began to take a special interest in me, and this attention, I suppose, marked the beginning of my commitment to geography. I learned much from him by staying within earshot on field trips into the English countryside. He must have realized my developing aptitude for the subject because he pressed me with extra reading and additional exercises, especially in map reading and air-photo interpretation, and in the physical geography that he loved. Because he pressed me, my performance improved, and as my performance improved, my enjoyment of what I was doing increased. I came to *like* geography.

Acton County School was a controversial place. The school's headmaster, G. C. T. Giles, was a Cambridge-educated Ph.D. in philosophy—shades of Burgess and Maclean, Philby and Blount—and a member of the British Communist Party's executive committee. When he retired several years later, it was to Maoist China. Half of the teachers were party members and made no bones about their beliefs. Local politicians kept up a drumbeat of criticism. The student body was not unaffected. Because it was part of the intellectual climate of the school, we paid attention to the contest between communism and capitalism as a new socialist government struggled to nationalize England and as the Cold War deepened. Political and economic debate became part of everyday fare, and even though we could not take classes in these subjects, an interest was implanted in me that bloomed when I got to university. There were always men outside the school, standing and occasionally asking political questions. They, too, reappeared later in my life.

In 1949, the family moved out of the Palm Toffee offices to a house around the corner that had been vacated by the death of one of the Walters family sisters. Aunt Mabel and Uncle Buck visited from Kent once they were able to buy a car, and despite the fact that I was underage, Buck taught me to drive. Dad also got a car—a four-cylinder, 1934 Ford that could reach sixty miles per hour on a long downhill—and on weekends we toured the south of England. I always tried to read about where we were going and became the tour guide, map reader, and way finder—especially when the main roads were clogged with traffic. But a critical turning point was approaching, the "O-Level" examinations. My eyes weakened, and with great heroics Mr. Urch and Dr. Giles persuaded the National Health Service to provide me with corrective glasses on an expedited basis. I sat for exams in English language and literature, mathematics, geography, history, French, German, and physics, with very satisfactory results. My scores in English and geography were particularly high.

I now was sixteen, and no one in the family had ever attended school past age fourteen. Should I now leave and enter the world of work? I earned my keep in the toffee factory all summer and into the fall, and, like a number of my classmates, I took the British civil service entrance examination and waited for the results.

Aunt Mabel stepped in. She and Buck lived at the home of Arthur James, a senior British civil servant, in Sevenoaks, Kent. They called him "Sir Arthur" because he expected to be knighted when he left governmental service. When he did retire, he moved to the nearby village of Westerham to become Winston Churchill's neighbor. Buck was a gardener and landscaper, and Mabel looked after the Jameses' disabled daughter. She told Sir Arthur what I was planning, and he advised otherwise, sending me a strong message: "Go back to school and complete the sixth form. Then you will have a choice of the executive-level civil service or even university." My mother was scared for me, fearing that I would ultimately face rejection because I was rising "above my station." Fortunately, Sir Arthur's logic prevailed, and—getting what I really wanted—in the middle of the autumn term of 1950, several months late, I knocked on Dr. Giles's door and asked if I could return to school. "Of course." There was no question of the civil service. It was the university that lay ahead.

"What are your interests?" Dr. Giles asked. "Well, politics and economics and . . ." "To study these subjects at the university you must first study English literature, mathematics, economic history, and geography. You *must* become literate and numerate." What important advice that turned out to be.

I had to catch up with my schoolmates, but then my teachers took over. Sixth-form classes were held as small tutorials with lots of reading and writing. Mr. Urch welcomed me back and laid out an imposing geography program: "The best book about physical geography is by de Martonne, and we will learn about climate from Köppen." One was in French and the other in German, and that's the way I had to acquire the materials, alongside reading Holmes's book on physical geology and some of Wooldridge's work. I liked the physical geography because it was systematic, something I could get my mind around rather than simply my memory. Mr. Urch made me work hard, insisted that I be organized, showed me how to prepare good notes that combined class work and reading, and how to write four or five intelligent essays in a three-hour period. First drafts were never good enough and always had to be redone: "What is the question? Did you answer it directly and straightforwardly? No. Try again." *"Inimical* is a very useful word. Learn to use it properly."

We organized younger school members in repeating Dudley Stamp's land-use survey in a rural area and in an urban area, and my map reading and air-photo interpretation skills were honed by comparing images, maps, and what was actually on the ground. I rewarded Mr. Urch by breaking the school's running and jumping records, by anchoring the winning sprint relays for Urch House as well as for the school in regional competition, by repeating as Middlesex County champion, and by becoming a regular Polytechnic Harrier in national events. I was a fast goal-scoring forward on the school's football team and a utility cricketer, sometimes bowling and sometimes wicket keeping. As the school sports captain, I organized after-school training sessions on the track for the younger boys. When we were required to write essays in connection with a major exhibition of Leonardo da Vinci's work, I wrote on his bird's-eye view renderings of Italian landscapes, describing them as effective three-dimensional perspective maps, and I won first prize. The music teacher enlisted me to teach my classmates ballroom dancing. Acton County was a boys' school, and the girls came over from their school for Friday night dances and became our Saturday night dates.

As the second year of sixth form ended, I sat for the advanced and scholarship levels of the General Certificate of Education (GCE) in English, history, and geography. I scored exceptionally well in all three, particularly in geography, and won a state scholarship that enabled me to pursue the university education that Dr. Giles and Mr. Urch had planned.

Where should I go? What should I study? Dr. Giles's influence took on a life of its own, although I doubt he would approve if he could see the consequences today. My interest in politics and economics had been sharpened by a new sixth-form classmate, Robert Dowse, a radical left-winger who shared all of my classes except mathematics. Bob and I became the closest of friends, and the Dowse household a place to hang out close to school. In addition to private meetings with Dr. Giles, who complained about being watched, his mail being read, and his telephone being tapped, we attended political gatherings of all flavors (although, I must admit, more often driven by hormones than ideology—the communist meetings, especially, were great places to meet girls who enjoyed a little revolutionary sparking) and debated the political issues of the day. The men who stood outside the school and watched became even more insistent with their questions. Once we reached eighteen, they could always be counted on to provide a pint at the pub across the street. After I considered the alternatives, a degree in economics and political science with—for me—a special subject in geography seemed the way to go. I applied to the London School of Economics (LSE), to University College, London, and, for safety, to redbrick Reading and Nottingham. Bob decided to get his military service out of the way first and ended up in the Suez Canal zone.

The LSE interview did not go well. I was scheduled to meet with the admissions committee in the morning and planned to go running in the afternoon, so I had my bag of gear with me and deposited it by my chair. One member of the committee, the geographer, spotted the bag. All that he wanted to talk about was my performance on the track for the Polytechnic Harriers and as a goal-scoring center forward. "Would I spend all my time trying to qualify for the British team?" "No." "Hadn't Queens Park Rangers [then a third-division team] been interested?" "Yes, but I wasn't interested in them." The committee's political scientist seemed offended that I had been counseled by the notorious communist Dr. Giles. The third committee member asked me about my career plans and seemed displeased when I said that I had none. All three appeared to have difficulty with my East Midlands accent, making it clear by their body language that they were "U" and I was not, belying the purposes for which LSE had been established. My application was rejected. Maybe mother was right.

But the next interview was at University College, London (UCL), one-on-one with economist A. C. Stonier. University College was "the godless college on Gower Street"—Jeremy Bentham's utilitarian alternative to Oxford and Cambridge. A mummified Bentham still presided over college council meetings. From where I sat in an armchair in the economics department, I could see that Stonier's eyes were off-puttingly askew, pointing thirty degrees port and starboard. You never knew whether he was looking at you. "What did I know about economics?" "Not much, but I wanted to learn." "What about politics?" "The same." "Did I want the bachelor's of science (economics) with geography as a special subject?" "Yes." "Why?" "Geography is a subject that I like, one in which I have performed very well." He liked the answers and offered me a glass of cider. I noticed that it was Merrydown, smooth, beautifully distilled, and very alcoholic. I sipped at the tumbler very slowly as the interview continued. I think it amused him, especially because I downed the balance only when the interview was over. We shook hands, and he told me that I was admitted. Joy! Later, Stonier always gave me a wicked grin whenever he invited the undergraduates to receptions at his flat. "Sherry or cider?" he would ask the girls. They usually chose cider, thinking it was the nonalcoholic stuff, and he would pour them big glasses of Merrydown. "Have fun," he would say, as we left. Usually, we did, although most of the girls were singularly lacking in any sense of humor.

I commuted each day between West Acton and UCL on the London Tube: *ta-dum da-dum, ta-dum da-dum* as I held an overhead strap with one hand and a book in the other, pressed tight to other commuters, and changing trains at Notting Hill Gate. I had classes in micro-, macro-, and applied economics, money and banking, international trade economic and political history, philosophy, government, statistics, and geography. Lunchtime discussions took place in the UCL bar over cheese rolls and half pints of bitter. Some of us walked to the London School of Economics to attend Michael Oakeshot's brilliant lectures on the history of political ideas, R. G. D. Allen's on mathematical economics, and an eminently forgettable series on sociology and anthropology. We also went to Birkbeck College for a lecture series that featured many of England's leading geographers. It was there that I met L. Dudley Stamp, Michael Wise, and Arthur Smailes. I regularly attended London's live theater and dance, usually paying only one shilling and six pence or two shillings to sit up "in the Gods," even at the Old Vic, Covent Garden, and Royal Albert Hall if I chose the right nights. There were rounds of cutthroat bridge at UCL and LSE, and we made it to the Houses of Parliament to observe debates in progress. I occasionally picked up pocket money teaching ballroom and "old-time" dancing to grateful matrons.

Over the years 1952 to 1954, the tempo of study picked up. I broke the

tension by running for UCL and the Polytechnic Harriers, and by playing soccer for UCL. Between terms, I continued to work in the toffee factory. I
needed the money to fund the sports activities and my first real romantic interest, the daughter of a Midlands shoe manufacturer. The English class system was to intervene in that affair, however. Her parents forbade her to see
me: I was "below their station" and could never be a "proper" match. She
obeyed them, but I learned that a year later she ran off with an American serv-
iceman. This romantic rejection once again raised all of the fears I had absorbed from my mother and provided a greater incentive to excel when the
Part I examinations came, after a wonderful European field trip led by Karl
Sinnhuber through the Netherlands and the Rhineland. It was on this trip
that I locked horns with Professor Troll. The B.A. geographers came along,
too, including longtime friend Akin Mabogunje. We were banned from the
Jugendheim in Bad Godesberg for rowdy behavior, and when we were required to "do something geographical" in Amsterdam, we surveyed the prices
of the oldest profession, plotted them on a map, and related the variations to
location, quality of street lighting, and other variables.

When the Part I results were posted, Stonier called me into his office and
poured me another tumbler of Merrydown: "You have a chance to get a
'first.' " Once again, I had done particularly well in the geography examinations, and, as planned, decided to concentrate on the subject in the third year.
But before I left for the summer, I asked the geography instructors what they
would expect me to read when classes resumed. Days, I worked in the factory.
Nights, after running in the nearby park, I did the reading, including Edgar
Hoover's *Location of Economic Activity*. On weekends, I traveled to athletic
events or occasionally into the Cotswolds or to the southern coast with the
family.

My final year at UCL was all geography: historical geography with H. C.
Darby, economic geography with B. R. Law and W. R. Mead, physical geography with Eric Brown (poor Eric: Mr. Urch had seen to it that I knew more
physical geography than Eric offered in his lectures, and I am sure that I was
too fidgety in his classes by far), Europe with D. Ricks and K. Sinnhuber, and
North America with Mead. Darby, a brilliant lecturer, provided us with systematic insights into the evolution of Britain's landscape, let us see how he ac-

complished his geographic reconstruction of the Domesday Book, and challenged us in a weekly seminar on the study of historical geography (my paper for him was on Frederick Jackson Turner's model of development). Law and Mead instructed us in economic geography via the theory of location—von Thünen, Weber, Hoover, and so on—and I took to it like a duck to water after two years of studying economics. Here was a systematic approach to human geography into which I could sink my teeth—theory, not memory, was what counted. Ricks reinforced my skepticism of the rote regional geography of the time. It bothered me that some UCL and Birkbeck geographers advised that the way to become a geographer was to take a part of the world and learn more about it than anyone else. Included in the year was a regular program of field trips because, in addition to being examined in five subjects, we were expected to submit field notebooks reflecting both the field trip program and independent fieldwork. My classmates complained, but I told them, *"Modo fac: cum tractu traducto"* (Just do it: it goes with the territory). I wrote about the Trent Valley area in which I had spent my early years, about the nearby Isle of Axholme, where the medieval open fields had never been enclosed, and about the West London industrial estate in which I worked.

I also began to make other plans. Derrick Sewell, with whom I played soccer, had graduated from UCL with a B.S. in economics in 1954, obtained a Fulbright travel grant to study geography in the United States, and landed a teaching assistantship at the University of Washington in Seattle. If he could do it, why not I? The alternative was military service, which I had deferred since I was eighteen. I knew I lacked the "right" accent to be considered "officer material" by observing those of my classmates who had or had not become officers during their national service. On the other hand, the men who still came around and asked questions told me that I seemed to be the right kind of person to be seconded for "special assignments." I did attend their weekend training sessions, but we agreed that a graduate education should come first. I could pay my dues later.

I sought the advice and help of Darby and Mead about the American option and began to fill out applications. Darby urged me on because, as he explained, Britain lacked the systematic graduate training that the best American universities provided. Brian Law had already introduced me to location the-

ory, and from my readings I knew such names as Harris at Chicago and Ullman at Washington. Darby added Wisconsin and Minnesota to my list because of the aptitude I had shown in his seminar in historical geography.

Chicago quickly rejected my application: my interests were "incompatible
with those of the department." I learned only much later that Robert Platt did
not see eye to eye with Darby. But suddenly, before Christmas, came the offer
of a teaching assistantship from G. Donald Hudson at the University of Washington. He requested a response by return telegram. I accepted, of course. I
did receive offers much later from Wisconsin and Minnesota, but by then the
die was cast. A Fulbright travel grant came through, and Darby also engi-
neered an invitation to participate in a month-long Fulbright summer school
on American studies held at University College, Oxford. There, former U.S.
naval intelligence officer and George Washington University geography professor Robert D. Campbell (who later helped me get my first U.S. research
grant from the Office of Naval Research and who crossed paths with me on
numerous occasions on special assignments in northern India) introduced me
to the field in the United States via the newly published *American Geography:
Inventory and Prospect*. I already knew I had my "first" before I arrived at Oxford. Bill Mead had called and located me in the toffee factory where I was
working after exams were over and had given me the good news. When I took
the call late in the day, tired, in grease-smeared overalls, the factory owner's effete son was there, elegantly suited. He had heard that morning that he had
received a "third" in geography from Oxford. But his accent was impeccable;
he had preceded Oxford with a term as second lieutenant in the British army;
and he was about to take over as deputy manager of the factory. Ah, England!
I relayed to him what I had just been told, but thought, "You twit. I'M OFF TO
AMERICA TO BECOME A GEOGRAPHER." My great adventure was about to
begin.

# A Journey of Discovery

## John R. Borchert

*John R. Borchert (1918–2001) received his bachelor's degree from DePauw University in 1941 and, after serving as a meteorological officer in the United States Air Force during World War II, obtained his doctorate from Wisconsin in 1949. Although he taught briefly at the University of Wisconsin (1947–49), he became indelibly associated with the University of Minnesota, rising to regent's professor before his retirement in 1988. He was a member of the National Academy of Sciences and the American Academy of Arts and Sciences since 1976, and received the Van Cleef Gold Medal from the Association of American Geographers after serving as president of the association in 1968. He devoted much of his professional life to his home state of Minnesota through his applied research and public service on such bodies as the Twin Cities Metropolitan Planning Commission, and the Minnesota Pollution Control Board, as well as Minnesota's Department of Education, Upper Midwest Council, Resources Commission, and State Planning Agency. His research, teaching, and production of meticulously detailed maps informed virtually every branch of state and local government to make Minnesota the most "geographically literate" state of the Union. At the national and international levels, his public service included a directorship of the Social Science Research Council as well as membership on the Environmental Studies Board, the Committee of Surface Mining and Reclamation, and the Transportation Research Board. He was also vice chairman of the Commission on Environmental Problems of the International Geographical Union from 1976 to 1980.*

One of the country's leading professional urban planners once remarked to me, "If I ever write my autobiography, the title is going to be *One Thing Led to Another.*" He quickly admitted that's a terrible thing for a planner to say, but repeated that it was true. Another time, a geographer who was a successful analyst and consultant on metropolitan growth and land development commented, "I really haven't had a career; I've simply careened through life." I heard those characterizations from others, but they are no less applicable to my own case.

For me, life as a geographer has been a succession of discoveries, beginning by chance at a particular time and place, and each leading to another. As a boy, by chance, I discovered the subject of geography, although I didn't assign a name to it. Then, by chance, I discovered the academic field of geography. A little later I discovered the discipline, and at last I could put together my accumulated knowledge of the subject matter and the field. Soon afterward, by good fortune, I discovered the need for geographical teaching and research. Meanwhile, I discovered the legion of curious, able, energetic students who themselves were discovering the subject matter, the field, and the discipline. And, alas, I discovered the chores to be done if one would conserve the institutions that provide the opportunity to teach and to study the field.

Too often among the hundreds of undergraduate advisees who came into my office, one of them would say, "I just don't see how I'm going to work out my course program until I figure out what I'm going to do with my life." We would then work something out, but my parting advice would always be, "Remember: as a student, you're a professional. Be the best, most professional student you can be, but don't be afraid to adapt to new interests and opportunities. Just be sure when you lay out each year's program that it makes the greatest possible use of the energy you've invested up to that time. Changes in direction are exciting and often inevitable, but continuity is extremely valuable. Save what you learn, invest it, and reinvest it."

### Discovering the Subject

As a boy in the 1920s and early 1930s, I lived, by chance, on the edge of one of the steepest geographical gradients in the world at that time. On one side of the gradient stood my hometown, Crown Point, Indiana. At that time, it was in most ways a typical Corn Belt county seat of twenty-five hundred. Its merchants served the farmers within a six-mile radius, most of whom farmed land that had been cleared by their immigrant fathers and uncles in the mid-1800s. Almost everyone had a northwest European background, the vast majority German. Many of the old folks still spoke German at home. The worth of a farmer was commonly measured by the size of his barns and manure piles and by the order of his fields, not by his formal education. Ethnic and religious tolerance did not have a high standing. There was not a single African American household and only two Jewish, both in menial trades. Even our annual family reunion crowd of two hundred or so divided along religious lines for the picnic dinner. The Roman Catholics ate at one set of tables; Missouri Synod Lutherans, plus a fringe of lesser Protestants and fallen away but otherwise honorable relatives, ate at another set.

Yet just ten miles north of my hometown was the southern edge of the new city of Gary, population one hundred thousand, laid out less than a decade earlier by the U.S. Steel Corporation on the marshes and sand dunes at the south end of Lake Michigan. Just five miles farther north were the gates of the largest steel mills in the world, the economic base of Gary. Twenty thousand of Gary's residents were African American. Easily 80 percent of the rest were recent immigrants from eastern and southern Europe, and a small contingent was from Asia. It was a linguistic and religious polyglot to match any in America at that time.

From Gary, a smoky, chaotic array of recent, modest residential neighborhoods, refineries, factories, and vast rail yards sprawled westward across the Calumet flats for fifteen miles into Chicago. Beyond that, just forty miles from my bucolic hometown, monumental office towers and hotels rose above the noise and bustle and soot of the Chicago Loop. The train ride from Crown Point to the heart of Chicago took fifty-nine minutes. Through the dirty daycoach windows I watched, on trip after trip, the quick, bewildering transition from my rural countryside—through a heavy industrial complex that matched the Ruhr and the Pittsburgh-Cleveland axis for world leadership, amid rail yards teeming with thousands of box cars emblazoned with system names that read like a gazetteer of North America—to the heart of what was then the fourth largest city on Earth.

The contrast was all the more striking because of my mother's connections with Chicago. Whereas my father was a country boy, she was a city girl. On completion of the eighth grade, she had gone to work for the next fourteen years as an office clerk at what was then reputedly the largest factory in the world, the giant Western Electric plant on Chicago's West Side. Close friends she had made in those years were now raising children my age in locations scattered in widely contrasting neighborhoods throughout the metropolis—from the sooty frame tenements near the industrial Chicago River to the elite North Shore and upper-middle-class western suburbs—housing everyone from inner-city, immigrant common laborers to corporate executive and suburban bank presidents. Unlike my cousins and many friends at home, most of the children from all these neighborhoods were college bound and knew it.

Then there was my maternal grandmother in Chicago. She was born the year the Golden Spike was driven to link Chicago with the Pacific Coast, and her earliest memories dated from the night of the Chicago fire. Self-educated, with good fortune, she had worked at a string of jobs, beginning as a seamstress stitching baseballs in Mr. A. G. Spaulding's upstairs workshop and ending as a housekeeper for the mayor of Chicago. Her neighborhood butcher had been the eventual meat-packing magnate, Oscar Mayer. Her recollections of the physical and social changes in her city, like everything else in the environment, were stimulating, often puzzling.

One of her nephews, briefly husband of the famous torch singer Sophie Tucker, was music director at the newly established National Broadcasting Corporation (NBC) radio network studios in Chicago. When he would arrange visits to the studios, I could go into the control room and confront on one entire wall a breath-taking relief map of the United States that showed the city location of every NBC affiliate, with a light where the local station was carrying a network program at the moment. I could look at that map and put my nearby, still-bucolic hometown in the context of the new coast-to-coast radio network communications system of the nation.

Other interesting contrasts highlighted my early life. As auto ownership spread in the 1920s and 1930s, people from Crown Point made the trip to the Calumet industrial cities more and more often for work, shopping, and entertainment. In my college years, I worked during Christmas holidays as a salesman in the men's furnishings department at the Sears department store in

Gary. I often helped appreciative black ladies select a shirt or tie that would look good on their sons or husbands, and met not only black men who labored in the mills but others who were professionals and small-business owners in the city's large black community. I learned much about their backgrounds and lives in the process—a thrilling and provocative experience for a boy from all-white Crown Point.

Three of my cousins, brothers from the same farm, were skilled, self-taught mechanics, like many of their neighbors. When model-T Ford commuting began, they found work at the steel mills in Gary; and when the worst of the Great Depression followed, they were laid off, along with 80 percent of the production force. One of them found work as a garage mechanic, another as a car salesman. The third worked for labor leader John L. Lewis in the effort to unionize the mills. In that work, he met a young Slavic immigrant woman who believed in the Soviet revolution, preached communism among her friends in Gary, and relayed propaganda literature from the Soviet cultural offices in New York, which gave me the opportunity to read vividly illustrated volumes on the goals and achievements of the Five-Year Plans. And I could try to match this version of the situation with the reporting of eastern European events in Colonel McCormick's *Chicago Tribune* and William Randolph Hearst's *Chicago American*.

The third brother persuaded the other two to join him as members of the Communist Party U.S.A. It seemed like a way to get the mills going full-tilt again. Later the three joined to buy the Chevrolet garage in a nearby small town that was becoming a residential suburb of the industrial cities. Thus, they were General Motors dealers and card-carrying members of the Communist Party at the same time! Meanwhile, back on the family farm, their father had resisted the township's plan to surface his dirt road because he did not believe automobiles were here to stay. He refused to get electricity on the farm because appliances would make everybody lazy, and he thought a telephone was unnecessary and would simply tempt his wife to waste the day talking.

So here was an adolescent in a remarkable and potentially very confusing transition zone in the country's—or the world's—historical geography. There was a rich array of inconsistencies and contrasts. There were plenty of questions. There was the bewildering initial problem of simply classifying and ordering the observations of places, people, and activities that tumbled in day after day. It was a feast of wonderful, challenging experiences, though nearly a decade would pass before I fully discovered that there was a field of study into which all of this feast would fit—geography.

Two important sources of help were at hand for me at the time. One was the small group of teachers, the minister, and the parents of some of my friends who were in the legal and land records professions associated with the local county courthouse. They guided me to the library early on and eventually helped me to understand that I should go to college. The other source was my dad, who was a railway postal clerk. For years, I watched him as he continuously thought about the routing of mail through the northeastern and central regions of the country and studying for examinations. His study habits were my introduction to intellectual discipline. Meanwhile, a map of railway mail routes provided me with a very useful system for arranging the welter of information about places beyond the horizon.

### Discovering the Field

It was a fitful, chancy journey to eventual full discovery of geography as an actual field of study. I guess the journey began in the first year after I finished high school. To be sure, I had two one-term geography courses in grade school, but the teacher was not known in the school or in the community as the geography teacher, let alone as a geographer. I surely did not understand that the books were actually written by people who worked in a college field by that name. To my knowledge, geography simply did not exist as a field of advanced study, let alone as an occupation. Therefore, to learn of the field I was going to have to go to college. And that was not a certainty.

I wanted to be a journalist. It seemed the obvious way to begin was to get a job with the local weekly paper and work up. I learned the rudiments of the printing trade and occasionally had to write the whole paper, from sports to card parties to weddings to funerals—with a deadline and with the knowledge that everyone around the courthouse square would soon let me know if I got something wrong. In later years, I envied my colleague James Parsons his longer and deeper journalistic experience before he took up geography at Berkeley. But during my first year as a journalist, the local Methodist parson intervened. He introduced me to a friend who was an executive at the *Chicago Tribune*. In his imposing tower office on Michigan Boulevard, Mr. Maxwell quickly showed me why I needed a college degree if I were going to get anywhere in that field. Then Reverend McFall introduced me, rather by chance, to the president of DePauw University, whose local high school commencement speech I was reporting. The following fall I was in college at DePauw, in Greencastle, Indiana.

By chance, I took a year of geology as a freshman. It took some time to get oriented before I decided on a geology major. I had two reasons for my decision. The study of historical geology—especially the Ice Age and recent—was my most liberating intellectual experience in college up to that time, and the study of economic geology could lead to employment. Like any youngster in any age, I was curious about my place in the scheme of things, and like any youngster in a working-class family during the Depression, I could not justify college if it did not lead to a good job.

The lone professor of geology did offer one course called "Geography," which served to meet a state requirement for students majoring in education. But neither he nor I had much interest in the course. The textbook seemed to me to be handicapped by what I now realize was an environmental-deterministic organization. It seemed either to garble or omit most of the geographical information I had accumulated informally. Meanwhile, however, Professor "Rock" Smith was a devoted and knowledgeable geologist and a stern taskmaster.

Rock saw the future of such fields as statistics, geophysics, and aerial photography in geology research and applications. He pushed his handful of majors (I was the nineteenth in twenty years) through a rigorous and well-rounded introduction to geology, the basic sciences, and mathematics unusual for the time. The college saw to the rest of a liberal program. Even more important, the campus lay astride the boundary between newer and older glaciation of the Midwest, and also astride the gradient between later settlement from northern Europe via New York and earlier settlement from the South via the Ohio Valley. Hence, Rock's love of fieldwork exposed his students to much informal observation of not only the physiographic but also the cultural changes across those boundaries.

Preceding graduate school in geology, a short stint in geophysical exploration for oil introduced me to the inspiring landscapes of the northern Great Plains and also to my future wife, Jane Willson, in Bismarck, North Dakota. Her unfailing support would help me at every point on the road ahead. A semester of graduate work at the University of Illinois—just long enough to discover that I enjoyed teaching—was punctuated by Pearl Harbor and U.S. entry into World War II. There was no way I could focus on traditional graduate work now. A job as a topographer with the U.S. Geological Survey (USGS) beckoned. Topographic mapping of the coastal areas had been accelerated in response to the fear of Nazi and Japanese attacks. My assignment was on the Tensaw Quadrangle, on the bayou-laced delta of the Alabama and Tombigbee Rivers, at the northeast edge of Mobile Bay. One result of this work is that my name appears as one of the surveyors on a USGS topographic sheet—the only member of the Association of American Geographers, I believe, to hold that obsolete distinction.

Far more important, at least to my education, were some of the features of the "culture" that we were locating on our field sheet. We kept encountering large mounds hidden deep in the piney woods and swamps, left by Spanish surveyors and their slaves when they had tried to locate the boundary of Florida in this remote wilderness, months removed in travel time from the royal palace in Barcelona. And there were the footpaths winding through the roadless pine forest—a mature forest growing where a previous mature forest had been logged early in the twentieth century and where cotton fields had grown half a century before that.

In clearings here and there along the footpaths were one-room, windowless, but still occupied log cabins. Adjoining small gardens were enclosed by crude stockades to keep out the deer and bear and wild hogs. The gardens supplied much of the food for the descendants of slaves who lived in the cabins. Trapping and hunting provided their cash. The people in the cabins were old. Their great-grandparents had worked in the cotton fields that were now under second-growth mature forest. The year was 1942. The younger folks had moved to Mobile in response to the wartime employment boom.

Where the paths converged in the middle of the forest, there was a larger cabin, with split logs for seats and desks. That was the community church. It had also doubled for the school when the young people were still there. A couple of miles away, out on the highway, stood a neat brick elementary school, built for white children with federal public works funds. Nearby, a tiny rough-sawn store building stood at the edge of the road, run by a friendly, toothless, grizzly white man who was the only remaining descendant of the one-time plantation owner. His store stocked crackers, canned beans, Coca Cola, and shotgun shells. He catered mainly to the people who lived in the cabins along the footpaths.

As my rodman and I went about our mapping, the families in the cabins always greeted us, and so did the preacher. We encountered him several times each day as he rode from cabin to cabin to minister to his flock and take his meals with them. His mule was the only transportation in the black community. We often visited with the pastor, but I had difficulty understanding the others. So did my rodman, a white sixteen-year-old dropout from the county seat. I was a stranger from nine hundred miles away. He was a stranger from only ten miles away. To those folks beside the cabins, I could not explain what we were doing. Most were not literate. Some had no certain surname, perhaps no legal existence. They had no context to understand a topographic survey. My mind so often ran to the urbanized, transplanted African American customers I had met in the Sears store in Gary.

My work on the Tensaw Quadrangle ended abruptly in early 1942 with a profound change in career direction—but definitely not a discontinuity. I changed my course of study from geology and geophysics to graduate work in meteorology at MIT, where Jane and I were married. As part of the Aviation Cadet Program, the course led to a commission in the U.S. Army Air Force. The most exciting part of the program to me was the work with synoptic weather maps. Foreshadowing events to come in my life, one course dealt with world regional climatology. That was my introduction to the Köppen classification of climates, which, unbeknownst to me, was a focus for much research and writing in American geography in those years.

After commissioning and a brief break-in period, I went to England. Experience at bomber bases in mid-1943 led to almost two years of work as an operational weather forecaster at the headquarters of the B-24 "Liberator" Bomber Division. There were similar weather centrals at each of the two B-17 "Flying Fortress" divisions and at the Eighth Air Force Bomber Command. On many days, a thousand planes and ten thousand men were flying on our forecast.

Briefing the crews and the generals and debriefing the crews after a mission were powerful and humbling teaching experiences. Analyzing and drawing the weather maps and preparing the forecasts were powerful experiences, too. The essence of our job was clear. We had very large array of numerical data on a map of a very large area of the world. We used isopleth analysis to locate highs, lows, gradients, air flows, and weather conditions generated by those flows as they diverged, converged, and crossed relief features and water bodies. Then we applied a mixture of fairly rigorous procedures and intuition to extrapolate the patterns through time.

Thus, we were doing four-dimensional cartographic analysis, which I would later come to believe is the heart of the geographic method. It was obviously the same method to which I had been exposed in the contouring of bedrock surfaces in search of possible oil-bearing domes on the northern Great Plains and in making topographic maps that could be used to follow changes in the shape of the delta at the head of Mobile Bay. The weather maps were far more exciting because they were so much more dynamic—changing by the hour rather than by the century or the eon. Hence, they were much suited to seeking and applying the understanding of change because change was going on while we watched.

Twice daily we analyzed weather maps that reached more than half way around the Northern Hemisphere, from the west coast of North America to central Siberia, and from near the North Pole to the Sahara. Four times each day we analyzed maps that reached from Greenland and the Azores to the Urals and North Africa. The map analysis and forecasting for routes and target areas were absorbing. We could never avoid reviewing and enlarging on what we had learned about the vast, ever-changing global atmospheric circulation system. But it was also impossible to range over all those lands without thinking about what people there were doing. Along with weather forecasts, from piecemeal intelligence photos and topographic maps, we conjured landscapes and little stories as our attention moved from Atlantic City to Aberdeen to Algiers to Astrakhan, all at the same hour on the same day.

During the preparations for the Allied invasion of the European continent, one of my former fellow geology students at DePauw appeared at the British Admiralty headquarters as a member of a small team assembled to forecast sea-and-swell conditions for D-Day. When we got together, I met one of his coworkers, Kenneth Hare, whose background turned out to be in geography—a fully recognized field of study in England, I learned. I hadn't even a premonition that our paths would cross again, in our mutual pursuit of geography, two decades later in North America. Nevertheless, my gradual, fitful discovery of the field—and certainly my widening observation of the subject matter—continued.

The end of the war set in motion a chain of chance events that led me quickly, once and for all, to the field of geography. My roommate on the base in East Anglia at that time was the division ordnance officer. He also mysteriously happened to be responsible for closing down the libraries at all the bases. Near our quarters he had a Quonset hut that happened to be filled to the roof with books from those libraries. A great many of them were introductory college textbooks. With combat missions discontinued, I found myself spending many evenings browsing through stacks of textbooks that some committee had once decided should be available to curious GIs while they carried on the war over Europe. As you might guess, the books were generally in excellent condition.

I became steeped in the fundamentals of practically everything. But one book caught my special attention—*The Elements of Geography,* by V. C. Finch and G. T. Trewartha, professors at the University of Wisconsin. Large parts of the book seemed to deal with mapping, geology, and meteorology or climatology. I had worked at the early graduate level and had some applied experience in all three fields. The authors seemed to be making an effort to relate the earth science material to human use of the land, and that connection had always interested me. The final section of the book tried to say something systematic about the morphology of human settlement, and although the effort was minimal and halting, the idea was intriguing. My thoughts also ran to the enjoyment I'd experienced in tutoring at DePauw, running labs and quiz sections at Illinois, and briefing crews and generals in England. I thought perhaps I should look into this.

When I returned to the States in September 1945, Jane and I, now with a three-year-old daughter, were impatient to settle in the civilian economy and get going. I soon rejected the possibility of continuing graduate work in geology or pursuing graduate work in meteorology. The outlook for satisfying work, advancement, and income seemed best in petroleum exploration or air transportation. A particularly exciting opportunity had appeared with Northwest Airlines, which was at the threshold of its major postwar commercial expansion coast to coast and across the Pacific. But I still had lingering thoughts about geography.

In late September 1945, Jane and I were visiting my home back in northern Indiana. I left for two days to go to Camp McCoy, in central Wisconsin, for separation from the military. As I was leaving, Jane suggested that I stop at Madison on the return trip and talk with those men, Finch and Trewartha, just to put the matter to rest. I said, "Aw, the train from Camp McCoy to Chicago doesn't go through Madison. Besides, I think we're pretty well committed to Northwest."

The next afternoon I was walking from the base headquarters back to my barracks to pick up my knapsack and head for the railway station to catch the train to Chicago. Overtaking two fellows on the walk, I asked the time. The conversation disclosed that they were about to drive to Detroit, and I was welcome to ride with them right to Crown Point, Indiana. Excellent. It occurred to me that this route would go through Madison. A couple of hours later, we got a red light at the corner of Park and University, on the edge of the University of Wisconsin campus. I thought, "If Jane learned I came this close and didn't stop, she'd be pretty upset." So I climbed out of the car and found a hotel room.

Next morning early I headed for the campus and easily found the geography department just where I expected it to be—relegated to the upper floor in a big, red brick Victorian building shared with geology. At the top of the stairs was a sign announcing Professor Finch's office. He received me graciously. We talked for some time about my background and interests and questions, his assessment of where the field had come from, and its postwar future. Presently he looked at his watch and observed that he had to give a lecture to the introductory physical geography class in a few minutes. He paused, then said, "The lecture today deals with the Marine West Coast climates in the Köppen system. You are certainly familiar with that climate and what it meant for our fliers in northwestern Europe. Would you like to give the lecture?"

Recklessly, I accepted the invitation. As we walked up Bascom Hill to the lecture hall, I thought about the chapter on Cfb climates in Haurwitz and Austin's climatology text at MIT and some memorable mission forecasts that illustrated both our location in the world wind system and what it meant to air force operations. I was alarmed to see two hundred students waiting in their seats. At the end of the hour, I was drained; the blackboard was filled with outline headings and map sketches; and the class applauded. I had found my field. I was hooked for good. A week later my family and I were living in Madison and beginning to burn alternative occupational bridges behind us.

## Discovering the Discipline

When Professor Finch and I returned to the department after the lecture, he introduced me to Professor Trewartha for much more conversation about the field, then I remained for a bag lunch with the half-dozen graduate students, including Allan Rodgers and Wilbur Zelinsky. By chance, the guest speaker at the lunch was Professor Wellington Jones from the University of Chicago. He reported on his research in the Punjab.

Jones's presentation was a revelation to someone at my particular level of preparation. His maps were simply work sheets, containing Indian census data on crops at successive time intervals. The data were overlaid with isopleths. He talked about areas of high and low production, intervening gradients, and changes in pattern from one time to another. He laid out his explanations for the patterns and the changes—explanations that were based on archival work, interviews in the field, and comparisons with other maps. He examined his subject at several scales. Behind him hung a couple of large wall maps on which he placed his study area within South Asia and the world. At the opposite end of the scale, he showed photographs of landscapes that were generalized on his maps. He discussed questions that still puzzled him, and he speculated about possible further questions the maps suggested.

I thought, "This is analogous in so many ways to what we meteorologists did with weather observations during the war—isopleth analysis, description, and classification of patterns; description at different scales from global to local; interpretation using both theories and simple, direct observations; discussions of the results with others who were interested." Here, once more, was a demonstration of what I would come to regard as the core of the geographic method. The data were for minor civil divisions rather than for specific weather stations. Wellington Jones was sampling an extensive surface using small areas rather than points. His time intervals were in years rather than hours, but there was plenty of opportunity to watch and map the change as it was actually taking place.

Graphically and intellectually he put his study area in its context of larger world patterns—just as we operational weather forecasters had placed the weather maps of the British Isles in the context of the Northern Hemisphere circulation when we briefed aircrews. Or just as my work in the oil company had placed the potential oil structures of North Dakota in the context of stratigraphic and structural patterns of the continent. Or as my work with the USGS had placed the Tensaw Quadrangle in the context of colonial boundary surveys and the U.S. Land Survey or the map of slave-holding agriculture. Or just as the route of my boyhood train rides from Crown Point to Chicago had placed my hometown in its context on atlas railroad maps.

The landscapes in Jones's slides represented regions on his maps. I thought of memorable landscapes in my own experience—the flat woods, the window-less cabins, and muddy bayous on the Tensaw Quadrangle; the contrasting scenes from the train window in the transition from the countryside to the nearby heart of Chicago. His interviews seemed analogous in a way to my conversations with the folks standing beside their stockades on the Tensaw; with the land agents leasing drilling rights from central North Dakota farmers on the land we were seismically surveying; with my white rural cousins and my urban black customers, all busy living their lives in the industrial northwestern corner of Indiana. Clearly, one good way to order these landscapes and con-versations was to put them on maps.

In short, this talk focused on an aspect of the human use of the earth. It classified selected features to describe a geographic structure. It described a geographic process of change in that structure. It looked into some of the me-chanics of the change process. And it depicted the structure and change at dif-ferent scales. The study suggested that maps proscribe a distinct geographic scale for the study of human settlement—from local to global. To appreciate the analysis I had to visualize myself in a place there in India and visualize my location on maps of the village, the region, the subcontinent, and the world. I had to be able to shift back and forth between images of the real landscape and images of my location at that place on maps at different scales, with graphic symbols at different levels of abstraction. Wellington Jones's maps also suggested the virtually endless range of possible observations about the human use of the earth that either are or could be collected and mapped, which, in turn, suggested the problem of selecting which data to work on and in what order. Not any easy subject. When I retired fifty years later, we were still far from understanding the cognitive aspects of the discipline.

Nevertheless, here was clearly a powerful intellectual tool. It would be difficult to imagine a more intellectually liberating experience than backing off and looking at one's self on the face of the earth in this way. Nor would it be easy to think of any more efficient way to understand the locations and interactions among a great variety of day-to-day activities while, at the same time, helping in the ultimate scientific quest to understand the role of humanity on the earth.

With the benefit of retrospect, it's easy to make too much of the brief encounter. Yet I'm sure that all of those seeds were planted at that time. The experience showed me a way to order the array of naïve observations of people's behavior on the land and the limited vocabulary of mental maps I had accumulated up to that day. At the same time, it set in motion my thinking about the discipline and practice of geography. To be sure, the thinking would be modified and reinforced from that time on, but it had begun.

The inspiration really never waned. It led to many rewarding discussions with fellow graduate students, including Rodgers and Zelinsky, as well as Cotton Mather, John Brush, and John Alexander. Richard Hartshorne added historical depth sorely missing in my peculiar and accidental background. Arthur Robinson brought so much insight into discussions of scale, generalization, and measurement. Glenn Trewartha added his penchant for viselike, orderly, unequivocal description. Reid Bryson's ideas about flows, gradients, boundary zones, and interactions between the earth and human settlement ranged far beyond his central interest in dynamic climatology. The same thinking that began in that lunch hour with Wellington Jones in 1945 carried through to later discussions with my new Minnesota colleagues, especially Jan Broek, John Weaver, Philip Porter, Joseph Schwartzberg, and Fred Lukermann, and with a procession of graduate students.

As a result, maps in time series used to analyze geographic processes were a hallmark of most of the research I did from the time I became a geographer. Let me give you a few major examples.

My first major publication, in 1949, was almost my doctoral dissertation. It compared the patterns of atmospheric circulation, rainfall, and temperature in different dry seasons and decades in central North America. A short time later, I did two studies of municipal water supplies of American cities that compared patterns of water use with available supplies in wet and dry periods.

A field class project in my early years at Minnesota returned to a one-time land-use and ownership survey that Professor Darrell Davis had done in a frontier Finnish settlement area on the Lake Superior north shore. His work dated from the 1930s, in the days of federal rural resettlement during the Great Depression. Mapping and interviewing in the same area at a second point in time enabled my students and me to describe a quarter-century of dramatic local changes that reflected global forces from the czar's Russification of Finland to Washington, D.C., and Kremlin policies in the Cold War.

A 1967 study, "American Metropolitan Evolution," depended on maps of the country's cities, using comparable-size classes, at successive dates in the evolution of transportation and industrial technologies. The study emphasized the importance of successive, unforeseen new rounds of initial advantage, reorganization, and adaptation, as well as new reasons to exploit new land and to abandon old. It mused over countervailing public pressures—either to speed up institutional adaptations to geographic change, notably in local government, or to try to retard the rate of change and thus reduce the need for difficult adaptation. Comparison of this paper with the grassland study shows my continued focus on mapping geographic processes, notwithstanding a big shift of application from resources to settlements.

A subsequent paper, "Major Control Points in American Economic Geography," dealt with one component of metropolitan evolution. It mapped a half-century of change in the location of the headquarters of large business organizations. The maps reflected the importance of entrepreneurship, instability, inertia, and drive for security. Another follow-up study in 1983, "Instability in American Metropolitan Evolution," described in more detail a century of increasing variability in local growth rates as the speed and capacity of intermetropolitan circulation increased. In light of my findings, chaos theory came as no surprise to me.

I based a 1987 regional monograph, *America's Northern Heartland,* on many maps comparing the settlement patterns of the Upper Midwest at successive times—at the beginning of railroading, the beginning of the auto/air age, and the beginning of the jet/satellite/fiber optic era. The study documented and interpreted dramatic changes in the way the region functioned. It also highlighted persistent features of the culture and circulation network of a busy part of the country, whose winters, most Americans think, make it basically unhabitable.

A short time later I had the opportunity to reflect on the changes since the 1960s. That paper was a chapter called "Futures of American Cities" in a 1991 volume titled *Our Changing Cities*. The book was conceived and edited by my colleague John Fraser Hart on the occasion of my retirement. My paper followed a stimulating, year-long series presented at the University of Minnesota by some of the country's finest urban geographers. In that study, I asserted that we have been in a new epoch since the 1970s and speculated on the settlement features that would be the hallmarks of the resulting new metropolitan "age rings."

I suppose I could not escape from impressions developed over the previous three decades of my life—the importance of evolving, pervasive technologies; unique local sites and histories; entrepreneurship; and increasing instability, complexity, and fragmentation. I added that an outpouring of atlases and interpretation would be more essential than ever as the inhabitants of these American cities seek to understand their options and their actions. There I had in mind two converging trends: society's growing need for geographic analysis and forecasting, on the one hand, and the potential power of geographic information systems on the other.

Both trends were foreshadowed, in a way, by a fairly massive study I carried out with the help of students in the late 1950s. The result was "The Twin Cities Urbanized Area: Past, Present, Future," published in 1961. Although that paper also rested on a time series of maps, it also had some added features.

For one thing, the goal of the study was to map a probable *future* geographic pattern of subdivision in the Twin Cities' metropolitan area. That goal demanded a historical series of data more consistent and detailed than the census. Computerized land records were still well in the future; hence, we had to devise a measure that could be obtained readily from both old and recent maps, and would be consistent through time.

From a large sample of mile-square sections in the land survey, we found that a count of public street and road intersections per square mile provided a virtually perfect indicator of the density of platted building lots and street mileage. Thus, we had a physical descriptor of the cultural landscape reminiscent of Finch and Trewartha's "cultural elements." We laid a grid of one-mile squares over the entire five-county study area, not unlike the process followed in the early works of Sauer and Finch. Then we added a simple index of relief and roughness in each square mile, and another index of water surface—an essential element in the Land of Ten Thousand Lakes. We found that when distributed on a graph, these measures in each of our several thousand mile-square sections fell naturally into classes of density, relief, and drainage. A resident or a developer looking at the land would probably put boundaries just where we were placing them. We added another descriptor for the quality of roads.

The resulting maps provided an unprecedented picture of the spatial growth of the Twin Cities from 1900 to the height of the post-World War II building boom in 1956. The next step was to extend the growth picture to 1980. We finally generated a map that accommodated the number of new persons in accepted gross population forecasts. The map also put all of the projected new people in places that developed logically from past decisions, terrain, and accessibility. The map showed unprecedented geographic detail. A quarter-century later it turned out to be approximately 80 percent accurate. Meanwhile, it had helped to plan major expansions of highways, parks, utilities, shopping facilities, schools, and subdivision locations.

Thus, the 1961 Twin Cities study went beyond the notion of a time series of synoptic maps unfolding from past to present. It attempted systematically to extend the series into the future. It also set an example locally for the use of fine grids and quantified descriptors of the landscape, looking ahead toward coming computerized geographic information systems. In that respect, it was part of the movement spearheaded by the "area transportation studies" stemming from the federal interstate highway program in the late 1950s. Our study also established the direction for two subsequent large-scale research projects: the Minnesota Lake Shore Development Study and the Minnesota Statewide Land Use Management Study—affectionately known to students and state legislators in those late years of the 1960s as the LSD and SLUM studies.

## Discovering the Need

In the preface to *Minnesota's Changing Geography* (1959), I asserted that the book's maps and narrative "reveal one of the most exciting facts which the human mind can discover—the fact that the varied landscapes all around us are parts of an orderly spatial pattern. That spatial pattern is the focus of the study of geography. And it is a fascinating, ever-changing composite expression of the combined works of men and nature." I also claimed that "Organized knowledge of the present is essential to give relevance to the historical past. Knowledge of the pattern of land and settlement provides the concrete framework upon which to build more abstract knowledge of human society. Knowledge of today's changing patterns provides the foundations from which plans for tomorrow must grow."

In subsequent decades of use of the book by hundreds of teachers and in the face of frequent restatements in classes and workshops, no one has ever denied those convictions. If they are true, there is little doubt about the importance of geography in liberal education, formal and informal, at every level. I am convinced that liberal education is by far society's most important need for geography. Other needs and the academy's ability to furnish students to meet them are important but subordinate.

Like most geography departments, we had several lower-division courses at Minnesota that gave us the opportunity to introduce large numbers of students to the field and the discipline. When I began teaching at Minnesota in 1949, I inherited, by chance, one of those courses—a long-established though weakly attended course on the geography of Minnesota. I assumed, somewhat naïvely, that students would come into the course with the common attitude that they would be expecting an unrewarding but easy three credits. I wanted to show clearly that they could gain new insights about their own territory, or about any other, by studying it as geographers—to show that though the place was familiar, the discipline was new to them, and that, as a result, they might not only enlarge their substantive knowledge and understanding but also learn useful skills and concepts.

I decided to organize the content of the course around major problems of public policy. It seemed best to select problems that not only have a major geographic dimension but also are important and persistent. The procedure was to state each problem in general terms, sort out the major dependent variables and study their geographic distribution, and then ask what principal independent variables account for the distribution. We then compared the resulting series of maps, attempted to explain the problem, and showed what variables would have to be changed in what ways to resolve that problem.

The problems, themselves, were not peculiarly geographical, nor were the answers. But the approach to them was. It used the vocabulary of regional patterns, place knowledge, and map legends, as well as concepts—location, scale, circulation, nodality—that are hallmarks of geography. It showed that geography is a way to clarify an issue, analyze a problem, and propose a solution. I selected five broad, interrelated issues. They were vital in Minnesota at that time and likely to be around for some years—the farm problem, industrial growth, metropolitan organization, the future of small towns, and the outlook for the "depressed" Northeast.

Rather than eliminate the need for traditional material, the new course framework demanded of the students more rigorous descriptions of the location and form of such features as moraines and summer drought, and it gave to their understanding an obvious urgency. It also demanded many new maps of cultural and economic features that had not been needed in the traditional approach and had never been prepared because the questions had not been asked. Students helped to do the research. The material turned out to be so timely and informative that the "instruction" soon spread far beyond the classroom to podiums, panels, and editorial pages. And there was no doubt that it was geography. People had to discuss the ideas from maps, compare and analyze patterns and locations. They had to know what was there, and they came out with place-specific statements about issues.

It's not stretching history to say that much of my direction for a great deal of the next fifty years' work tumbled from the experience of teaching the Minnesota course. In the first place, the course reached several thousand students in subsequent years. Moreover, spurred by that experience, I completely reshuffled the large, introductory physical geography course offered by the department. We reordered traditional material, using as the text a world atlas and specially prepared supplementary maps; added information on population and technology; and organized student work around the creation of thematic world maps and accompanying essays. Class means on the final examination rose fifteen percentage points over previous performances. Ten thousand undergraduate students took the course over a dozen years. With an equal number in Jan Broek's introductory human geography course, we had a strong underpinning for the undergraduate major and graduate programs.

The need for material for the Minnesota course motivated the first atlas of the state, which Neil Salisbury and I produced when he was a senior undergraduate major in the early 1950s. The first edition emphasized the state's agricultural geography and helped to spawn a lasting relationship with the agricultural extension service. That relationship, in turn, led to research funded jointly by agricultural economics and geography in the late 1950s. Faculty and graduate students investigated freeway impacts on land use and land value. The geographical studies used time series of maps to separate freeway influences from independent, long-term trends in both rural and urban settings, and thus to show the complexity of many changes that had been assumed to be caused simply by freeway building. Meanwhile, subsequent issues of the atlas in the 1960s, 1970s, and 1980s engaged an extended family of graduate students who went on to positions of leadership in public agencies.

Another set of new materials for the course generated a year-long series of articles for school classroom use in the Minnesota educational journal in 1954. Again, that research used time series of maps not only to show the geographic evolution of the state's surprising array of manufacturing, but also to suggest the overwhelming role of local entrepreneurship in explaining it. The series stirred enough interest to lead to publication as a separate booklet. Besides its original purpose, this research encouraged changes in state industrial development policy and alerted the governor, other officials, and local business to the importance of geography.

Development of material for the metropolitan unit in the course led eventually to the project "The Twin Cities Urbanized Area: Past, Present, Future." Students who worked on that project fanned out to positions in the metropolitan area's pioneer urban-planning enterprise. Work with both public and private planning opened the opportunity to organize an urban research program under the Upper Midwest Economic Study in the early 1960s. Now the subject was the changing geography of towns and cities all across the Upper Midwest. The ostensible goal was to encourage more urban planning in the changing economy. But our studies resulted in a much deeper understanding of the irreversible geographic trends the auto era had brought to every part of the settlement system. Moreover, my experience with the Upper Midwest study led to the development of a new year-long course sequence on the geography of American cities. There, it was my good fortune to intersect the emerging careers of quite a few dozen of the most enthusiastic, talented students I would ever know.

Also back in the late 1950s, the visibility of the atlas and industry studies led to an opportunity for me to work with state legislators on a new program that would respond to the federal Outdoor Recreational Resources Act. In the Minnesota setting, of course, attention went directly to lakes and forests— money for fisheries, for public access, for tourism, for control of polluted runoff, for exchange of public and private forest lands, and so on. We badly needed centerpiece studies of the basic geography of those topics, and by the mid-1960s geographers were involved with virtually all of these issues.

The most urgent need was for a study of the state's thousands of recreational lakes—their physical properties and status, and trends in their development. That project soon consumed us. We brought together data from sources scattered through state agencies and local courthouses, and supplemented them with survey research. And we joined all the data in a grid of forty-acre cells in the basic land survey, covering twelve thousand miles of inland lake shore. The study had many applications to public policy, lakeshore property development, and the recreational business. It also provided a context for contemporaneous research in the basic sciences, which was necessarily localized.

Widespread interest led to the expansion of the lakeshore study to a statewide land inventory covering more than a million forty-acre cells. By 1972, the project had produced a land-use map of the entire state and files that became the basis for the state planning agency's pioneering land-management information system. This big colored map might well have been the first such computer-generated civilian work in the country. In any case, it soon hung in hundreds of state and local offices and libraries, and it certainly raised many aspects of geographic awareness to a new level.

Four more applied studies from the 1970s and 1980s come especially to my mind as I remember my career. They dealt with such seemingly disparate topics as higher education enrollments, historic preservation of buildings, origins and destinations of redistributed tax revenue, and the market value of land and buildings. All had in common certain traditional features of work I had done over the years. They included fieldwork and analyzed time series of maps. They focused on features of the settlement pattern. They brought geographic detail on topics that were otherwise being dealt with only in generalities that had much more limited value in policymaking. But, also, all these studies were collaborative efforts; all depended on managing large amounts of computerized data; and all were perhaps more narrowly specialized than most of my earlier work. Through the 1980s, more of my projects tended to be retrospective "overviews."

These projects more and more reflected my aging, I suppose. And times were changing. In 1990–91, I had the opportunity to work in video and TV. Our team produced a ten-part course called "Minnesota on the Map." It was an updated combination of the time-honored undergraduate Minnesota course and the "hard copy" book on the Upper Midwest region. As of this writing, the broadcasts are long since lost in the ether, but I'm still grading papers from the interesting students who use the videotapes in the "distance learning" course (I still call it "correspondence").

### Discovering the Chores

By the late 1960s, I was growing acutely aware of the institutional obligations that go along with the opportunity to teach and to study. I had already served as department chair, filled in briefly as associate dean and assistant to a vice president of the university, and served on the Association of American Geographers (AAG) Council and on miscellaneous National Research Council (NRC) committees. But at this time there was a full-time center directorship and chairmanship of the NRC's Earth Sciences Division. Then the presidency of the AAG came along at the moment of controversy over moving the meetings from Chicago, in the midst of the Vietnam years, and over needs to raise money to support the national office. And there was more to come for me in the 1980s in the stimulating affairs of the NRC and the International Geographical Union. To be sure, assignments like these are not merely obligations. They are opportunities to know personally many outstanding individuals. Often, too, they are honors one can only humbly accept from colleagues.

In retrospect, these excerpts from what I might call a career have been to a very large degree a succession of fortunate, really wonderful opportunities to observe human settlements, to study, teach, and learn, and sometimes to serve institutions on which I had depended. Each opportunity was peculiar to the place and to the time I was there. At many critical points, I depended on support from individual friends, colleagues, and students. By chance, I have mentioned only a few among the many whom I would dare not even try to list. The journey evolved. It's all explainable, but it was not predictable. Nor could I ever replicate it.

## Selected Works by John R. Borchert

"American Metropolitan Evolution." *Geographical Review* 57 (1967): 301–32.

*America's Northern Heartland.* Minneapolis: Univ. of Minnesota Press, 1987. 250 pp., maps, photographs.

*Atlas of Minnesota Resources and Settlement,* with Donald P. Yaeger. St. Paul: Minnesota State Planning Agency, 1969. 626 pp., map folders. Earlier editions, 1954, 1958; later edition, with Neil C. Gustafson, 1980.

"The Climate of the Central North American Grassland." *Annals of the Association of American Geographers* 40 (1950): 1–49.

"Instability in American Metropolitan Growth." *Geographical Review* 73 (April 1983): 124–46.

*Legacy of Minneapolis,* with David Gebhard, David Lanegram, and Judith Martin. Minneapolis: Voyageur Press, 1984. 195 pp., maps, photographs.

"Major Control Points in American Economic Geography." *Annals of the Association of American Geographers* 68 (June 1978): 214–32.

*Minnesota's Changing Geography,* Minneapolis: Univ. of Minnesota Press, 1959. 191 pp., maps, illustrations.

*Minnesota's Lakeshore,* with George W. Orning et al. Minneapolis: Univ. of Minnesota Department of Geography, 1970. 72 pp.

*Projected Urban Growth in the Upper Midwest,* with Russell B. Adams. Minneapolis: Upper Midwest Economic Study, Univ. of Minnesota, 1964. 34 pp.

*Public College Enrollment in Minnesota's Changing Population Pattern,* with Thomas Mortenson and Arnold Alanen. Minneapolis: Univ. of Minnesota Center for Urban and Regional Affairs, 1973. 95 pp., maps, tables.

*Real Property Value in the Heart of the Upper Midwest,* with William Casey. Minneapolis: Univ. of Minnesota Center for Urban and Regional Affairs, 1994. 41 pp.

*State of Minnesota: Land Use,* with George Orning et al. Minneapolis: Univ. of Minnesota Center for Urban and Regional Affairs and Minnesota State Planning Agency, 1972. Computerized nine-color land-use map, showing data for 1.3 million forty-acre parcels.

"The Surface Water Supply of American Municipalities." *Annals of the Association of American Geographers* 44 (1954): 15–32.

*Taxes and the Minnesota Community.* Minneapolis: Univ. of Minnesota Center for Urban and Regional Affairs, 1979. 34 pp.

*Trade Centers and Trade Areas of the Upper Midwest,* with Russell B. Adams. Minneapolis: Upper Midwest Economic Study, Univ. of Minnesota, 1963. 44 pp.

*Urban Dispersal in the Upper Midwest,* with Thomas Anding et al. Minneapolis: Upper Midwest Economic Study, Univ. of Minnesota, 1964. 24 pp.

*The Urbanization of the Upper Midwest, 1930–1960.* Minneapolis: Upper Midwest Economic Study, Univ. of Minnesota, 1963. 56 pp.

# Center, Periphery, and Back

## Karl W. Butzer

Courtesy of Karl W. Butzer

*Karl W. Butzer (b. 1934) was born in Germany but escaped with his parents at the age of three as a Catholic refugee, residing first in Britain and then in Canada. In 1954, he received his B.S. in mathematics from McGill University and his M.S. in meteorology and geography the following year. He earned his D.Sc. from the University of Bonn in 1957. He taught at the Universities of Wisconsin, Zurich, and Chicago before his appointment as Centennial Professor of Geography and Anthropology at the University of Texas. His research has opened up the fields of paleoclimatology, ecology, and archaeology, and his fourteen books and more than two hundred research articles document his original field-based studies. He is the recipient of many awards across several disciplines, including honors from the Association of American Geographers and medals from the Royal Geographical Society, the Society of American Archaeology, the Geologists' Association of London, the Geological Society of America, and the Archaeological Institute of America. In 1984, he was elected a fellow of the American Academy of Arts and Sciences and in 1996 a fellow of the National Academy of Sciences. His extensive fieldwork includes research in Egypt and Nubia, South Africa and Namibia, Spain and Mexico. He is currently the R. C. Dickson Centennial Professor of Liberal Arts at the University of Texas (Austin).*

## With Open Eyes

My memory bank begins before age three, when I ran away from a nunnery in Belgium to navigate to the church, a half-mile away, where I knew I would find my grandmother. This was in 1937, when my family emigrated illegally from Nazi Germany. But that goes back to another story, when Prussia united Germany in the 1870s. It promptly began a culture war against the Catholic minority that mobilized Catholics as a political party, working to remove discriminatory laws. My father's family were Catholic activists who saw themselves as Rheinlanders and Europeans as much as Germans, and they resented the centralizing policies of Prussia as well as its militaristic and nationalist ideology. During the early years of the Great Depression (1929–33), German Catholics recognized the parallels between Prussian and Nazi goals or ideals, and studies in electoral geography verify that few Catholics ever voted for the National Socialist Party. After the Nazis took power in 1933, with only 35 percent of the vote, both my parents were implacably opposed to the increasingly totalitarian state.

Working as an engineer for a large industrial company, Father refused to make the Nazi salute or march in the political parades, until he was warned to take part on May 1, 1937, "or else." Shortly before that point, my eight-year-old brother, reflecting attitudes at home, had loudly refused to go along with a Nazi ceremony in the schoolyard. Non-Jewish Germans could not emigrate or even visit abroad without special permission, but my parents decided to leave—illegally. In March 1937, Father was allowed to attend an engineering conference in the Netherlands with nothing but a briefcase, a clean shirt, and ten Marks. He cabled to two Jewish German colleagues then in London, whom he had urged to emigrate two years earlier. They wired back, offering him a job, the paperwork for immigration to England, and money for the trip. A few weeks later my mother crossed into Belgium, ostensibly on a shopping trip, but went on to London. My brother and I were smuggled across the border to stay at the nunnery until our parents had found a place to live in England.

We were fortunate that my father had a marketable profession and connections, so he could escape. Many Jewish Germans could not obtain work permits for England or the United States and were arrested by the Nazis in the

Netherlands or France in 1940, thereafter to be shipped to concentration camps. We had some wonderful years in England, but after Dunkirk in 1940 my father was picked up early one morning by plainclothes police and sent off to an "internment camp." A few weeks later there was another knock on the door, and this time my mother, brother, and I were similarly rounded up. We spent six weeks on the Isle of Man with a minimum of necessities before we were released because my father's boss had connections and insisted that father was vital for the war effort. Again we were lucky because most of the other internees, wives, and children separated from fathers spent the duration of the war in camps in Canada or Australia—even though the great majority were Jewish German refugees.

In January 1941, our family was translocated to Canada, where my awareness of and exposure to ethnic dissonance was magnified. Because I have written about this elsewhere,[1] I will simply suggest that emigration, dislocation, and repeated changes of social environment made me aware of the world outside at an unusually early age.

I enjoyed my first "geographical" experience during the summer I was turning four. Accompanied by one of my father's colleagues, I flew in an airplane from London to Brussels. From a window seat in one of those small and wonderful passenger planes of the 1930s, I first saw the world from above. The images are still vivid in my memory. All the way to the horizon there were fields, lines of trees along winding roads, and villages. All the shapes and colors were different, as light clouds drifted by below us. It was simply beautiful! I could actually recognize human figures while we were ascending, and after we reached cruising altitude, I could still see the cars slowly moving on the roads. But everything seemed tiny, as small as toys. That miniature world from above amazed me most. It was the first time that I was actually aware of how things became smaller and smaller at a distance.

A few years later, when my brother introduced me to an atlas for the first time, I had no trouble transferring those images of the world from above to the different scales of maps or grasping the abstract reality that they repre-

1. K. W. Butzer, "Coming Full Circle: Learning from the Experience of Emigration and Ethnic Prejudice," in *Light from the Ashes,* edited by P. Suedfeld, 361–98 (Ann Arbor: Univ. of Michigan Press, 2001).

sented. At age eight, I counted that atlas one of my favorite "books," and at nine I could identify every country that my brother pointed to on our globe. In retrospect, I understood at how early an age a child could learn to love the beautiful world in which we live, grasping spatial relationships and differences of scale, and shifting from the reality "outside" to its symbolic abstraction on a colored sheet of paper.

I tried it out when my granddaughter, Maddie, was three, showing her a road atlas of Texas and pointing out Dallas (where she lived), Austin (where her grandparents lived), and Houston (where her cousins lived). I had her trace the roads from one to another with her small index finger, telling her that those red lines were the roads that the car traveled down. A few months later she was back visiting, and I got out the same atlas, pointed out Dallas, and asked her how she had come down to see us. That same pudgy finger traced the red road down to Austin, and then she smiled shyly at me. Next, my five-year-old grandson, Sebastian. He had been playing a "world" board game with his older brother that involved moving around tiny tanks, planes, and ships from country to country. Although he couldn't read yet, he could show me where Texas, Africa, and Asia were, and where his other grandparents lived in Germany. His attitude was, it's no big deal. China? Here. Russia? There. Our current lack of geographical literacy is a problem of parents who barely know about other countries and who fail to stimulate their children's minds. All children should grow up with an atlas and with a parent or sibling who will show them how to use it.

But back to my putative interest in geography as a child. When I was eight, mother bought me an incredible book, Richard Halliburton's *Complete Book of Marvels*. Halliburton was an adventuring pilot in the early days of flying, and he flew to all kinds of faraway places, talking to people, photographing them, and writing about them. He always combined some history with a vivid description of the country, and his *Complete Book* was written for youngsters and printed in big type for beginning readers. It also had a special part, namely a chapter on each of the "seven wonders of the ancient world." I don't know how many times I reread those descriptions of romantic sites in Egypt, Mesopotamia, and Greece, a sort of introduction to ancient history. There were large but slightly grainy photos of pyramids, temples, and monuments, always linked by maps showing how a person could travel from one to the other. That same book is now displayed on the shelf of my younger son, an architect: he, too, grew up with it.

I was a late starter and still couldn't read near the end of the first grade. At that point, my teacher sat down with me after class every day for two weeks and patiently explained the "system" to me in a low-pressure situation. I finally grasped the code and soon became an avid reader, especially when I discovered that there were more interesting things to read than Jack and Jill primers. Halliburton greatly improved my reading speed and comprehension, in addition to illustrating the logical linkage of geography and history that stayed with me throughout my career.

In seventh grade, we spent the year studying the geography of Canada with the help of a prescribed book of that name. The descriptive text was a bit dull, with too much detail, but the older teacher was very good, and I learned how contour lines show topography. I also had a new atlas that included continental maps in three colors showing industrial zones, agricultural areas, and "nonproductive" land. That latter concept intrigued me, and with a pencil I superimposed the boundaries between those categories on a number of the political maps. It drove the point home that large parts of the continents were "nonproductive," and that the solid colors of a political map were misleading as to how much land there was that was either used or developed. By now, I was drawing maps of imaginary countries that had mountain ranges, hills, lakes, and deserts on them.

At this formative stage of my life, one other book was important. When I was thirteen, my mother picked up a remaindered copy of John Henry Breasted's *Ancient Times* for me. It was a 1913 high school text on the civilizations of the Near East, a quite rigorous study that was not dumbed down like many of today's schoolbooks. I had to push myself to get through it, but it introduced me to the Stone and Bronze Ages of prehistory, which were quite novel for me and made a lasting impression. Breasted also used the size and elaboration of burials or tombs to identify levels of civilization, my first introduction to social evolution. I remember being a bit puzzled, however, that burial goods and tomb chambers should be a measure of civilization.

But not all I learned was from books; my father taught me how to plant a vegetable garden. During World War II, the United States supplied less produce to Canada, and by 1943 people were encouraged to plant Victory Gardens. There was a large tract of unused land behind our back fence, and father began to grow vegetables. It plainly was something he took great pride in, and as I later understood, he had learned as a boy from his widowed mother, who grew flowers to sell at market for a living. Before that, her father and grandfather had been in charge of the flower gardens at the *Schloss* in Benrath since the 1830s. Father used a spade to dig up a ten-inch tilth that was then built up into raised beds, with a six-inch ditch around. On top of the bed, he made three-inch ridges in which to plant the seeds. Beans were planted three together at intervals of four inches. He weeded regularly and, when necessary, dusted for bugs with some deterrent compound. In the fall, when everything was harvested, he would turn the soil over loosely, dismantling the mounds. The results were spectacular, the plant rows coming up thick and strong, rainwater collecting in the ditches but not washing out the seed rows. Our Scottish Canadian neighbor was visibly covetous; he used a hoe instead of a spade, simply working up a few inches of soil, and his rows of beans were thin and had many gaps. Four years later he bought the lot we used and ejected us, presumably thinking there was some hidden charm under our garden, but the results of his efforts in our old plot were no better.

Father had always explained to me why he did things the way he did. Now he told me with a wry smile, "It was time to apply fertilizer." At about ages eight to eleven, I often worked with Father in the garden or went to faraway nurseries with him to find transplant stock for ethnic veggies such as cauliflower and red cabbage. It was our first bonding experience because we didn't engage in organized sports. Years later, when my interests drifted into cultural ecology, I recognized that I had little more to learn about ditched fields or the intensive horticulture of market gardening. My own children didn't have this privilege, but they did know their Opa's rose garden, which was his joy. But my older son, an attorney weekdays, loves to spend a Saturday working in the garden, with the help of his little children, who are also learning about the life that springs from the soil.

Another thing I learned about in the garden was soil texture and how it affects fertility. We had bought an old country house in the Laurentian Mountains in 1945, and my brother and I tried unsuccessfully to garden there. The tomatoes were still small and green when they froze in mid-September. But our farm friends, the St. Pierres, had a model vegetable garden. The difference was that they piled on manure from their barns. The explanation became intuitively clear to me as a builder of elaborate mud castles: in the Laurentians, I found that the soil was insufficiently cohesive to build a two-foot tower with battlements. The next spring I brought a bag of Montreal soil to mix with the Laurentian soil, and it worked. In my first year of graduate school, I would learn that Montreal Island was part of a late-glacial lake plain (with clay loam texture), whereas the Laurentians had a mantle of stoney glacial drift (a sandy loam), covering infertile granites.

As a child I had learned a variety of practical things about the world through direct experience, but it was still largely an imagined world, as if seen from a window. As an adolescent I became conscious of my love for the outdoors and of a strong interest to learn to know and understand the natural world in which I lived for three months each summer. The transition was gradual because I had frequently worked with the St. Pierre boys, cutting hay, milking cows, and forking manure. In fact, before I went to college, I had wanted to become a dairy farmer when I grew up. Back from the gravel road front and away from Echo Lake, the land still belonged to a half-dozen French Canadian farm families, getting caught up in the shift to service industries or moving to the city. The extensive hardwood forests were interrupted by occasional pastures, but I had found old apple trees smothering in the secondary growth. As I eventually surmised, they were testimony of abandoned farming ventures.

The incentive to begin exploring my environment came from the library. At fourteen, I had become addicted to reading Westerns when I was in Montreal, and I soon focused on the novels of Zane Grey. Apart from the need for better character development and historical context, his novels gave evocative and remarkably "geographical" descriptions of southwestern landscapes. In later years, I would become familiar with the mesquite brush beyond the Nueces River, the pine forests below the Mogollon Rim, and the purple sage of the mesas of southern Utah, and these landscapes turned out to be exactly like I remembered from Zane Grey's descriptions. During the summer of 1948, my self-confidence bolstered by my wire-hair terrier, a .22-caliber rifle, and a Stetson hat, I was out in the "bush" most of the time. An ornithologist had told me about the mysterious Mud Lake, and I eventually found it, locked in between some low hills and a perimeter of dense forest. It was more like a large swamp, with a wide ring of aquatic plants, while the water surface was almost covered with lily pads and pond scum. But to me the lake seemed a big achievement, and it would be my first nonoutlet lake of many, although the others would be amid open and much drier environments.

I therefore attribute my avocation for field geography and my sensitivity to the nuances of the biophysical environment not to some academic icon, but to roughly fifty westerns by Zane Grey. Indeed, even before I finally found my Mud Lake, I had written a short, landscape poem about a lovely spot that I called "Seven Springs"; I submitted it for an English assignment, and the teacher sent it in to a poetry contest—it took second prize. Not surprisingly, I have never accepted the stereotypical put-down that physical geographers are less humane. It's just that we see the horizon as well as the foreground.

During winter months, before I started shoveling the snow off the front lawn in April, I would read. The greatest library discovery of my teen years was the twenty-volume family encyclopedia of 1894 we had at home, but it was in German and, to make matters worse, in that impossible Gothic typeface, so I couldn't use it. But I cajoled my father to read to me from it. The first article I selected was on Oman, a country I had not been able to find any information on. Each item that followed was the description and history of a place, but with a time-warp of more than fifty years. Without realizing it, I was becoming fascinated with an unusual kind of historical geography, namely the nineteenth-century Near East and Africa. Father's patience fizzled out after a few weeks, so that I had to learn to read the encyclopedia myself, first with his help, then on my own. That task proved to be quite important down the road because I began to master German, perfecting my reading skills over the next few years. Mother now gave me two incredible atlases, dating to just before the First World War, that had belonged to her favorite brother, Karl Hansen, who had fallen at Verdun in 1916. It was a symbolic act because I had been named after him. One atlas was topographic and economic, the other historical, and I used both intensively until I bought new editions in Germany a decade later. It is fair to say that those historical maps, still imprinted in my memory, significantly influenced my later interests and breadth of historical comprehension.

During the summer of 1953, I spent months reading the abridged edition of Arnold Toynbee's six-volume *Study of History*. It was my first exposure to

macrohistory, but presented and interpreted in a normative fashion. Because I
was in a hard-science track at college, I was intrigued by the notion of patterns, such as the rise and fall of "civilizations," which I would later revisit in
the framework of systems theory.

That fall I had to choose an elective for my last undergraduate year at
McGill University. The chairman of the mathematics department suggested I
try the two-semester course on human geography. I balked a little: What was
that? But within weeks I knew that *geography* was what I had been devoting
most of my spare time to for the previous twelve years. Moreover, it was an academic discipline with distinguished origins. That course proved a joy, as
taught by two young and enthusiastic geographers. Theo Hills was a New
Zealander, working on a dissertation in South America, and his tropical experience was just what I was looking for. My term paper was on development
in the African savannas, an environment I so much wanted to see for myself.
The other instructor, Hugh Thompson, had just obtained his doctorate at
Oxford on subarctic glaciers. Although I didn't much care for the Arctic, he
impressed the systematics of physical geography on me in a way that I could
really appreciate.

After two months I told my parents that I wanted to go on to graduate
school in geography. They were appalled, as if I had suggested I wanted to become a musician. How could I earn a living that way? Mother never quite believed that geography was a respectable field, but in the end both my parents
were persuaded. If I completed my honors degree in mathematics, I could
switch to McGill's master's program in meteorology and geography during
the following year. I lived up to the bargain and was accepted—thrilled at the
prospects as much as I was relieved to leave mathematics behind. But I could
not so easily discard the logical structures and the abstract thinking inherent
to non-Euclidean geometry, which had appealed to me.

### The Best of All Possible Worlds

Yet another fortuitous event took me to Europe with no advance notice.
Shortly after my last exams in the spring of 1954, we were visited by a doctor
from a German ship that had just docked in Montreal. There was some brain-
storming at home, and forty-eight hours later I was on a small freighter bound

for Germany, wondering how I had ever gotten myself into this situation. Any doubts I had quickly evaporated when I climbed up on deck next morning. We were sailing along the coast of the Gaspé Peninsula, a string of lonely villages in view, on an almost treeless landscape. Church bells calling to Sunday mass were the only signs of life from our distance. I was exhilarated.

The next two days we inched through a fog bank to the mournful tunes of our foghorn and then rode out a three-day Atlantic gale. I stumbled around on the decks, photographing waves breaking across the bow, enjoying every bit of it. The officers and crew of the *Franziska Hendrik Visser* were all German Frisians and very laid back. Because there were only two other passengers, they allowed me every reasonable freedom and let me spend hours in the control room. I was shown how the meteorological equipment worked and was logged, and how diurnal positions were measured and drawn in on the navigation charts. By the end of the trip, I could estimate Beaufort wind speeds from wave development to the satisfaction of the second officer, who became a good friend. I also copied the weather log for our voyage and later plotted those data on graphs. The experience proved to be a most useful workshop on weather recording as well as an opportunity to improve my conversational German before landing in Emden.

Traveling by train south to Düsseldorf in a prewar carriage with wooden benches gave me a small cross-section of changing landscapes outside and of passengers inside. At first there was picture-postcard scenery looking stereotypically Dutch, with quiet but friendly people. By the time we approached the Ruhr, the train had filled up with a more varied crowd that dressed, talked, and interacted differently. In a distance of three hundred kilometers, equivalent to that from Houston to Dallas, there was more cultural difference than between Texas and Minnesota. This was Europe.

I was puzzled by unfamiliar items of vocabulary. My parents used *billet* for a train or bus ticket, one of many French loan words current in 1930. A ticket was now called a *fahrausweis* (a neologism equivalent to "travel identification card"), however. Here was top-down language change in the style of the French Academy, but, instead of Americanisms, the Nazis had eliminated French loan words. My father always spoke standard German, but his family in Benrath (on the southern edge of Düsseldorf), where I stayed for three weeks, used only Rhenish dialect, which is close to Plattdeutsch or Low German. It's a slow, rhythmic, and comfortable way of speaking, analogous to the contrast between southern and standard American English. But German dialects don't just differ in pitch and pronunciation; they also vary somewhat in grammar and vocabulary. At the scale of major dialect groups, Rhenish and Bavarian are as different as Dutch and standard German, or Russian and Serbo-Croatian. In picking up a working knowledge of Rhenish, I was beginning to understand the use of language and its nuances in social negotiation.

That summer I also spent time getting to know my mother's family in Aachen, some of whose suburban communities lay beyond the border in Belgium or Holland. My cousins crossed the borders regularly on shopping trips, and I began to grasp the rituals of ethnic behavior that come with such regular interaction. In fact, Aachen, a very old imperial city and the residence of Charlemagne's court in A.D. 800, was very much a border town with a transnational outlook. Even today, busloads of Dutch and Belgian visitors daily swarm inside Charlemagne's cathedral, which they see as part of a shared heritage. The national boundaries of much of Europe are more a matter of historical accident than they are delimitations of difference, particularly where they are drawn in the countryside. But each city has an intangible but distinctive ambiance, grounded primarily in the temporal and economic context within which it had once commanded wealth. That distinctiveness became intelligible to me only little by little, over the years, but I remember how my curiosity was first piqued during that incredible summer in Europe.

Next on the agenda was a trip to Spain. The most unforgettable moment was waking up in the first light of dawn as the train rumbled across the *meseta* of Old Castile. Wheat fields stretched flatly from horizon to horizon, glowing pink in the rising sun. Then I saw the silhouette of the great wall of Avila on the horizon. My brother, a cousin, and I got out there, spent hours walking around that huge wall, and then went inside, through the maze of narrow streets with their small plazas and medieval church facades. All the stone was colored buff and set off against a crystalline sky with a stark, understated beauty. It was my first image of an arid environment, one that would capture my lifelong professional interest. On the rickety local train from Avila to Madrid, we sat in third class with a crowd of chatting, sociable Spaniards, being made to feel right at home even though we didn't speak a word of the language. The warmth of those simple country people activated the powerful, historical landscape that drew me back to Spain time and again across some forty years.

Italy, which I now visited with a tour group, was somehow less dramatic. But it was here that I first saw Mediterranean orchards and began to take an interest in church architecture, conscious of my total ignorance in the subject. The Germans in my group were a delightfully wacky bunch, with a good sense of camaraderie—my first taste of living and working with research teams much farther afield and under more trying conditions, both complicated by a higher ratio of thorny personalities. Field-camp behavior quickly diverges from that at a university, sometimes for the better, often for the worse.

These formative impressions of that halcyon summer in Europe were rounded out by a "vacation" with the family of one of my uncles in the Alps. I insert qualifying quotation marks because my uncle was either hiking or negotiating mountain passes in a minibus every single day. I also learned a great deal about haymaking and the pungent application of liquid manure. It was an incredible introduction to southern Germany, Austria, and high mountain landscapes. When I returned to Canada to begin the graduate program at McGill, I had sampled much of the agenda of geography with open eyes, but still in the dark as to how one moved from travel experience to professional observation and understanding.

My master's advisor at McGill was Ken Hare, the departmental chair. He was British and had worked with the Meteorological Office during the war. A fine process climatologist, he had a broad interest in climatic change and in the interplay between climate, vegetation, and soils in boreal and subarctic Quebec. He inspired by emphasizing the importance of the question and the quest, and avoided simple answers. As a supervisor, he stimulated and facilitated rather than dictated. For my proclivities and needs, he proved the ideal choice, and his relaxed style impressed me. I owe my interest in environmental change to him above all others.

The course work was preceded by a week-long field trip to the glaciated landscape of southern Ontario, led by Brian Bird, an arctic geomorphologist. The trip was fast-paced and rigorous, an almost overwhelming eye-opener for a novice. Also aboard was a young Swiss glaciologist, Fritz Müller, whose company I enjoyed. Twenty-five years later he invited me to join him at his institute in Zurich.

After I tried out some sterile courses in botany, my revised curriculum for the year included three key pieces. The first was Hare's inspirational seminar on arctic ecology and Quaternary paleoclimatology. One of my two papers examined the work of Carl Troll and his students on ecozonation in the Andes, including the issue of ecological convergence of unrelated biota above the treeline in various tropical mountain environments. The second piece was a brilliant lecture course on modern European history by an English historian, Noel Fieldhouse, for which I did a paper on the ethnicities of the Austro-Hungarian Empire. The third piece was a self-directed reading course with Hare, which began with Ellsworth Huntington's books on climatic change and "climate and civilization," the latter a disappointment compared with Toynbee's sophisticated, nondeterministic approach. From there, I sampled the putative literature about climate and disease on human physiology before finding the trajectory for my master's thesis. I decided to explore the literature on historical climatic variation in the Near East in relation to nomadic migrations. Visiting people in other departments, I encountered my first polemic ideologue, an anti-Marxist anthropologist who assumed I was a Marxist because I had read Gordon Childe. My meetings with the orientalist Bible scholar on campus were more congenial, and he became my second reader.

I would be completing the thesis during the summer, so I was looking into options for the Ph.D. The McGill department had an obligatory Friday afternoon colloquium that almost without exception included speakers on arctic themes such as calving glaciers and permafrost coring. After six weeks of such a cold diet, it was clear to me that I had no interest in an arctic dissertation. An assistantship at McGill would have required me to spend months at the permafrost laboratory atop some self-heaving sphagnum moss in Ungava. Faced with that prospect, I instead wrote to Carl Sauer at Berkeley and explained what I'd like to do; he replied promptly and encouraged me to apply for a teaching assistantship. But then a Sri Lankan classmate told me about German government scholarships, which required an invitation from a German university. I wrote to Carl Troll, an obvious choice, and my letter elicited an enthusiastic, two-page, hand-written response. After my application for a Canadian government fellowship was denied—almost by return mail—I interviewed with the German consulate. Two weeks later the Germans offered me an exchange scholarship. The choice was clear: Troll, Bonn, Germany.

In September 1955, after a somewhat breathless year, I was aboard ship for Le Havre, France. On the train ride from Paris through Belgium, I became engaged in a conversation with a young passenger who I thought was Belgian because he spoke fluent French, without an accent. After half an hour of exercising my Canadian French, it turned out that he was a German and that we could actually converse with less effort. That would be a preview of the vibrantly international flavor I found at the University of Bonn. The Geographical Institute was housed in a renovated eighteenth-century palace, and the three spacious rooms for doctoral students had eighteen desks; of these at least five were assigned to foreigners like myself—a Spaniard, a Japanese, an Iranian woman, the Sri Lankan from McGill, who had also received an exchange fellowship, and a scholarly Englishman who spoke in such a strong regional accent that we had to communicate in German. At the university cafeteria, several dozen Norwegian students always sat in a closed group, while a hundred or more Iranians fluttered around any unattached women. At get-togethers organized by the Academic Exchange Service, one could also meet a variety of Commonwealth students.

The doctoral students, some of whom had served in the military, were enthusiasts for a united Europe without borders. The foreign students were all given warm welcomes and genuine support. Up to two hundred students from all over the world faithfully attended several long slide shows on Japanese rural landscapes by our Japanese friend. Six to nine hundred citizens of Bonn could be expected to attend any open lecture on a foreign country. African Americans, who were unable to sing in U.S. opera houses, starred in the town opera, and were enthusiastically applauded. Whatever unreconstructed thoughts lurked somewhere among an older generation, those were heady times not only to be in Germany but anywhere in continental Europe. The era of obsessive consumerism and guest workers had not yet dawned, and it seemed as if everyone had just come up for pure oxygen after the years of repression, war, and occupation. All this stood in contrast with the grim parochialism of Canadian society at the time.

My first course with Troll was on regional Germany, a subject in which I had little interest in principle, but I was impressed by the way he integrated physical and human geography. Most days, before class, he drew a detailed chalk map of the geomorphology of the area he would discuss. Later he would talk about the settlement history and craft or industrial change in that context. The full power of his seamless integration of physical and cultural geography became apparent on the related field trips. The second course covered regional South America, but Troll never got beyond the Andes, and it became a systematic exposure to tropical mountain ecology, a subfield that he had pioneered and dominated for decades. From the autobiographical bits he occasionally threw in, I began to appreciate his grasp of Andean environments, a knowledge based on many years of working up that mountain chain from Chile to Colombia, mainly on mule back and sometimes on foot, climbing over Pleistocene moraines or contemporary glaciers.

Troll always carried thermographs in the Andes to record various temperature changes at the ground and at the regulation level of 1.5 meters. Against that microclimatic data, he explored changing associations of vegetation and small-scale geomorphic processes, such as needle-ice displacement of soil particles. Over the course of his career, he used this approach from Spitsbergen to Antarctica, emphasizing the importance of diurnal versus seasonal temperature regimes as sensitive diagnostics of both zonal and vertical environmental types. I followed up this classroom exposure by carefully studying his many papers on periglacial processes and geomorphology, a subject that did not enter English-language texts until a generation later, after several of Troll's major reviews had been translated into English by U.S. government agencies. Eventually I was able to develop a corresponding field familiarity that was quite helpful when "born again" periglacial enthusiasts were wont to make spurious claims for Pleistocene periglacial activity in unlikely subtropical mountain settings, but in the 1950s I declined to take up this topic for the Ph.D.

The reason that Troll did his fieldwork in South America was that it was the only foreign venue where German scientists were welcome to work in 1925—German geographers were not even allowed to attend International Geographic Union meetings at the time. Troll always remained grateful for this opportunity, and he respected and was genuinely fond of South Americans. I remember when the Paraguayan embassy invited him over to award him with some medal of distinction. He was quite excited by the prospect, taking the honor at face value and enjoying the opportunity to expand his network of cordial relationships, even though he already had a drawerful of Latin American medals. His attitude contrasted with the cynicism of former intelligence service geographers who became engaged in the U.S. study of Latin America during the 1940s. Troll's field-oriented students learned from him that one doesn't patronize people of good will from less-developed countries. By extension, he also didn't question the quality of my Canadian training, however green I actually judged myself to be. He supported and encouraged me to concentrate on my dissertation without any preliminary hurdles.

During the semester break, I went to Egypt in the winter of 1956, traveling with a German student group as a matter or organizational convenience and minimal expense. After docking in Alexandria with a Turkish steamer, we ate Arab cuisine the first night. The next morning I got out of the train at El Alamein and set off on foot for the coast, walking through the not-yet deactivated mine fields with a roll of toilet paper in hand, following the narrow trails of local goatherds. Later on, in 1960, I wrote a paper on the area that continues to be cited because I first recognized a major fault zone interrupting the Pleistocene beach ridges at the western edge of the Nile Delta. While the other students were visiting mosques in Cairo, I was walking up and down dry desert valleys or wadis, observing arid-zone geomorphology that looked rather different than that described in the textbooks I had read.

I also had an opportunity to fly to Jerusalem via Jordan, and the Egyptian pilot let me sit in the cockpit to observe the panorama of desert landscapes over the Sinai and Dead Sea, which I was able to photograph from the air through the open window of the DC-3. In the Jordan Valley, traveling in an old taxi, I visited the high lake beds of the Dead Sea described in the geological literature. By now I was hooked on arid lands. Still later, after a train wreck, we, the student group, reached Luxor in Upper Egypt. There I walked all the way up the Valley of the Kings, examining alluvial sediments for the first time and finding Paleolithic artifacts in place within them. Afterward I climbed across the mountains above the tomb shafts, stumbling on the mass of talus debris and learning to appreciate the role of mechanical weathering in such an environment. I paid the price for my explorations with a slight case of sunstroke, but after a day in bed I was out again. The beauty of the total desert in Egypt, with its luminous limestone scenery and its deep blue sky, fascinated me. It was a strong stimulus to move on with the dissertation.

A year later, a paper that I had developed from my master's thesis but that

was informed by the Egyptian exploration was accepted for publication—my
first publication—in the same week that I defended my dissertation. I felt I
still had a great deal to learn, however, and decided to stay on at the University of Bonn for a while, taking courses that I knew I needed, especially in geology, and auditing others in history that I was interested in for less-tangible
reasons. In 1958, Troll arranged a meeting for me with an Egyptologist who
wanted to carry out a survey in the Nile Valley for late prehistoric sites. Just
as Troll had found support for me as a research associate of the German
Academy of Sciences and Literature, he now now steered me to a grant for
fieldwork.

The idea in Egypt was that prehistoric sites had been reported from the
desert edge of the floodplain in some places but not in others. The Egyptologist and I studied the sites that were known in order to understand in what
geoarchaeological contexts they were found and preserved. With that in
mind, we walked some five hundred kilometers up and down the desert edges
of Middle Egypt, looking for new sites or developing an explanatory rationale
as to why they were absent, which was usually the case. My associate was high
strung and cantankerous, but he was genuinely interested, which helped in a
strenuous undertaking, with poor local food and dirty hotels, sometimes
without a lock on the door. Midway through the project, I moved out for two
nights to an American mission in Asyut for a chance to take a bath and catch
the fleas that insisted on traveling with me. There I heard the wavering strains
of Puccini on short-wave radio, and my eyes grew moist. I can therefore relate
to the soldiers listening to the military radio station in Belgrade signing off
with "Lili Marlene" during World War II. Something you otherwise take for
granted can become your last sane link to reality.

We came within a hair of being lynched at an isolated town called Dalga,
west of Mallawi. After we were run down by an armed horseman, a growing
hostile crowd dragged us to the cemetery, stood us up against a wall and
picked up stones to throw at us. They desisted when we called over and over
again for the *omdeh,* usually a respected civil and religious leader in rural
towns. Instead they brought us to a small military post where the soldiers
couldn't read our Arabic identification papers, but insisted on beating us up a
bit before taking us in to the commandant. He could read and was upset, if
only because several hundred people were shouting for our blood outside. We

eventually left under military escort. His explanation was that the townspeople thought we were British spies, a reasonable notion given that the 1957 war was still fresh in people's minds—British fighter-bombers had destroyed the entire Egyptian air force on the ground.

A year later, when I told a British geologist at Oxford about this adventure, I received a very different reading. In 1919, a British troop train with wounded soldiers had been boarded by rebelling Egyptians, fifteen kilometers away from Dalga, resulting in somewhat of a massacre. Dalga had been the scene of one of several British reprisals, and a decade later the Oxford geologist had been attacked here and was forced to "use" his shotgun. Middle Egypt has been a rough place. Some rival villages were still settling blood feuds with AK-47s in the 1970s.

By comparison, Spain was just this side of heaven. I had been there on various long excursions during an international congress in 1957. In 1959, I returned, this time on my honeymoon. The upbeat schoolteacher from Bonn whom I married knew me well enough to expect that it would become a working vacation. Elisabeth and I both loved Majorca and its deeply incised marine inlets, and I started a project on fossil beaches, dunes, and paleosols that continued until 1962. My bride wrote down the numbers derived from gravel-rounding measurements and giggled when I tried to use olive oil to avoid a sunburn. We also established what became lifelong friendships with two Majorcan scientists, one of whom later invited my wife and me to start our Spanish village study in 1979.

In 1959, Troll showed me a letter from the University of Wisconsin asking him if he knew a good physical geographer to recommend for a new position in Madison. Was I interested? Of course I was. Without an interview, I was hired, presumably because I had already published ten papers and my dissertation, "Quaternary Stratigraphy and Climate in the Near East." Little did I suspect that my years at the University of Wisconsin would be the most tumultuous of my career.

### "On Wisconsin," or Whither Wisconsin?

At McGill, critical discussion had been something that graduate students could engage in even if they argued with a faculty member, but at a German conference in 1958 I found out that such argument wouldn't go over as well when I contested a claim by a senior geomorphologist with a counterexample from Canada. He promptly wrote a letter complaining of my behavior to Troll, who was genuinely embarrassed because it wasn't the first such letter he had received. A year earlier I had made some offhand comment at supper to some German professors on one of the Spanish excursions about the unattractive lodgings we were in. That comment prompted a letter to Troll noting that if the accommodations were good enough for German professors, they should be good enough for me. It had become apparent that I didn't want to stay at a German university permanently because I was temperamentally disposed to call a spade a spade. I also was under the illusion that Wisconsin would be much like McGill, only bigger. Wisconsin, as a state, and Madison, as a city, were friendly and indeed welcoming to newcomers, but the Department of Geography at the time was smug and authoritarian.

One problem was that I was only twenty-five, younger than most of the graduate students, and the next youngest faculty member, Fred Simoons, was thirty-seven. Most of the senior faculty had worked for the intelligence services during the war and had known each other well for at least fifteen years. They formed a good old boys' club, with a strong military flavor and redolent of rank. The chairman, Andrew Clark, was a Canadian. At the institute in Bonn, there had been an explicit hierarchy that comfortably structured formal relationships, but did not inhibit periodic fraternization at Carnival parties in the institute or all-night story-telling at Troll's home, with the participation of a number of his nine grown children. His authority was based more on his charisma and on the affection of students and staff than on the institution. Not once did I see him act overbearing or wield his authority to patronize a subordinate; his critical mode was paternal and constructive, never ad hominem. The Madison department worked the other way around, with a fictive egalitarianism and all decisions made in private by only three to five power brokers, despite obligatory weekly faculty meetings that ran all of each Tuesday afternoon. Subordinate full professors either remained inconspicuous or were sarcastically put down. The graduate students were nervous and uptight, and displayed no spontaneous adulation for faculty.

The worst human traits appeared each winter when everybody had to "vote" on everybody else's salary increase. For weeks before the vote, faculty would go around lobbying from office to office, bemoaning how hard up they were. Salary-increase increments were $100 for assistant professors, $200 for associate professors, and $300 for full professors, and one could recommend anywhere from zero to three increments. The year I arrived I had a salary of $6,250, whereas the four senior professors were paid three or more times that amount. The inequities were self-perpetuating because, throughout my seven years at Wisconsin, Andrew Clark and Arthur Robinson constituted the salary committee and thus set their own salary increases, which moved in tandem by three increments each year. Just how fictitious the process was became clear the year a new instructor turned in his recommendations to Clark and was forced to explain and argue each case, with encouragement to criticize his colleagues. When the ordeal was over, we were told that the salary committee didn't count the votes of untenured faculty anyway, but it did want to know how that faculty voted. That process strengthened the position of the ruling oligarchy by making advancement conditional on "loyalty" and by dividing the lower-status faculty, with the unfortunate result that it took months before the department regained some sort of equilibrium.

By refusing to let myself be bullied, I had opted for a painfully slow advancement. My first confrontation with Andrew Clark came within a month of arrival. He had written a letter to all faculty for ideas or recommendations for a possible new position in Soviet geography. After again being asked directly, I wrote a note mentioning a Czech colleague I knew from McGill and who had a doctorate from Prague. Clark thereupon sent my Czech friend a condescending letter, questioning his qualifications and even the validity of his Ph.D., but concluding that he could apply if he wanted. I protested this cruel and unnecessary letter because if we didn't think him good enough, we didn't need to write him to tell him so. Clark exploded. My voice was just as loud when I responded.

Three months later I received an invitation to give some talks at Minnesota, where several geography faculty pressed me relentlessly to say something negative about Wisconsin. Inexperienced as I was at such games, I eventually did make some cautious but limited negative comments. The telephone was faster than my return flight, and I was accused of "running down the department."

At Bonn, there was a free market for ideas and approaches to geography, and I never sensed a prescribed way of thinking about anything. Not so at

Wisconsin, where I was told, reminded, and eventually warned that my class presentation of physical geography was to be descriptive rather than processual. I had to sit in on Clark's and Simoons's presentations of the elementary physical (bread-and-butter) course, as well as Glenn Trewartha's climatology courses, so as to see how they were done. I also *had* to use the departmental textbooks. I wasn't necessarily being difficult about these points, and I was still experimenting with teaching methods and had no fixed ideas, but the departmental ideology was gratuitously expounded to me as if I were some sort of subversive force. I was first allowed to teach an advanced climatology course in 1962, but I didn't use Trewartha's books. It was 1965 before I could teach an advanced course in "landforms" (*geomorphology* being a taboo word), after Edwin Hammond left the department.

It bears noting that all the established Wisconsin faculty either explicitly or implicitly adhered to the maxims of Richard Hartshorne's *Perspectives on the Nature of Geography* (1959), even though Hartshorne himself never questioned whether what I did was "geography."[2] That spirit of a bounded geography, retreating behind disciplinary fences to focus on global categorization and regional description, had deep roots at the state universities of the midcontinent, but it was disseminated and championed in the postwar years mainly by the geographers who had once worked in the Office of Strategic Services (OSS, the forerunner of the CIA).

Although I was told to my face that I could either teach well or do good research, but not both, my response to the challenge was to try to do both. I wrote four new papers, which were accepted for publication in the first nine months I was at Wisconsin, but I received no salary increase, although the dean gave me the semester research leave originally promised. I spent that time working in Majorca and later doing the backup soil-and-sediment laboratory work in Bonn under the initial guidance of one of my old graduate classmates. At about this time, I learned that my doctoral defense three years earlier had become enshrined in legend. It was the first time that a Ph.D. candidate had sat down on top of the desk while answering questions in Rhenish

---

2. K. W. Butzer, "Hartshorne, Hettner, and the Nature of Geography," in *Reflections on Richard Hartshorne's* The Nature of Geography, edited by J. N. Entrikin and S. D. Brunn, 35–52 (Washington, D.C.: Association of American Geographers, Occasional Paper 1, 1989).

with an American accent. My strategies for scientific networking were more mundane, but were beginning to pay off. In the 1950s, there were no photo-copying machines, but reprints were still affordable, and so I had started off writing for reprints of others' work even before I had the degree. When my first papers came out, I reciprocated by sending out reprints of my own. By the time my dissertation was published, I sent out notices of its publication to the 110 people on my reprint exchange list. The edition of eight hundred copies was sold out in two years and was then reprinted in New York.

Shortly after arriving in Madison, I was contacted by the Chicago archaeol-ogist Robert Braidwood, whom I had already met at a conference in Ham-burg. He invited me to submit a paper on my geoarchaeological work on prehistoric sites in Egypt at the American Association for the Advancement of Science (AAAS) meeting in Chicago that December. A reception was held for me in Chicago, where I met F. Clark Howell, a rising star in paleoanthropol-ogy. We had many common interests, and he invited me to a prestigious Wen-ner-Gren Symposium at Burg Wartenstein, Austria, in July 1960. The intense ten-day workshop was devoted to stratigraphy and environmental change in the Mediterranean Basin as a background to the archaeological record. More than half of the twenty-four participants were French scientists, again expand-ing my network of acquaintances in this interdisciplinary arena. In May 1961, Braidwood invited me to an informal symposium in Indiana, where just about everyone working on agricultural origins in the Near East was present—from palynologists and zoologists to archaeologists. Two months later Howell co-organized another Wartenstein symposium, this time on hominid evolu-tion, ecology, and Africa, and I was aboard again. It was an unrivaled learning experience, from watching films on baboon behavior to seeing Louis Leakey and Desmond Clark explain the making and use of stone tools to hearing well-known geologists describe ancient lake beds and cave fillings. The people I met at the AAAS and the two Wartenstein affairs were excited by their re-search, eager to learn more about anything that might be pertinent, and most were pleasant and personable. They indeed were world-class professionals, driven by their quest for knowledge, irrespective of disciplinary boundaries or training. By comparison, the Wisconsin geographers seemed more and more like burned-out teachers at a second-category British public school.

Andrew Clark did inadvertently do me one good favor. After I arrived in

Madison, he informed me that I would never be able to teach geomorphology but that I could have his course "Introduction to Historical Geography," which I could teach as a sort of prehistoric geography. Beginning in 1960, I offered it each spring, and it became the crucial link between my teaching and my research. Its topics ranged from methods of paleoenvironmental research to reconstruction of Pleistocene environments, and from human-environmental interrelationships to the beginnings of agriculture and floodplain settlement in the Near East. As my understanding and sophistication grew, the mimeographed lecture notes that I circulated in my classes were gradually transformed into a manuscript on which I sometimes worked until three o'clock in the morning. The resulting book appeared as *Environment and Archeology: An Introduction to Pleistocene Geography* in 1964.

According to Fred Simoons, the department met a half-dozen times during 1961–62 to debate whether or not to renew my appointment for another three years. In the end, they promoted me to associate professor instead, which completely surprised me because I expected to be terminated. In retrospect, now that I understand better how mobile the job market had become by 1962, my surmise is that they decided they couldn't afford to lose me.

My outside professional work continued to evolve and intensify. In 1961, I first joined Clark Howell to decipher the burial circumstances of extinct elephants at Torralba and Ambrona in central Spain. They had been killed and dismembered approximately four hundred thousand years earlier by Paleolithic hunters at the foot of a periglacial slope at the one site and in a valley-margin swamp at the other. It was a most productive intellectual exchange from one day to the next, and the work continued during the following two summers, then later resumed in 1980–81. In 1962–63, I was on leave as co-principal investigator on a National Science Foundation grant, spending seven months in Egyptian Nubia, where I and my Ph.D. student Carl Hansen mapped the Quaternary record and examined the geoarchaeological contexts of prehistoric settlement.

When *Environment and Archeology* appeared, Andrew Clark began to lobby the Wisconsin faculty that it was an eccentric book, more an embarrassment than an asset to the department. In April 1965, I put copies of a 1,300-word lead review from *Science* into everyone's mailbox, thus ending this speculation but not the machinations. In the fall of 1965, I visited Louisiana

State University, where I was made an offer as full professor. During the interview, the former director of the Coastal Studies Institute, Richard Russell, told me that he had already received an unsolicited letter about me from the University of Wisconsin. That letter apparently described me as a troublemaker, among other things, but he had given it no credence.

The chair at Louisiana State, Jesse Walker, agreed to give me three weeks to digest and consider the offer, but after only a week he was on the phone, pressing for a decision. Because I don't like to be pushed in such fundamental decisions, I declined the offer with regrets after the third call. The dean at Wisconsin promptly authorized my promotion to full professor, with a substantial increase in salary. Unfortunately, it wasn't quite that simple. At a February 1966 faculty meeting, the figurehead chairman informed us that there would be little of a salary increase the next year because most of the money would go to *my* salary. Some of my colleagues looked down at the table with discomfort, others stared or glared at me. I didn't say a word, although his statement was a lie; the money for counteroffers came from the dean's discretionary funds, not the department's existing allocations. After the meeting, I walked down the hall to my telephone and called Clark Howell in Chicago. He had already hinted to me that Chicago was interested and now promised to get the gears in motion.

In March, I interviewed at the Department of Anthropology at Chicago, where I talked at a formal dinner attended by the whole department. Several sociocultural anthropologists expressed their delight that I was joining them. By contrast, when I visited the geography chair the next day, Wesley Calef smiled at me and said, "You know, Karl, we aren't really interested in you in this department, but if the anthropologists want you, that's certainly all right with us." I smiled, too. I really did understand.

The terms were generous: teaching only two quarters per year with a total of three courses; a huge, combined laboratory and office in the paleoanthropology wing; moving expenses; and a 50 percent salary increase over Wisconsin. I accepted without hesitation. At Wisconsin, I was removed from all my committees and "shunned" by the senior faculty. My only regret was my initial perception that all the effort I had put into that department had been a total loss. But of course it wasn't because the students and not the institution are what matters. I had four Ph.D. students, two finishing after I left, and one joining me in Chicago. There also were other students who had taken or audited my prehistoric geography course and who later told me that I had made a difference: Alfred Siemens made his reputation as an innovative geoarchaeologist; Cole Harris moved on to make a second career in ethnohistory; and Stanley Brunn, as editor of the *Annals of the Association of American Geographers,* chose me as editor of the special *1492* issue. Wisconsin colleagues such as Fred Simoons, Bill Denevan, and Jonathan Sauer remained friends for life, and the map librarian, Mary Galneder, continued to help me find items years after my departure.

The underlying ironies of the Wisconsin experience crystallized for me some years later at a meeting in Milwaukee when I overheard an Andrew Clark student sarcastically say to one of his peers, "There goes the world's only Pleistocene geographer." It was Wisconsin that was out of step, bemired in a conservative, postwar paradigm of geography, unable to grasp the ongoing theoretical revolution and to open up to more versatile social sciences beyond the fence. Yet the "national departmental ratings" system continued to reward the strongholds of orthodoxy and stasis with billings among the top three or four. That lagging ability to move from derivative teaching to engagement in an open, intellectual discourse would eventually cost the discipline of geography dearly.

### Chicago, Africa, and Beyond

My first years in Chicago in the late 1960s were an exhilarating experience as I interacted with the incoming graduate classes of anthropologists (in the team-taught "Human Career" seminar) and geographers (in my physical geography course), some thirty in each group. The anthropology faculty were interactive, and there were regular faculty-student lunches at a local pub, at which good discussions were common. I sat in on some of Clifford Geertz's seminars in cultural ecology, held in his home. In the spring, I would take excursions up to Wisconsin, visiting many of the physical features to which I had taken my Wisconsin classes. In the meantime, Clark Howell was preparing a multiyear expedition to the Omo River of southwestern Ethiopia to search for early hominids, and I would be a participant.

East Africa turned out to be everything I thought it would be and more, and my teen experiences in Canada proved to be the right preparation for bush bashing, whether in the riverine forest or the adjacent tall-grass savannas. The tribal peoples along the Ethiopia, Kenya, and Sudan border were fascinating. I made my first, unexpected encounter on a woodland path not far from the Chicago camp, running into a six-foot man wearing little or nothing, with an equally tall spear in his hand, walking with his mate, who was decorated with a bark skirt. When I stopped abruptly and my jaw dropped, he grinned. We both started laughing and shook hands. Africans, I soon discovered, have an incredible sense of humor. Eventually I learned to work comfortably at sedimentary exposures with a half-dozen armed warriors watching me with great solemnity.

Camp life was pleasantly convivial, and I spent some time with Richard Leakey in the Kenya camp on the other side of the river. Richard had found an anatomically modern cranium, eventually dated at 130,000 years, by far the oldest such specimen known. I mapped and interpreted the sediments that secured the context for this find and a more primitive fossil (Kibish I and II) that set in train the "out of Africa" hypothesis for modern humans. The hypothesis would be supported by biomolecular evidence that contemporary people are all descended from an African "Eve," very roughly 150,000 years ago.

In 1968, I was back again at mapping, by helicopter, the surficial sediments of an area the size of Rhode Island. This project included a study of the meander belt and distributaries of the Omo Delta, as analog data for the Pliocene fossil beds being concurrently explored by others. Lake Rudolf (Turkana) had been sixty to eighty meters higher 10,000 to 3,000 years ago, at times overflowing into the Nile drainage through the swampy flats of southeastern Sudan. In fact, the Omo River in flood and indigenous cultivation next to its banks provided a useful model for the primeval Nile in Egypt because it derived its waters from a watershed abutting that of the Sobat and Blue Nile. Upriver reconnaissance by light aircraft was equally exciting, and we regularly flew in and out of the study area via the Rift Valley from Nairobi.

Even after I began to work in southern Africa in 1969, I continued to extend my East African experience. In the Serengeti, I studied savanna landform evolution, and in the old center of Aksumite civilization I was able to show

that soil erosion had seriously impaired the productivity of this sector of up-
land Ethiopia. Despite a helicopter crash and shooting scrapes, my experiences in East Africa remain among the most memorable of my life.

Then I began research in South Africa that would eventually cover nine
seasons, spanning thirteen years. My interest in the area was based on its potential parallels with the arid and semiarid environments of the Near East and
Mediterranean Basin, and comparatively few researchers were qualified for
such work. I studied ancient cave breccias in the Transvaal, younger cave sediments and beach accumulations on the spectacular south coast, old playa lake
beds of the High Veld planation surface, waterfall tufas on ancient dolomite
escarpments, and alluvial fills in the Vaal River Valley and in several of its tributaries. This work was focused on some three dozen key archaeological sites
and on a variety of rock-engraving clusters, and was complemented by a season of archaeological surveying and excavation.

Here, even more than in East Africa, I worked in a wide range of topographic and environmental settings, with sites that ranged from fossil beds to
stone artifact horizons, rock art, and architectural features. Observational
skills are best honed not by endless repetition under similar parameters, but by
comparison and contrast. Equally so, different kinds of sites record change
with respect to different thresholds to provide a more complex view of the nuances and amplitudes of environmental change, while helping to identify different patterns of climatic deviation in specific parts of a region. In South
Africa, I was also able to show that early modern people were present before
100,000 years ago.

Given the regional data base that I had established, it was apparent that
later Pleistocene groups left extensive areas of marginally productive environments unoccupied for tens of millennia at a time.[3] I interpreted this phenomenon as a risk-avoidance strategy in drought-prone and low-predictability
environments because highly dispersed bands of foragers would intersect with
other groups at intervals too long to provide current information on animal
migrations or where plant foods were available. In isolation, a lack of information is as hazardous as an inadequate gene pool. Such a "marginality model"

---

3. K. W. Butzer, "A 'Marginality' Model to Explain Major Spatial and Temporal Gaps in the
Old and New World Pleistocene Settlement Records," *Geoarchaeology* 3 (1988): 193–203.

implies long periods with discontinuous occupation of the African continent, which would accentuate human evolution through alternations of gene flow and genetic drift,[4] which in turn could lead to genetic "modernization" without invoking directed migration.

I had always been interested in the interface between the environmental and social sciences, beginning with Toynbee's notion of challenge and response as a prime mover. To avoid the trap of determinism I knew that I would need a great deal of sophistication in the social sciences, which I acquired during years of interaction with anthropologists of all kinds who commonly were on the cutting edge of social theory. I had discovered early on that environmental history, whether short or long term, had to be generated inductively and deductively through primary research. By the 1970s, I felt comfortable enough to begin examining those relationships, starting with seminar courses on settlement geography and archaeology. I wrote a small book on irrigation ecology in ancient Egypt, based heavily on historical sources, and continued to pursue the subject of Nile failures and political devolution in successive approximations. My lectures at this time explored unusual combinations of positivistic and humanistic themes, borrowing heavily now from modern geography.

My departmental homes at the University of Chicago were in decline by 1970, when the anthropology department lost Clark Howell and a chain of other brilliant men, replacing them with vulgar poststructuralists. Geography lost three of eight full professors in 1968–69, without even attempting to replace them. After 1971, endowments and scholarships dried up, as did the influx of top-caliber students. But I had learned at Wisconsin not to try to change the academic establishment. Eventually I sought most of my stimulus and feedback off-campus or generated it in experimental courses. In 1981, I took a cautionary leave and accepted the chair of human geography at the Swiss Federal Institute of Technology in Zurich, Switzerland. That caution proved wise because within days of my arrival I tangled with the university president there in regard to his unexpected plans for the Geographical Institute. I stayed for only two semesters but developed a course sequence that em-

4. K. W. Butzer, "Environment, Culture, and Human Evolution," *American Scientist* 65 (1977): 572–84.

phasized historical human ecology, including such themes as demography, long-term population cycles, famine, the Industrial Revolution, social justice, war and genocide, and Third World exploitation. Such an agenda was perceived as radical in the conservative framework of Swiss geography.

Meanwhile, in 1980 my wife and I had begun a totally new, long-term project in Spain, studying a group of mountain villages across the last nine hundred years, from the Muslim past to the Christian present. Elisabeth worked in the archives and did ethnographic work in "our" village, while I excavated Islamic sites and studied traditional land use. This intensive historical and cultural ecology marked a new career shift for me that incorporated ethnicity and decision making. But I saw it as the logical closing of a long circle, during which I acquired the necessary skills, experience, and understanding to work comfortably at the interface between physical and human geography.

After I returned to Chicago, it was only a matter of time before I made an alternative career move. It came in 1984 when I accepted a university professorship at Texas. In Austin, I was once more able to interact with superior students, and my teaching evolved with my research, increasingly focused on ethnicity and on what I call the intellectual encounter of the Old and New Worlds in colonial Mexico. Some of my Ph.D. students at Texas have been geomorphologists, others have been human geographers, but none are narrowly subdisciplinary.

Just as my old Wisconsin department was later reconstituted anew, American geography has moved from a closed to an open system. Perhaps it now qualifies as one of the least inhibited of the social sciences, so much so that the Sturm und Drang of my own generation will seem like a tale from the Iron Age to the irrepressible baby boomers who simply ignored the military uniforms in the closet. In their wake, I, too, found freedom. Not that I rediscovered geography. It seems, though, that I have been reclaimed, and that is a good feeling.

# A Random Walk in Terra Incognita

## Leslie Curry

Courtesy of Leslie Curry

*Leslie Curry (b. 1922) served in the Royal Navy during World War II, and afterward obtained his bachelor's degree in geography and economics from King's College, the University of Durham, in 1949. His master's degree is from Johns Hopkins (1951) and his Ph.D. from the University of Auckland, New Zealand. He taught at the Universities of Washington, Auckland, Maryland, and Arizona before becoming a professor at the University of Toronto. While at Johns Hopkins, he was introduced to probabilistic models of operations research, in particular queuing and storage theory, theoretical approaches that have marked his research over four decades, culminating in his foundational theoretical book* The Random Spatial Economy *(1998). Never content with explanations offered by case studies, although drawing on them for theoretical insights, he has searched for deeper probabilistic models that have the capacity to generate multiple geographical outcomes from more fundamental spatial and temporal structures. He has received awards from both the American and Canadian Associations of Geographers and was the Connaught Senior Fellow in the Social Sciences during his residence at the Rockefeller Foundation's center at Bellagio, Italy. He has held visiting professorships at the universities of Minnesota, Australian National, Cambridge, and California (Santa Barbara), and has served on the editorial boards of geographical journals emphasizing analytical approaches. He is currently professor of geography emeritus at the University of Toronto.*

## School and Afloat

I know practically nothing about my forebears other than they were of rural Northumbrian stock. My mother's name was Hedley. In "The Steel Bonnets," George MacDonald Fraser (of Flashman fame) has them domiciled in the valley of the Rede, a tributary of the Tyne, in the middle March. In the sixteenth century, the Scots and the English of this country were "Border Rievers," raiding each other's farms and villages at night to take horses, cattle and anything else. This had little to do with politics and everything to do with "gainful" employment as a way of life. All is peaceful in the Cheviot Hills nowadays, just as it is farther south across the wall built by Hadrian, perhaps to repel the Scots, perhaps to define the limits of empire—the latter notion suggested by Owen Lattimore in analogy to the Great Wall of China. The "Geordie" accent is endemic, hardly understandable in its rural version. I still have some traces of it, but my father's was perfect. He visited us in Maryland, and I wonder yet at the content of a conversation he was having with a neighboring black tobacco farmer of the richest southern drawl.

My earliest home until I was about fourteen years old was a downstairs flat in a terraced house in Gosforth, Newcastle-on-Tyne. England in those days was not exactly Dickensian, but it was class-ridden. Fortunately, I was bright enough to pass the eleven-plus exam to go to secondary school—the other twenty-odd kids were cut off right there at age eleven. My elder brother by four years obtained a half scholarship, and the family raised the rest by my father growing vegetables on his allotment, which my brother and I sold around the houses early on a Saturday morning. I was shocked to find, when visiting England for a year in 1969, that my children were faced with this same exam.

When I was fourteen, we moved from Gosforth to the west of Newcastle, where my father became head gardener of Hodgkin Park, and I changed schools to Rutherford College. In addition to its being a much better school academically, Rutherford gave my athletic abilities greater scope in that our football (soccer) team won the city, county, and two-county championships. I was on the school cricket team, but found it almost as boring as baseball. I also escaped from the schoolmaster who had chosen me as his star boxer. Following high school, I went into a bank for a year. This was the "Peace in Our Time" era of Chamberlain. My friends and I used to go cycling at the weekends. In later days, the road signs were taken down to hamper any German parachutists, so a good map and sense of direction were necessary. I shudder now to contemplate our journeys to the Lake District across the Pennines on a Saturday afternoon (we worked in the mornings), returning on Sunday evenings, and also to remember that most of my companions were killed soon after. I was in a first-aid unit in this period, serving overnight duty once a week and having to report when bombs dropped. This happened only once, but I slept soundly through it.

With war, I volunteered, age eighteen, for the navy. My imperfect eyesight had me turned down for the Fleet Air Arm, so I became a radio direction finder (RDF, i.e., radar) mechanic. My first draft was to *H.M.S. Heythrop* in the Mediterranean, but the fact that the BBC announced its sinking did not halt my inexorable posting around the Cape of Good Hope, with both warm—and cold-weather clothing, on the *Mauritania*. There was a hilarious occasion when I was put in charge of a party to holystone the decks—sand and water everywhere. In Alexandria, I was posted to *H.M.S. Janus,* but because it had been beat up by the French and was going home to Britain, I transferred to the *Jervis.* There was an ignominious flight to Port Said as Rommel advanced, but after that things turned round. We were captain—that is, leader of the Fourteenth Destroyer Flotilla—with shipmates having incredible histories, sunk twice in one day in the Crete evacuation and so on. We did a couple of Malta convoys and then night sweeps from various ports as they became open—Alexandria, Malta, Bizerta, Tripoli, Famagusta, and others, shooting up anything that moved on sea or land. Then there were invasions to support: Leros in the Aegean, Pantelleria, Lampedusa, Sicily, Salerno, Anzio. The *Janus* had returned, but was sunk at Anzio when we had our bows blown off. I was lucky; many people did not come through unscathed in this period. We returned to the English Channel for the Normandy invasion, basically to bombard special targets. One day we were off a tiny island sprouting large antennas among the Channel Islands, and it was decided we should capture its radar-radio equipment. I was given a revolver, which I had never handled, so there I was, a bag of tools in one hand, the gun in the other, in a whaleboat with other heavily armed sailors, approaching the shore. Suddenly there was a *tat-tat-tat,* and we all flinched and cowered low. But it was only the boat's engine. The island was deserted, the equipment taken or blown up, so we had a swim and returned to the ship. Such were the hazards of hand-to-hand combat in the navy.

At this point, I transferred to submarines, going through the whole training routine. The European war ended, and I was on a boat going to the Far East when the A-bombs were dropped. There was no crisis of conscience with us that I discerned. Again, there were crew members with daunting histories—for example, those who had escaped from a sunken boat, and a first lieutenant with a Victoria Cross earned on a midget submarine when it damaged the battleship *Tirpitz* in Norwegian waters. I was high on the list for demob and scurried frantically toward that end before the ship sailed.

## Under- to Postgraduate

All this background has little to do with being a geographer. I had had a standard and interesting introduction to world knowledge at a British school, had cycled around Scotland and Northern England, and had seen much of the Mediterranean—romance in people and landscapes perhaps. Also, on reading Henry Williamson's *Story of My Heart,* I felt a mystic bonding with nature. Too bad, as I learned later, he turned out to be a fascist. After the war, I enrolled in King's College, the University of Durham—in my hometown—in a joint honors program in geography and economics. The latter choice no doubt stemmed from the socialist sentiments I shared with all the thinking members of my class and age brought up on Bernard Shaw and H. G. Wells. I had formed a deep resentment of unequal opportunities, a disgust at the range of income levels, while squirming at the patronizing posture of privilege and, above all, at the assumption that this structure was right and proper. I hope that things have improved, but in today's paper there is the grandson of the queen going to Eton in his morning coat.

Economics in those days was exciting: John Maynard Keynes, the roads to serfdom and reaction, Beveridge's welfare state; while in geography physical planning loomed large. The analytic power of economics was more appealing to me, but the subject matter of geography more interesting, and the challenge was to bring the two together. Fortunately, in those days students were not afflicted by television so that I had time for athletics (cross-country and track), student affairs (vice president of the Students' Representative Council), politics (Labor Party), and a social life as well as academics. On graduation, I luckily did not take up the chance to continue to the Ph.D. in England, but spent a year doing rather mundane research for a regional development association. People in the United Kingdom had then, and probably still have, the belief that one can move directly from an undergraduate education to choosing a doctoral topic. Specialization begins even earlier in the schools now.

The following year, 1951, thanks to a Fulbright and a fellowship, I was ensconced in the Isaiah Bowman School of Geography at Johns Hopkins, taking a self-chosen and weird array of courses: Owen Lattimore on Inner Asia, Vladimir Sokoloff on soils, Douglas Lee on physiological climatology, while Clarence Glacken, a fellow student, was giving a seminar on his "nature and culture" thesis. Ernest Penrose, my tutor, emphasized population and institutions, and I also took a couple of courses in economics—and no doubt in other subjects. At a departmental talk, I met briefly the geographer whom I have admired most, John Wellington, who in those early days was battling apartheid from within South Africa, alongside, as I recall, an Episcopal bishop. There was an opportunity to read widely in the location literature for my thesis, "The Regional Concept in Economic Studies"—Bertil Ohlin, Heinrich von Thünen, Walter Christaller, August Lösch, Lloyd Metzler, and others I have forgotten. Linear programming was just coming in with Tjalling Koopmans scheduling tramp steamers around the globe—I went up to the New York publishers for an early copy of his *Activity Analysis of Production and Allocation*. The only paper I wrote from this time I sent to the Institute of British Geographers. It arrived scorched around the edges, survivor of an airline crash, but the referee talked about regions as unitary objects, so I gave up. American geography was not an intellectual hothouse, but at least there was the idea that things were not set, that they could change. Interestingly, the ordinary engineering schools were teaching the latest technologies, such as jet engines; the good ones were teaching much mathematics and pure science so that their graduates could follow through on all new innovations. This approach mirrored my aspiration to know how things worked in some fundamental way.

Although I was not personally affected, as this was a happy time for me, geography at Johns Hopkins was a political sewer. Isaiah Bowman had evidently wanted to establish a top-notch department, but, as I found out, eminent scholars who had been approached had a hate on for the school. Then came George Carter's denunciation of Owen Lattimore as a top Soviet agent to McCarthy and the House Un-American Activities Committee. At least one graduate student also suffered, and the aftermath of Chiang Kai-shek witch-hunting in U.S. politics using U.S. funds was an eye-opener to an innocent abroad. I joined the Baltimore Olympic Club, which trained at Hopkins, and we traveled to various places to compete against other clubs. I shall never forget being with a black team member when the bus stopped at a coffee shop, and he did not go in. One needed to have been brought up in the South to have entered without him. In those days, mixed marriages were illegal in Maryland.

Then followed six months as an economic affairs officer at the United Nations, at a salary that was my peak for many years to come, until I realized it was true that to destroy anyone you only have to have him dig holes and then fill them up again. In the evenings, as well as dating my future wife, Jean, I pondered how to bring time (anyway, recurrent time) into the space with which I had been dealing. I wrote to Warren Thornthwaite, at his lab in Seabrook, New Jersey, whom I knew because of his connection with Hopkins. He invited me down to stay, but regrettably we could not get funding for our proposed research. Nevertheless, I stayed for nine months working on timing activities, learning some climatology, and often busing up to New York to see my fiancée. During this time, I almost ended up working for a rainmaker who wanted to move into crop scheduling. Thornthwaite was a devotee of Gilbert and Sullivan, buying their records at bargain prices at a Left Book Club store, thus qualifying himself for questioning on his political affiliations.

### Migration Down Under and Over

At an International Geographical Union (IGU) meeting in Washington, D.C., I had met Marion Marts and Ed Ullman, who invited me to Seattle to replace Bill Garrison for a year. Just as my bride and I settled in, we were uprooted after three months to move to Auckland, New Zealand, for a permanent appointment. Obviously, knowing a little about climate, I was assigned that course: in fact, I learned a great deal and ended up publishing two papers on the atmospheric circulation of the tropical and southern South Pacific. As a faculty member, I could submit a doctoral thesis, and I began my research on the relationship between livestock-farming practices and climate in the different areas of New Zealand. Two lifelong interests emerged from this thesis: probability theory—here in terms of rainfall and of grass growth—and the treatment of storage operations—here in measuring soil moisture, in transferring surplus feed via hay and silage, in scheduling calving and lambing to fit in with grass growth, and so on. I tried to develop an analytic queuing model rather than simulating to obtain actual evapotranspiration, a topic I have returned to, but without success.

I spent one summer in Western Samoa attempting to record its physical geography as part of a departmental project. Considerable diplomacy was necessary to disarm local suspicions on the island of Upolu that I was part of a scheme to plant an airfield on village lands. Seaplanes still operated there in those days. Another summer I spent in New Caledonia on a mainly cultural topic, irrigated taro cultivation. Those were the days of genesis, form, and function in geography—debates that, thankfully, are no longer with us, replaced perhaps by even more stupid ones. Kenneth Cumberland, the professor at Auckland, was a dedicated Hartshornian. Genesis, conceived as history, was forbidden by Richard Hartshorne. Form was sufficient in its own right and certainly a necessary forerunner of function. Even the latter was essentially the extent of correspondence of different phenomena in areas. There was little concept of process, for example, or operations. Geomorphology could be (and in Wisconsin was) reduced to study of elements of landforms to give regional differentiation. As an undergraduate in Newcastle, I could not fit geomorphology into my schedule so that I had Wooldridge's book assigned to me. My main impression of the subject was that it was no more than a vocabulary for naming physical features. Genesis was largely tracing present features, terraces, and so on back through the history of the Ice Ages. Of course, a few brave souls were concerned with process—John Leighly, for instance—and thank goodness these much more stimulating studies are fairly general these days. My Samoan physical geography concentrated on form in Auckland fashion, informative but hardly breathtaking.

In 1957, I paid a three-months visit to the University of Washington again, this time with Bill Garrison and his coterie of students in residence, Waldo Tobler, John Nystuen, Dick Morrill, Brian Berry—and I think Michael Dacey was there, too. It was nice to see some changes happening. Once we were back in New Zealand, our twin boy and girl, Bill and Claudia, were born, and in 1959, with Union ticket (Ph.D.) in hand, we left for the University of Maryland. En route we visited the Lund Symposium on Urban Geography in Sweden, and I gave a paper that again resurrected the twin themes of probability functions and storage systems. Unfortunately, Maryland had a long-established permanent chairman with all the faults of such an arrangement: extreme conservatism and a partiality entrenched to the point that graduate courses were taught in the evenings by civil servants whose departments consulted with the chairman. I started taking stochastic processes seriously as a way of life: this was the way the world worked. As I saw it, climatic change as a random series again involved storage. My first paper in the random spatial economy series was a nerve-racking experience. I had never been so abstract; I was not sure that I knew what I was doing. Another excitement was the birth of our second daughter Ann in 1961.

Frank Ahnert and I took FORTRAN classes at Maryland, and I solidified this preparation with more programming at a summer institute at the University of Chicago, as well as learning a good deal of Bayesian statistics there. Those sub—and superscripts from Raiffa and Schlaiffer! Then once again we were on our way, this time to Arizona State University, which was still in its adolescent stage. I was placed on an intellectual climate committee, presumably because I taught meteorology, and we served as an audience for the dean's complaining about students' coffee cups outside his office. The computer had a random number generator within, but its call sign had been forgotten. I finally found it by judicious guesswork. The university has grown immeasurably, and I am sure has entered adult status. It was around this time I got into the essentially geographer's problem of spatial autocorrelation and its effect on spatial association, a problem that still worried Dan Griffith, Eric Sheppard, and myself in 1976. We stayed in Arizona only one year, driven out by the climate and the infections it produced. In nearby Scottsdale, the city had just completed a street of shops with something of an Old West decor. When we visited this sprawling city a couple of years ago, we noticed that the area was now labeled "Historical District."

## Looking for a Framework

We thus came to Toronto, where we were to stay until retirement and where I did most of my work. My first essay here was in a volume honoring my former professor Henry Daysh, "Chance and Landscape." Most of my papers are usually in the works for one or two years or more; I finished this one in a week or so, and it is the most reprinted. It showed my dedication to viewing natural and social processes as stochastic. Then came my Lund paper of 1960, an attempt to generalize the arguments about inventory management. This is interesting because it shows the convoluted steps often preceding a publication. Given my interest in climatology, I had read Lewis Richardson on the dispersion of two floating particles by atmospheric turbulence. Initially together, the velocity of their separation is constantly increasing as they move into larger and larger distinct eddies. In other words, the mechanism is not spatially invariant as in heat diffusion. To retain spatial invariance, and to regard the process as independent of structure, we use another marked particle as datum, noting only the particles' distance apart and so the efficiency of separation. Time is irrelevant in describing structure. To exploit the analogy with the central-place hierarchy, I had to learn dimensional analysis, learn from John von Neumann how turbulence theory used this analysis, and then use it in turn for modeling purchases from central places. Fortunately, I was acquainted with autocovariance and variance spectrum statistics. A number of years later I picked up a book on fuzzy sets and found a paper on settlements by a couple of Italians who referenced two writers who had preceded them in the topic. One was the founder of the mathematics of fuzzy sets, and the other was Leslie Curry. Because I had and have no knowledge of them, I stand chastened on my fuzziness.

The last paper on climate I wrote was in 1967 for a Kansas conference on time series. I have always had a strong belief that a generating model for climate needs to be stochastic, and I attempted to derive temporal probability density functions and spectra of several climatic elements from moving and standing waves. Indeed, I later tried a general circulation model from random walks, which Hans Panofsky said was ingenious because it had no physics in it. However, these excursions remind me of a cartoon showing two scientists, surrounded by complicated apparatus, contemplating a little pile of sand and saying, "Nobody really wanted a dehydrated elephant, but it's nice to see what can be done." I used to respond to invitations from departments to speak, and one year I kept bumping into Don Deskins, who had just finished his Ph.D. Wherever I went, there he was. I loved his attitude. He would have obtained a good university post whatever the circumstances, but he was black, and black was beautiful to administrators at that time. He understood all this and joked about it.

My incursion into remote sensing was instigated by my attending a summer course at the University of Michigan run by its natural resources program. A number of my interests came together at that point: radio work in the Royal Navy having to do with wave frequencies, the fitting of daily and seasonal waves by periodic regression in climatology, and my probability and statistics work already mentioned. Bruce MacDougall and I obtained a contract with the U.S. Geological Survey Earth Orbiter Project for a study entitled "Autocorrelation and Spectral Density Functions in the Interpretation of Remotely Sensed Data," a rather wide-ranging topic. One thing we (and especially Bruce) did was to construct a laser processor. As we learned to explain it, if you shine ordinary white light through a diffraction grating, you will see all the colors of the rainbow; the uniform spectrum of wavelengths is being split up into its constituents. Here, the input is a variable, and the grating is fixed. Now do the reverse: pass a laser beam (of single wavelength) through an airphoto transparency (of variable spatial wavelengths). The output is the constituent wavelengths of the photo specified according to direction. One can now filter this output by wavelength or direction and reconstitute the amended image. This is possible because the light beam retains phase on passing through the processor—that is, it is deterministic, unlike its statistical counterpart, the variance spectrum. We could remove the underlying hills from their forest growth and vice versa. Fascinating stuff! Our final report was late, 1972, and comprised a 250-page step-by-step exposition of statistical spatial analysis: point processes, polynomials, Fourier forms, autocovariance and spectrum, Gaussian surface, convolution, Z transform, impulse response and transfer function, some theoretical derivations, cascaded averaging and differencing and its uses, spatial entropy, and forecasting. We really should have done more about having it published.

I've published only two purely statistical papers since then. One was a general exposition of spatial systems analysis that I was to take to the IGU meeting in Japan, but my wife's illness prevented this. The other applied the Rice and Longuett-Higgins approach to the three-dimensional statistics of wave surfaces, amended for air photos of land forms lit by a high sun. Around this time, the Toronto department became involved in a Bell Telephone study aimed at forecasting Ontario population growth. I first surveyed all the types of approach—empirical and theoretical, deterministic and stochastic—to spatial forecasting. In particular, I emphasized the method of filtering the scale components of a map by cascaded averaging and differencing in analogy with the Fourier spectrum. Geoff Bannister and I produced "Township Populations of Ontario, 1971–2001, from Time-Space Covariances" in 1972. The most interesting parts were the very many experiments we performed after rejecting polynomial trend fitting: we preferred the nonstationary autoregression with weighting depending on a compromise between flexibility and stability because we had only a short series. The way in which the scale components interacted in the forecast was a revelation. Geoff wrote his doctoral thesis on a related topic, and we both submitted papers for the Association of American Geographers (AAG) meetings in North Carolina. We were both turned down. Because I could imagine the standard of reviewing, I did not mind for myself, but I was furious for a graduate student—he subsequently was published in the *Annals* with much commendation. Consequently, at the next AAG General Meeting I proposed, quite seriously, that there should be a *Salon des Refusés*. What did happen was that all reviewing ended, all papers are accepted, and some of the blame for the twenty-minute papers in twenty-five parallel sessions presently devolves on me. However, I do think that student papers should be in the 8:00 A.M. sessions, not those of distinguished scholars from abroad.

In 1969–70, my family spent a year at Reading, England, notable mainly because of the strong planning research people I found there, such as Michael Batty and Peter Hall. I did a lot of lecturing around the country and was totally exasperated by the amateur attitudes to research then prevalent. Scientists just did not want to talk about what they were doing—it was their private thing. Compared to the interactions that Waldo Tobler and I had at all our meetings—we seemed to be venturing into the same subjects for a number of years—the lack of mutual stimulation in Britain was ridiculous.

For the next several years, I appear to have been on the same kick, exploiting the areal scale decomposition as a Z transform or the spatial spectrum as a transfer function. I also investigated spatial regression, misspecification of the gravity parameters, and geographical specialization and trade (location of industry is as much an effect as a cause of trade). The next string to my bow was the notion of potentials. I happened to be a member of a Ph.D. examining committee in economics for a dissertation titled "Models of Optimum Currency Areas," in which the author used potential theory concepts. I was excited by the potential theory approach and persuaded Eric Sheppard that it was an ideal topic for his thesis, and it turned out to be an education for me. In 1976–77, I was in Cambridge and took up the study of demand in terms of potential. Strangely enough, Roy Allen's text *The Foundation of a Mathematical Theory of Exchange* in 1932 used this approach, but it disappeared from his 1962 edition and did not seem to have surfaced again. One of my papers covered the deterministic version, showing how map patterns and flows are complementary, and the similarity between the ordering of geographic space and of utilities in the mind. The stochastic version was even more apposite because it allowed changes in utilities and aggregation of different people. Now the transition probabilities between states of mind were equivalent to spatial gradients in demand. Later, in 1981, Eric Sheppard and I examined the notion of spatial price equilibrium from a potential-theory standpoint, actually solving a market situation using it. Fortunately, our referee picked up some faux pas we had committed. Another purely accidental common feature of some of my future papers was the harmonic oscillator. This was first used in my trade and specialization paper already mentioned. Next, in trying, uncharacteristically, to be practical, I conceived inflation to be a large-scale fluctuation in the central area of a line on which a dollar bill was random walking. This idea has appeared incidentally in other papers, and I am currently trying to see its significance.

At this time, my wife Jean died. She had been ill a long time, but this had not stopped us from spending our summers in Greece. We tried a different island or group of islands each year, Naxos, Samos, Paros, Rhodes, Corfu, Cos, and others. My routine on those trips was well established. Waking before dawn and going for a run; working until breakfast; quitting at 3:00 to 4:00 P.M., with perhaps a break for lunch; meeting with my family on the beach;

then dining out with retsina, and so to the next day. Not every day, of course. I had a large briefcase full of photocopied articles and chapters organized around my topic, which had accumulated over the winter. All that is left after thirty years of living together is to retreat within one's research horizons. This I did in examining the ecological literature toward an understanding of the role of externalities in geographical structures—I had always been leery of prices as explanatory in any basic sense. First, I chose vacancy chains and showed the close analogy between two-by-two community matrices and the external interaction between agents. Thus, the Lotka-Volterra predator-prey (-,+) model was relevant, but was disappointing in that it could exhibit only stability and not different configurations. The second subject was the (-,-) matrix—that is, competition, looked at from the point of view of the geographical division of labor. It provided much insight. I introduced, at the microlevel, the utilization function, the fitness set, and the adaptive function and, as we shall see, used them in a later paper as well as here. The third subject surveyed the field of diffusion models, regarding recruitment as diffusion and the resulting spatial structure of occupations. Previously, diffusion in geography usually concentrated on flows and not on the resulting patterns, which is odd.

## Getting It Together

At this time, I was struggling with relating notions of entropy, potential, and various spatial statistics. I actually produced a paper for a conference at Umeå, Sweden, in 1981, fortunately unpublished, which mirrored my anguish. Finally, I realized that thermodynamics had been there before me, and its formalism was available for interpretation. I have often wondered at the remarkable formal resemblance between two widely disparate phenomena. My conclusion is that the physicists were simply inventing the simplest equation to deal with a linear system, a coefficient of efficiency (temperature) relating order of position and flows to disorder (entropy), which is exactly what I required. A string of papers followed. I defined the inefficiency of spatial prices in these terms and discussed the value of information and the contribution of externalities to rents. Then I analyzed the inefficiency of trade patterns using Ilya Prigogine's "Brusselator" model showing how a large fluctuation can drive a system to a new steady state. For example, a single world system can be broken up into regional blocs by a shock such as the OPEC price rise. Again I developed the inefficiency of labor markets following the Ising model, showing how a coherent market could break up into stable separate zones of employment and unemployment. This was a highly complex paper in its writing, and I was fortunate to be doing it as a resident at the Rockefeller Foundation Study Center at Bellagio, Lake Como, in Italy. I followed my Greek routine, carrying a briefcase of photocopies, rising before dawn, putting in the equivalent of a day's work before breakfast, but working through till late afternoon; then running or perhaps playing tennis (I had not played in many years). Somewhere in there I fitted a walk around the beautiful estate. There was charming hospitality and congenial colleagues. Salt mines perhaps, but certainly gilded. Wives were invited, and I met Caryl, who was to become my wife a decade later after I had met her again in Washington, D.C., when she was divorcing. It so happens that she was born in New York City, graduated in Romance languages, is Jewish and brunette, has a beautiful smile, had a back problem for which she swam, and is a lover of opera—all features common with Jean. Incidentally, I like my music simple, primarily baroque, and, beyond that, later chamber music and Satchmo.

The last topic in the "thermodynamic" series was the examination of trade as spatial interaction. A theme that became recurrent in my work later was the difficulty of reconciling homogeneous regions and heterogeneous continuous space. I examined autocorrelation as a description of resource content. Then I looked at the ambiguity of the regional demand and supply curves because they have already taken into account the interregional structure of trade they are used to derive. This modeling of cause and effect applies particularly to the substitutability of goods: I treated a matrix of substitution coefficients like chemical potentials and called them the constituent potentials. Then I handled the fact that imports need to equal exports as an electrical potential, named the balancing potential, so that an equation could be written to include analogies to the physical, chemical, and electrical potentials. I also made an attempt to introduce inefficiency into the central places in the random spatial economy theory. Back in 1977, I used a paper by Peter Whittle to tackle the interesting problem of how the spatial distribution of settlement develops in response to the interactions between their elements, here both inhibiting (competition) and promoting growth according to their distance separation. I had the temerity to "correct" Whittle. In a somewhat related paper in 1984, I used communications theory to discuss the evolution of the economy. Structure must be maintained, but variety must be allowed. Freedom of choice in activity is given up for freedom of choice in consumption. An interesting parallel development concerning information and evolution was Manfred Eigin's hypercycle notion to model prebiotic evolution. I used it to try to push past my dissatisfaction with the Lotka-Volterra predator-prey model for vacancy chains. It provided for the village economy a much modified evolutionary approach that honed in on the stability of early interacting forms and their coherent growth. While conserving information, the system seizes on advantageous chance changes, which it fixes.

My major criticism of economic theory from a geographical viewpoint was titled "Factor Returns and Geography" and involved the issue of heterogeneity-homogeneity of resources. Concomitant features are the additive supply curve with its derived marginal productivity curve, and the receipt of rent in all earnings owing to the spatial variation in bargaining power based on the range and distribution of alternative opportunities. Bargaining has a fair degree of indeterminacy, so that productivity and agents returns are only loosely related, and, because of heterogeneity, it is wrong to have marginal revenue equal marginal cost. I used the graphs from the ecological competition paper of 1982 to show heterogeneous and flexible factors in partial competition with each other and in spatial conjunction with alternative opportunities to produce a bargaining relationship by which their prices are set. Technologies and products are variables in this adjustment. A 1987 paper emphasizes some of the important points. Density is arbitrary and needs to be related to mobility. Pareto optimality as a criterion for a project is useless for welfare because the area is arbitrary. Again, the topic was predetermined regions versus commuter models, labor demand being assumed known for a given technology and product prices, but these prices are variable in the trade problem. In teaching a course called "Information in Geography" for a few years, I tried to cover much ground. A paper, "Formalized Behavior in Agents' Markets," picked up some elements addressing factor supplies and demands necessary to a theory of trade. Types of uncertainty, criteria in choosing, flexibility with distance, search practices, mobility, distance operators, keeping options open—all are involved and provide better geographical bases than the initially given regions of Ohlin.

## On the Beach

In recent years, since moving to the Chesapeake Bay and observing the seasonal round of the osprey and Tundra swans, I have slowed down considerably but have not stopped completely. I wrote a plea for systematic theory in geography, my only methodological article after more than forty years in the field. Then in 1998 I published a compilation of many of my published articles on the random spatial economy. I had to add three extra chapters to fill gaps. One of these chapters, on spatial trade and factor markets, had been published in 1989, denying the possibility of integrating spatial labor markets, spatial trade, and spatial demand except by simulation. Another chapter was necessary to formalize a dynamic formulation of the "thermodynamic" approach to the economy, especially in its entropy balance. Last, but certainly not least, was the final chapter, "Endogenous Geographical Evolution," in which I made an interesting excursion into the mathematical philosophy of autonomous dynamics, likened MacKinder's "going concern" with technical change to a continual crossword puzzle, and conceived spatial development as interaction between the scale components of maps.

My current major project has gone and will still go through a series of changes. It is intended to be a considerable generalization of geographic theory. Initially, my aim was to model economic operations within a few basic forms. The idea was that geographic structures evolved and elements changed, but that these elements were different manifestations of the basic adaptive mechanisms. They were to be conceived as affecting single units operating in both space and time and also structures in space and in time. In the next phase, the way in which the harmonic oscillator equation kept reappearing in my modeling began needling me and instigated my using it in a purely mechanistic manner on the basic forms. I conceived these forms in electrical terms, with capacitance, inductance, and resistance affecting the current caused by a potential difference. The economic analogies were, in turn, integration (i.e., drawing on the surrounding environment and on the past);

phase shift (i.e., here versus there and now versus later); and limits on encounters (i.e., the ability to move and to consult the past). After pursuing several special situations in these terms, I felt I was missing the boat and began looking for a more basic interpretation of the harmonic oscillator. After all, this equation comprises in distance terms a direct response, a first derivative, and a second-order derivative response to inputs and is not limited to electrical circuits. We shall see.

Geography has matured considerably in my lifetime. I remember when I could read any article in any of our journals and absorb all the theory it might contain. It would have been difficult to find a score of books that could be recommended for their intellectual content. Things have changed immeasurably in this regard, not necessarily to my liking but at least in the right direction. It seems I have always had a desire for generalization, for systematizing. My first paper back in 1952 carefully chose quotes from Jean Gottmann, even from Jean Brunhes and Isaiah Bowman, urging a search for principles, for process. In 1981, I have Sir Joshua Reynolds's "Theory is the knowledge of what is truly Nature," which emphasizes the close relation between formalism and aesthetics; Fyodor Dostoevsky's "Realism, limited to the tip of its nose, is more dangerous than the most insane phantasmagoria because it is blind"; and Sir Arthur Eddington's "It also is good rule not to place too overmuch confidence in observational results . . . until they are confirmed by theory." In 1985, I quoted Thomas Huxley, "Those who refuse to go beyond fact rarely get as far as fact," and Oliver Wendell Holmes, "Even for practical purposes theory generally turns out the most important thing in the end." Things have not changed by 1998, when I am quoting Schopenhauer, who could not "avoid seeing in all history nothing but a repetition of the same things, as when a kaleidoscope is turned, you see all the same things in different configurations."

Who has had the greatest influence on my research? In the very early years, I mention, surprisingly, Robert Buchanan, writing on plantation agriculture; this was a theoretical analytical view of the scheduling of harvest and labor relative to climate, which showed what could be done. Bill Garrison's attack was certainly important, but then he faded from view. Michael Dacey's point processes and enthusiasm contributed to my work, but his refusal to clothe his mathematics in substantive meaning left me bereft. From a philosophical point of view, Fred Lukermann was crucial. He quotes David Hume, "Cause and effect are neither empirically perceived nor inductively testable." Only through experience can we gain knowledge of the real world, yet only through logic will we accept the "truth" of that experience. The single empirical event is the result of multiple causation, the convergence of many contingent series at a spatial-temporal point. It is necessary to make statements about many similar events, and these statements are necessarily probabilistic. Only in accepting process as an integral part of empirical investigation in geography do we finally make possible the relation of the particular to the general. In combining the object-level description and model-level explanation in a speculative but probabilistic schema, geography achieves a discourse-level narrative. Unfortunately, being somewhat short on models and rather long on descriptions, geography is more possibilistic than probabilistic at the moment. The history of my academic and research pursuits is, I now realize, one of opportunistic adventures, a feeling that I know something about this so that I can, after further education, tackle that. This is a recurring cycle, later reaching topics and methods I could not conceive of earlier. The fact that I was involved in climatology and economic geography, both individually and jointly, certainly guided the journey. Another feature that I observe on scanning references in my papers is the incredible variety of sources I tapped, comprising most physical and social sciences. Is this just me, or is it most geographers?

# Lessons from the Design of a Life

## William L. Garrison

Courtesy of William L. Garrison

*William Garrison (b. 1924), shown here receiving instruction in the finer points of computing from his grandson Riley, received his B.S. and M.A. from Peabody College, and his Ph.D. in 1950 from Northwestern University. During the Second World War, he served in the United States Air Force as a meteorology officer in the South Pacific. Prior to his retirement in 1991, he taught at the Universities of Washington, Pennsylvania, Northwestern, Illinois, Pittsburgh, and California (Berkeley), frequently holding the directorships of the Centers of Transportation, Urban Development, and Environmental Engineering. He is known as a major catalyst for the developments in geography during the late 1950s and early 1960s that became known as the "quantitative revolution," a sea change that was also strongly influenced by his students at the University of Washington. His professional life has been characterized by an intense concern to make insights from the academic world relevant to and informing of concrete problems in transportation and urban development. He has served on numerous committees at the local, state, and national levels, including the Department of Energy, Department of Housing and Urban Development, and the Department of Health, Education, and Welfare, as well as on many advisory committees of the National Research Council. The author of more than two hundred professional articles and reports, his greatest pride is reserved for the many students he has helped toward professional careers in geography, engineering, environmental studies, and closely related areas. He is professor of civil engineering emeritus at the University of California (Berkeley).*

O ne lesson is that when you come to a fork in the road, take it. And that wisdom about life's path, most often attributed to Yogi Berra, sums up my career strategy. It worked for me—kept me from being bored and supported the search for interesting-to-me things to do that matched my talents and limitations. Taking the next fork also provided an escape hatch when signs said, "Back Up, You Are Going the Wrong Way." It was a route *through* geography but never away from it, in spite of the labels with which I have been plastered, some not very flattering.

I didn't find geography until I looked for it. I certainly wasn't looking for it when I had my first brush with it in 1940. I remember the classroom and the teacher, Russell Whitaker, about whom I tell more later, but the course topic escapes me. No matter, that encounter didn't take.

That was at Peabody College in Nashville, Tennessee, and I had entered the summer after graduating from the demonstration school associated with the college. No career planning there—I just moved up the hill from high school to college courses. World War II had been in the cards for several years for anyone who looked, and the grip of the Great Depression was only beginning to weaken. Those were not very favorable times, and this sixteen-year-old was just waiting, for I felt I had no control over my future.

My waiting was not very graceful. I drifted along with a why-bother feeling about schoolwork and skimmed the surface of a liberal education. Two decades later, when I worked with Herman Porter, he once remarked that his previous employer had been a thirty-five-year-old Nobel Prize aspirant who had had that goal since he was in high school. For me, on the other hand, a number of career tracks started in high school, especially in the sciences. Though never a Nobel aspirant, my surface skimming was a mistake, and I would be better off if I had paid more attention to language and literature. That is hindsight, and it may be fallible. Yet I am certain that awaiting an uncertain future gave me empathy for young people of the Vietnam and similar eras.

I found myself an aviation cadet two years later in a nine-months-long meteorology course. The dynamic meteorology course was taught by Harry Wexler from MIT, an excellent teacher who gave his students a very tough mixture of fluid mechanics and thermodynamics. Armed with some working calculus but rather superficial physics and chemistry, I had to run very hard to keep up with the subject. The alternative was washout, a transfer to Jefferson Barracks in St. Louis, and the loss of my enamored cadet status.

Nevertheless I managed, and in spite of playing catch-up I tied for top score on the first exam with a chap who had an advanced degree in physics. (Most of the credit for that achievement went to my skills in taking tests, skills that had been honed by the many tests inflicted on me while in a demonstration school). With that high score and more hard work, I managed the dynamic and synoptic meteorology courses, as well as climatology and a mixture of other things in the curriculum. Climatology was taught by John Leighly from Berkeley. Decades later I would visit with Leighly on the Berkeley campus, and I found him warm and stimulating, in marked contrast to his descriptive and boring lectures before a class of 250 cadets.

I said above that "I found myself an aviation cadet." I could have just as well said that "I found myself when an aviation cadet." I found that by trying I could do what I had to do and do it very competitively. I also began to understand the design-of-a-life task. That task, as I began to understand it, involves matching what one can do with enjoyable and constructive things to do. But already I could see that meteorology and its intellectual relatives didn't match what I could do with what I liked to do.

The remainder of World War II for me involved two years overseas in the backwaters of the Pacific theater and long periods of boredom. If I hadn't been there, someone else would have been in my place. That was my contribution to victory—freeing up someone else who may have made a difference.

After being discharged from the service, my unfinished education at Peabody was first priority. Receiving credit for meteorology study shortened my undergraduate stint to a year, and I stayed on for another year to receive a master's degree, concentrating in geography. I found Russell Whitaker's course on historical geography fascinating, as was his course on conservation. Visitors gave regional courses, but these courses were presented as classification and fact listing in character and were not very interesting.

By that time, I had pretty much made the choice between a career built on opportunities in Nashville or going on in geography. Although there were many more steps to be taken, I had decided what I could do and firmly put my feet on the ladder of a career in geography.

There was still teaching versus doing something else, so before going on to Northwestern in the fall of 1947, I spent the summer at Ellensburg State in Washington State. Whitaker had recommended Northwestern and had also located my summer employment in Washington. The longtime geography professor at Ellensburg was away for the summer, so I was the geography program.

That worked just fine. I enjoyed teaching, the students, and the supportive social studies faculty. I was reading the textbooks more quickly than the students and teaching from them, as well as swimming and hiking. Offering a regional course, I began to appreciate Preston James's concept of settlement. Being an academic geographer looked to be a pretty good deal, and the only decision to be taken was whether to buy golf clubs right away or later.

Then came Northwestern. The incoming graduate students numbered about twenty, and there were, say, thirty to forty folk who overlapped in the program that I think of as being in "my class." I made many good friends at the time, and although we later scattered and did very different things, the warmness sticks. We did a fair amount of off-campus socializing, including get-togethers at the home of the chairman, Donald Hudson.

Before the first quarter, the incoming class spent a week at field camp. Learning how to classify soils, vegetation, and land uses required an hour or so, and another hour or so of walking around gave enough practice. I recall the remainder of the week-long camp as a pleasant walk along the Milwaukee railroad tracks in Wisconsin. It wasn't all Mickey Mouse work; I did learn a little about glaciated topography and got to know my fellow graduate students.

Course work at Northwestern? Except for cartography, which I somehow missed, it was one each of everything offered. Someone, Clyde Kohn, I think, used Edgar Hoover's book on the location of economic activity, and we read and discussed it. There was Hudson's seminar, in which I read and wrote a critique of Hartshorne's *The Nature of Geography*. I attended a University of Chicago-hosted urban-planning course by Harold Mayer, who was then research director of the Chicago Department of Planning. And then there was Ed Espenshade's China course, which involved reading J. L. Buck on Chinese agriculture.

Northwestern's offerings were pretty much par for the course for graduate programs of the time. The graduate students emerged ready to teach regional or systematic courses, with not much depth beyond textbook learning from several professors. Nor had they learned much about research, by either involvement, emulation, or just conversation about it.

Except for Mayer's course, I took only one course outside the department and had no depth in languages beyond the reading skills required for the Ph.D. I did very little writing. How could I have written a critique of Hartshorne by reading nothing but Hartshorne?

Eventually I did make use of the course on China. In 1983, I toured and lectured in China as a guest of the Chinese National Railways, and my escort was Tang Jian, one of Buck's former students. An economist by training, Tang Jian headed the English department at the Jiao Tong University in Beijing—economics being out of fashion in 1983. He was much taken that I knew of Pearl Buck and that I had her husband's picture of Chinese agriculture in my head. As a result, we had very open conversations, and I learned a great deal about life beyond the bamboo and other curtains.

Then there was the hurdle of the thesis. At Malcolm Proudfoot's suggestion, I read his article and monograph on shopping centers in Philadelphia and elected to gather data and do something similar for the northern suburbs of Chicago. I did a lot of walking around and field mapping, classification, and description. I despaired because Proudfoot wouldn't accept my writing. I solved that problem by borrowing Proudfoot's thesis from the University of Chicago and writing in a style imitating his. His view of me turned around the next time he reviewed the text.

I learned something from the Proudfoot experience: in order to be accepted, do what folk in the peer group do. As a result, my thesis looked and read like a Proudfoot-era geography thesis. It had many maps from field data, photographs, and tables. There was a little on how the spatial pattern evolved, and the text was mainly ad hoc classification and description.

At some point in that time period, I proofread autopsies from Evanston Hospital as a favor for a friend. How could someone knowing almost no anatomy do that? But read one and you have read 'em all. Just description of this and that usually followed by a laundry list of causes of death, with the leap from description to causes implicit. I have always thought of my thesis as autopsy-like.

Looking back, I regret not learning more about Proudfoot's work on

World War II-driven population redistribution in Europe. Also, I began to appreciate that I was to be faulted for the disappointing thesis experience. I should have pushed for something better, peer-group acceptance be damned, rather than something adequate. Several years later I was happy to say "No" when the Northwestern Library reported my thesis as lost and asked if I had a replacement copy.

The thesis was done, it was 1950, and I was off to my first choice of position as assistant professor at the University of Washington. It was a fine place to be, with a pleasant campus of good size, good faculty, and relaxed living. I bought and began to use some golf clubs, but I didn't actually use them very much. I began to realize that I felt inadequate and needed to learn how to do more interesting and useful things.

It took a while for that unease to emerge. I began teaching a large course in economic geography, lecturing and having teaching assistants. I'd already had two years of doing the same thing while lecturing in parallel with Clarence F. Jones at Northwestern. Using Jones's text at Washington was a bit radical, for previously the course had been organized regionally. I also saw no need for an expensive world atlas and asked only that students acquire an inexpensive one. There was some frowning about those changes because it wasn't "the way we have been doing things." If there was a lesson there, I didn't learn it; I just kept blundering along as I still do.

The question arose of other courses that I might teach. With what I had learned from Harold Mayer in mind, I thought about urban geography, but I eventually settled on transportation geography. Transportation seemed more of an untilled field. In hindsight, my selection of it was just plain good luck.

The year after I arrived in Seattle, Donald Hudson arrived from Northwestern to become department head, with promises of slots to fill. The department began building, recruiting very good young faculty from Michigan, Toronto, Chicago, and Northwestern, as well as Edward Ullman, who was already well established at Harvard. Hudson undertook to strengthen the cartography program already developed by John Sherman, did academic planning, and recruited students. He saw to it that links were strengthened within the university, in particular with the area programs that were becoming all the rage at the time.

All of us felt that we belonged, students and faculty old and new. But, inside, I had nagging concerns. What was my "thing?" True, I had responsibility for large classes and made the dean's list of excellent teachers from time to time. Hudson decided that an entering field camp for graduate students was needed, and I organized and led the camp during its first session. Modeled on my Northwestern experience, we did mapping, land and land-use classification, and such. The site was delightful. We headquartered at the university's Fisheries Laboratory at Friday Harbor in the San Juan Islands and walked around and mapped San Juan Island.

I had teaching niches, but I wasn't excited about them, and I didn't have anything to say that was intellectually mine. I just did out-of-the-textbook stuff.

A fork in the road came in the fall of 1952, when I accepted a one-year appointment at the University of Pennsylvania. I became a lecturer in the Department of Industry and did some teaching. Lodged in that department, the geography program was atrophying, mainly because of poor teaching and a failure to develop supporting relations on the campus. My major assignment was Project Big Ben, a classified Defense Department endeavor for which I was well paid, but for which I did no work because Eisenhower had just been elected president and his appointees were lined up before me in the queue for security clearances.

As it happened, I lucked out. Coming in every day and sitting in an office got stale pretty quickly, so I began to learn about the marvelous resources held by university libraries. Others in my underemployed situation included a well-established microbiologist, several young economic and statistics types, and an applied mathematician. We interacted and played with ideas, including imagining research and papers on a variety of topics. The microbiologist tried to talk me into becoming a microbiologist: "You can do it in a year or so." But I didn't take that fork in the road.

The project included travel money, and I remember very well a paper I wrote on the spatial organization of markets given by Karl Fox at the 1952 Econometric Society annual meeting. It was a real eye-opener, for it stimulated my imagination for analytic work on interesting economic geography-like topics. A decade or so later I got to know Fox well, for we shared many interests.

Deciding that I needed to learn something about statistics, I sat in a course at the Wharton School, which consisted of a few weeks worth of content stretched out over several months. The instructor presented the topic in a "this is it" style, and we students went away convinced that we knew it all. A terrible course: no unanswered questions, no philosophical issues, just a cookbook.

Through a contact at Project Big Ben, I was asked to review the fourth edition of L. C. Tippett's *Methods of Statistics* for a physics journal. That was easy because Tippett's work tracked closely on R. A. Fisher's, and I could refer to reviews of previous editions. The review was fun because I became interested in where Tippett and Fisher were coming from, so to speak.

Doing this review turned me on to why people do what they do, questions that I have greatly enjoyed. It stimulated my looking into the history of statistics, and I've since paid attention to the history of academic fields, technology, transportation, and many other areas.

During the year at Penn, I learned the fun of playing with ideas from my underemployed colleagues at Project Big Ben and had a larger world than geography opened for me by those folk, as well by my retreating to the library, going to meetings, and such. From the instructor in that boring statistics course and the reading I did to put Tippett's book in context, I learned to be skeptical about cookbook knowledge and to think of background and context as everything.

It may not seem like much, but the Penn experience was a fork in the road, a career-changing year. It gave me ideas to build on.

Returning to the University of Washington, I built into my transportation course context and the issue of how things came to be and played with such notions in the undergraduate economic geography course. Several incoming graduate students worked as teaching assistants in that course, and the approach in the course interested them in topics that interested me. Brian Berry and I talked of writing a textbook on the basis of the course, but nothing came of it. I put much of what nowadays we might call historic path dependence into my transportation course and developed courses in location theory and statistical analysis of geographic problems. I like to think of the latter as an analysis course, but it used a statistics textbook and students thought of it as a statistics course. I thought of the location theory as a spatial theory course because it built from Isard's *Location and Space Economy* as a text.

I leaned on textbooks for only a year or so while beginning to develop course notes, outlines, reading lists, illustrations, and other things that hewed to the ideas that I wanted to develop. As time went by, I developed such supporting materials for all my courses, materials that had a work-in-progress character—though they never matured into final textbook products. I probably should have published these materials in course-notes formats. They stressed processes because they were steered by transportation and communications, and emphasized ways to estimate the shape of those processes. That was my focus, and I never felt that the label *quantitative geography* or its diminutive *quantifier* was appropriate.

As my courses changed, so did the students' evaluations of my teaching. Then as now I began to hear complaints about my nonuse of textbooks, the unfinished character of my course notes, and remarks along the lines of "too much history" and "I couldn't understand what he was trying to say." As my average teaching-quality ratings slipped, I stopped appearing on the dean's list of excellent teachers. Average became the norm as I steered away from cookbooklike teaching, and the evaluation scores took on a bimodal distribution. So be it.

The department purchased some mechanical computers that students used in the analysis course, and their use increased as we began the highway-oriented work described below. Duane Marble and I began to use an IBM computer housed in the university's accounting office. Interested in computing possibilities, I attended short courses on mercury delay tube UNIVACs. Similar to the situation elsewhere, academic computing took hold at about 1960. The university purchased an IBM 650, and Douglas Newton, the IBM representative, offered courses on FORTRAN. Newton was a Washington Ph.D. in mathematics, and he had previously helped us by writing a special-purpose matrix inversion program.

By the end of the 1930s, the states had pretty much finished the network of federal aid primary highways, and planning during the 1940s centered on a limited access interstate system. During the early 1950s, several states had begun to build interstate-like facilities. Legislative interest in Washington State was in where to build and how to fund. Objectives of economic development and integration of the state's economy pointed to additional routes across the Cascades, as well as to bridge-highway-ferry combinations crossing

Puget Sound. Consultants did project-feasibility studies and, as is the case today, found ways—sometimes rather implausible ways—to find projects feasible.

Facility tolls, or the capture of increased land values, were possible answers to the how-to-fund question. What's the thinking behind value capture? Simply that improved highways will increase land value. The goal was to find ways to estimate how much and how to fund road building from value increases.

Robert Hennes in civil engineering began to assist the legislature, and I was on the team he put together to study financing possibilities. Edgar Horwood from civil engineering and folk from the real estate program in the Business School focused on a case study in Seattle. Using the word *study* liberally, I worked with graduate students on topics such as travel patterns through what might be called organization of the space economy: shopping centers and shopping patterns, rural cities and their tributary areas, and the delivery of medical services.

Geography, business administration, and mathematics students were involved, and from time to time my geography faculty colleague Marion Marts pitched in. We spent perhaps thirty thousand dollars per year, which went a long way in those days. We produced theses, papers, and other publications. Bob Hennes had the task of making our work meaningful to the legislature, and he did that very well.

The university had a policy of one trip per year to a professional meeting, so every year I was off to the Association of American Geographers (AAG) annual meeting. At one of these meetings, I gave a paper on the optimal size of the city. The concept was good, but the paper never got beyond the draft and suitable-for-reading stage. In December 1955, I attended the first and founding meeting of the Regional Science Association (RSA). Edward Ullman had suggested to Walter Isard that I give a paper. And toward the end of the decade I began to attend the annual meetings of the Highway Research Board in Washington, D.C. Beginning to know different kinds of people and ideas was not only fun, but stimulating.

Edward Ullman used to refer to a chap who belonged to both the economists' club and the explorers' club; he wore his economist clothes to explorer meetings and explorer clothes to economist meetings. As an outsider, he was nonthreatening and had something different to say. And he could listen as an outsider, too. I learned that I was advantaged by being an outsider. I liked that, although it took a while for me to learn that when the chips are down, outsiders are just that—outsiders.

The scenario changed in the highway work that I had been doing when the federal interstate program was funded in 1956. The Washington State-oriented work we had been doing was pushed aside, and the federal interstate locations and their urban extensions answered the question of where improvements were to be made because 90 percent of the money was federal.

But the basic problem remained. Although how to pay was no longer a state question, it remained to be answered at the federal level. Congress asked that cost studies be made, and the Bureau of Public Roads (now the Federal Highway Administration) embarked on a test-road program and a study that supported allocating user fees to the cost of providing and using facilities. How much should trucks of differing sizes and weights pay? What about automobiles?

In addition, Congress had asked that nonuser benefits be examined. The idea seemed to be that development benefits occur but may not be closely tied to uses. Then as now, policy and political folk talk development, whereas transportation folk count vehicles, calculate travel cost savings, and look for demand elasticity. That was a mismatch about which I became more and more concerned. The mismatch and my concern continue.

Our work for the state of Washington had received some attention, and I soon found myself on the advisory committee for the bureau's Highway Cost Allocation Study—the first of many advisory or report-preparing committees I served on in Washington. It was one of my better experiences, and I learned much engineering economics from a fine professional, G. P. St Clair, the leader of the study.

I worked hard on the reports from the study, did some research, monitored the work of others, and provided text on nonuser benefits concepts and empirical findings. Interim and final reports of the cost-allocation study were published as congressional documents, but very little of my writing made it through the review process.

The review process was very simple. The reports went to the White House and the Bureau of the Budget (BOB, now Office of Management and Budget), where an economist applied scissors to large parts of my verbiage. The

BOB reviewer, whose authority stemmed from his position rather than from the merit of his canonical ideas, treated my ideas with disdain. He didn't even bother to confront my heretical thought by arguing his superior wisdom. Needless to say, I had my nose put very much out of joint from that experience. I learned a bitter lesson, and I continue to give short shrift to those who argue ideas from position rather than from merit.

I began to learn and eventually realized that the simple and plausible wins every time, especially if it is a canonical view. I also learned to pick situations where there is a chance of not being thwarted by minds already made up and to be prepared to invest time to educate clients about ideas. I also learned to accept a low batting average—say, positive contributions in one out of ten situations. What's more, I became sensitive to the possibility of doing harm by applying dogma where it is inappropriate. Those were lessons I learned not just with the one experience, but over a period of a good many years.

I began to be concerned with the question, "If I'm not being effective, should I walk away?" One particular experience may explain this question. I recall very well attending a meeting to discuss studies needed to estimate the impact of the interstate system. That was in about 1956 and just after the highway legislation passed. My question was, "What's the interstate going to do for society?" and I had some thoughts about how to approach it. But other folk saw social impact having to do with construction contractors' purchases of tons of steel, cement, and asphalt; purchases of equipment; and the dollar value of the hours of labor to be used. Those are reasonable questions, but surely we need to invest for reasons that extend beyond purchasing materials and labor?

In this situation, I said what I thought and chalked up being ignored as a not-invented-here reaction. Back at the drawing board, I worked to develop more cogent ways to present my thoughts, thinking that better communication of my ideas might help. It doesn't seem to have helped in some situations. Just two years ago I was unable to convince an airport commissioner that the increased money paid to airport employees was part of the cost and not part of the benefit of airport expansion.

By the end of the 1950s, I was involved in Washington, D.C., in AAG matters, and in several professional organizations. Travel was a pain. True, the Boeing Commercial Airplane Company had ignored the naysayers (some of whom were my colleagues at the university), who knew the planned aircraft would be too expensive for too small a market, and so had begun to market the Boeing 707. But Seattle didn't have such a service. About the best we could do was a direct flight to O'Hare on a DC-6. When Northwestern University offered me a position in 1960, the opportunity to reduce isolation beckoned, as did my desire that my children be able to take advantage of family ties in the Southeast.

Was I a job-hopping, academic bum? One could say that. But my crimes as I see them are that I tend to bore easily and that I like changes in venues. Those crimes help explain the job hopping and taking forks in the road. They also help explain why, after having resolved the highway benefits question to my satisfaction but not having convinced many others of my insights, I began to look for entertainment elsewhere—another fork in the road.

Only Ed Espenshade remained of the faculty that had been at Northwestern ten years earlier when I received my degree. He had done a good job of recruiting new faculty, and things seemed to be developing reasonably well. I began to teach the transportation geography and analysis courses I had given at Washington and to participate in college committee affairs. The Transportation Center at Northwestern had provided funding for my appointment, so I tried to find ways to participate in its work, at first without much success.

I was busy, very busy, in the early 1960s. I had some support from the Office of Naval Research (ONR) before coming to Northwestern, and with ONR help I organized work at Northwestern on quantitative earth science analysis with Bill Krumbein in geology and John Logan in civil engineering. Krumbein and I sketched a model of beach equilibrium process that took account of the supply and characteristics of sand and of fetch and wave characteristics. As part of that model, we had a mapping program written for us, but we didn't develop it beyond a few mapping exercises. I became bored too quickly again and took another fork in the road.

ONR support enabled us to start thinking about the promise of remote sensing, and Duane Marble, Michael Dacey, and I pressed NASA to develop a remote-sensing capability. Complicated by security and diplomatic concerns, this project moved at a snail's pace.

In spite of the glacial pace of NASA program development, we made some progress, and perhaps our work deserves some credit for today's remote-sensing programs. Looking back, I recall very supportive relations with ONR, visits to NASA facilities, and a paper I gave at an astronautical meeting in which I showed a view of the earth from outer space with the caption, "World's Largest Information System."

While I was at Northwestern, my contacts with the Bureau of Public Roads continued. The bureau supported some work, and I served as a member of its Research Advisory Committee. It also wanted to strengthen its research program, and our committee helped. I learned from committee chairman K. B. Woods the importance of good committee management as well as of having the ear of the sponsor—in this case Rex Whitten, the chief of the bureau. I also learned a good bit about the fragility of intergovernmental relations as the bureau sought increasingly to steer state-federal cooperative research programs.

The Army Transportation Corps provided support for work on efficient logistics, which enabled my thinking about optimal configurations for transportation networks. Duane Marble, Martin Beckmann, and I—along with Waldo Tobler, David Boyce, Karel Kansky, and Nick Boukitis—tried several approaches to optimal synthesis and comparative network analysis. We compared the interstate and the rail network, did factor-analytic studies of air networks, and simulated the growth of the Irish rail network. Many students made stabs at this aspect or that, including one student who worked on optimal warehouse location questions. Except for the latter type of problem and Transportation Corps interest in optimal route locations, our work didn't seem to excite much interest. I was satisfied that I knew the shapes of optimal network problems. Today, our Berkeley librarian likes to remind me that one of the publications from the work has the word *prolegomenon* in the title, and I occasionally see references to the index numbers we used when comparing networks.

Donald Berry headed the transportation program in civil engineering at Northwestern, and with the buildup of interstate planning and construction his program had an influx of students, most of whom began to take my courses. Don Berry and Ben Gottas, the Technological Institute dean, arranged for me to be appointed half-time in civil engineering in order to develop planning aspects of transportation engineering and environmental engineering generally. My work was soon augmented by George Peterson, a full time civil engineering faculty member. Peterson did his Ph.D. with me, but otherwise we didn't do much work together.

Because of my previous experience of working with the civil engineering department at Washington, I was very comfortable in the same department at Northwestern. Don Berry was a fine, generous person with whom to work, and Ben Gottas and I became great friends. Gottas had worked to orient the Technological Institute away from practical training to theory and research. Having myself made a good start on that, I was his sounding board for his ideas about the institute's development paths, technology and society, and many other things. He would drop by with a question on his mind, and we would settle it and all of the world's problems before the day was over.

Gottas had been on the Rockefeller Commission (Nixon administration) review of Latin American development needs. He could enter into conversations about capital formation and such, but when the chips were down, he would say clean water was number one on the list of needs. I thought, "Once a sanitary engineer always a sanitary engineer." It was years later before I realized the value of Gottas's insight—certain things are needed before development can begin.

In the 1960s, the National Science Foundation (NSF) was actively sponsoring summer institutes for academic professionals, so I gave lectures in a regional science summer institute at Berkeley. At about that time, Duane Marble was organizing institutes in geography, while I was organizing program sessions for transportation planning and engineering. Besides the substantive values of these institutes, they were great for networking. They created peer groups that gave folk a sense of belonging as they were pressing ahead in exciting areas. I continued to be active in Washington, D.C., and in the late 1960s Grosvenor Plowman, undersecretary of commerce for transportation, asked me to look into Senator Pell's proposal for high-speed trains in the Boston-Washington corridor. From that experience, I learned something of the ways of Congress and how programs take on a life of their own. Plowman had one eye on Congress and the White House and another on potential stakeholders. He arranged for meetings up and down the corridor to get a feel for the interests of potential stakeholders, including the Pennsylvania Railroad. When the Department of Transportation (DOT) was created, the Northeast Corridor Program found a home, where it has limped along ever since.

As did Ben Gottas, Plowman used me as a sounding board for all kinds of things. Deregulation of railroads was not yet thinkable, but mergers and acquisitions were, and they were a favorite topic. Although coming to Washington from the U.S. Steel Company, Plowman was from Portland, Maine. He dreamed of the reorganization of New England railroads, and I can still see the map of those railroads in my mind's eye and hear his musing about possibilities.

A year or so later I learned even more about interest groups and Congress when serving on the congressionally mandated Independent Study Board. We had some data and thoughtful things to say about the impact of government expenditures on regional economic development. I learned a lesson about timing when our report fell on the deaf ears of the new Nixon administration.

Around 1970 I served on a combined National Research Council and National Academy of Sciences (NAS) committee on the uses of science and technology in regional economic development. Dan Alpert, a physicist from the University of Illinois, chaired that committee. Largely because of his leadership, we produced a good report. Deaf ears again. The report was for the Economic Development Administration at the Commerce Department, where regional economics dominated thinking, and science and technology wasn't their bag, so to speak. Another lesson about clients for ideas and work.

In the early 1960s, the founder of Northwestern's Transportation Center had moved to a fund-raising position in campus administration, and a new director had arrived. As an academic outsider, the new director let problems build without resolution, and as I look back, it is clear to me that he needed a mentor or guide through the rapids of academia. I should have tried to help.

A seemingly minor crisis occurred when the director resigned rather abruptly. I say "minor" because there were managers in the center's short-course program who could have taken leadership, which would have held things together but would not have overcome the center's detachment from the campus.

Unfortunately, it was soon found that the crisis was major. Although the center was committed to funding students and faculty, the director had made little effort to report on its activities to its major industry sponsors (mainly airlines, railroads, and large shippers). Also, reports from a large study funded by a sponsor were overdue. Indeed, the money had been spent on rounds of summer academic salaries, travel, and lots of working lunches, yet no product of the study could be found.

I accepted the directorship in part from the pull of new venues, but I was also pushed by the opportunity to reduce my overcommitment to two departments and pulled by loyalty to the university, with thoughts that research centers were good ideas and such. Did I act in haste and regret at leisure? Not really.

With the help of Joe Schoefer, we were able to put together a proxy for the overdue research report. Actually, the proxy was quite thoughtful, but it fell far short of the product expected from a study costing several hundred thousand dollars that was to have included case studies and empirical analyses. I can remember to this day facing the client (a committee) in a walnut-paneled board room. I just leveled with them, successfully. I came away with their respect and without their demanding their money back, a demand that would have killed the center.

I had thought that I would spend a year at the center and then return either to geography or civil engineering, but not to both. That plan didn't pan out. It took more than a year to mend fences with center sponsors, and I was surprised and pleased that I did not find that work onerous at all. Typically, I would make two calls at a corporation: one to the person interested in transportation and the other to the head of the corporate foundation. I would tell them what we had been doing and discuss their activities. Doing that, I learned a great deal about transportation modes and shippers.

The center's advisory committee was chaired by Downing Jenks, president of the Missouri Pacific Railroad, and he was very helpful. He gave me a railroading education as I went along on his annual trip over the road. A decade later Robbie Pfeiffer of the Matson Lines gave me an education on maritime and port affairs as we flew back and forth across the country to committee meetings.

I learned what any geographer should know—one learns by talking with people, kicking the tires. Over the years, I've talked with agency and industry folk of all stripes, and it has helped me understand what they do. I helped the American Public Works Association with its oral history of the builders of the interstate highway system, and I found things that I didn't know even though I had participated in some of the activities.

With no preliminary discussions at all, Lyle Lanier, provost of the University of Illinois system, called and offered me the directorship of their universitywide Center for Urban Studies. The center was to be housed at the new Chicago Circle Campus and would have funding from the university on the order of that for Northwestern's Transportation Center. I talked to Dan Alpert, dean of the Graduate School at Illinois, and to a few other people and decided, why not?

I'd been lecturing in the Brookings Institution's Urban Policy Program, which involved visiting cities and talking with stakeholders of various stripes. I heard lectures from others in the program and felt comfortable with their ideas. My thought was that the urban emphasis acknowledged that we had become an urban society and that the time was ripe for new thinking in many fields. Why not have fun being involved in that development in a university that was busy expanding?

At the Circle Campus, we had fine quarters in a brand-new building, and I was at home immediately because Dorothy Plant, who had been my administrative assistant at Northwestern, came along. During the next year, Ed Thomas, Joe Schoefer, and others joined the faculty, and they, along with folks already there, gave an instant critical mass. Mathematics, three social science, and two engineering departments were involved.

There were as many ideas about what the urban center should do as there were faculty, and even faculty at the Chicago-based Medical School and Urbana campuses had their claims on where center resources should be spent. To respond to the hands-on, activist urges to cure social problems, Maurice Larry, a one-time boxer and a public administration graduate from Urbana, joined our program. There were complaints that he gave short shrift to wishing-would-make-it-so notions, yet he had a good nose for legitimate ideas and requests for aid and was very effective. I certainly learned a great deal from him because he was in touch with the communities that I didn't know. I remember very well the day he said he would be out for a bit. "A Blackstone Ranger [a South Side gang] snatched my wife's purse." He was back with the purse, contents intact, in about an hour.

Close to my interests, there was the thought of *urban* as a metaphor for modern conditions of all sorts and an interesting context for research. People were developing as intellectual resources for such research, and psychology was a bright spot in the social sciences, while sociology showed promise. The dean of engineering had good ideas, but he moved on to the Brooklyn Polytechnic Institute. No matter that things were spotty, with new hires being co-opted, there seemed to be promise in the urban center, given enough time.

At about this time, I served on a committee to look into the establishment of a School of Public Health in the University of Illinois Medical School. We did the usual things. After surveying schools elsewhere, we projected a school that mirrored what we found. The chair of the committee suggested that we should think of a School for the Health of the Public, and this sent us back to work. I felt much better about what we were undertaking. The lesson here is that the flags in which one is wrapped do matter, for they can change the direction in which folk march. Course changes take a while longer, of course.

Having served for several years on the NSF Advisory Committee on Economics (at that time the committee was handling geography and regional science proposals and doing so in a very evenhanded way), I was asked to serve on a panel of the National Science Board looking into the contributions of the social sciences to science and engineering. The agenda was increased social science funding, of course. It was another one of those efforts that didn't lead to much, yet it was an education. I learned about how social scientists think of themselves and how others think of them. Howard Hines headed the economics program and later the social and behavioral science program. He was an unsung hero for geography and the social sciences.

Preparing for the 1970 Census of Population, the Bureau of the Census had in mind a mail-out/mail-in questionnaire to reduce costs (as well as the political hassle of hiring congressmen-recommended enumerators). Address lists were available, but once a form comes back, how can the address be matched to block and other census-reporting areas? Easy enough, just give postal addresses $x, y$ coordinates.

But where would we get the money for the straightforward job of assigning addresses to geographic coordinates? And there had to be much thought and effort given to developing computer programs. Some technical problems had to be faced, such as the preciseness of the coordinates on block faces.

Wilbur Steger of CONSAD (a Pittsburgh-based consulting firm) made a marriage. He brought the Census of Population and Research Division managers together with potential data users—academics (from geography, urban economics, and sociology), business people, and government agencies (the Department of Housing and Urban Development; the DOT and the Federal Highway Administration; and the Department of Health, Education, and Welfare). A venue was created, the Advisory Committee on Small Area Data, and I chaired that committee for about three years. The census needed money, and we thought that agencies would make money available if small-area data could be shown to be valuable.

For example, the census asks a "where do you work" question on its long form, and the answer provides data useful for urban transportation planning. The DOT was willing to spend money to assure that data were appropriately collected and that computer programs were written to aid its use.

Two legs of the stool were there—the census and federal agencies interested in funding the census. The third leg was demand by users, especially urban and state agencies. Adding that leg was a close call.

In late 1968 or early 1969, the Census Bureau went to New Haven to conduct a trial. We, the advisory committee and the census folk, sat at a large table in a room at the Yale University computing center with people from the local agency. Census said what it wanted to do. We then went around the table as the locals responded. One after another, education, highway-planning, parks, police, and zoning administrators said roughly the same thing. "We are very pleased that the Census Bureau is here and that you are interested in what we do in New Haven. But, to be frank, we are doing very well, have exactly the data we need, and are perfect in every way. Give us a ring if we can be of help to you."

The committee had long faces at dinner that night.

The next morning started out as a continuation of the previous day. The last of the New Haven participants was on my right—a not-so-young, fragile-looking person. She sat up straight and said with great presence, roughly, "I am Dr. Smith from the Child Health Division of the State Health Department. Unlike my colleagues in other lines of work, we are doing a terrible job in New Haven, and we will clutch at any straw to improve what we do." Going on, she had a list of things that she thought it would be useful to do, mainly map-matching things—matching areal patterns of disease, crime, income, and more.

With those words, the small-area data program was off the starting block. With local demand, the vast Health, Education, and Welfare money bags opened, and the department sponsored much of the programming and data-management technique development. The highway folk at the federal level came in with money from the pot of planning money, but after the 1970 census was taken, things didn't go smoothly. There were long delays in making data available, for example. It has taken quite a while for the capability to develop, but today it is there. Right now, an e-mail to the Bureau of Transportation Statistics of the federal DOT will bring you a CD-ROM with more census data than you wanted to know about your city, its transportation facilities, and travel.

Lessons? One needs clients and sponsors, and things may take a long time to reach fruition.

In addition to stimulating the small-area data work at the Census Bureau, Will Steger helped Edgar Horwood create the Urban and Regional Information System Society (URISA), which has played a major role in developing small-area data capabilities. I was able to play a small part in this project and look back on it with pleasure.

Unable to say "No" to things that seemed interesting, I got somewhat bogged down again in transportation work. Serving on the executive committee of the NAS Highway Research Board, I headed a committee to ease its transition to the Transportation Research Board. Problems within the Highway Research Board were small because most members recognized the desirability of expanded scope. But then there was within-academy concern about transportation work being such a large part of the budget and activities (a tail-wags-the-dog issue), about the mismatch between the academy folks' self-image of brilliant people advising on important things, and about the pecking-order notion that transportation questions were trivial—"I know solutions, why won't people do what I say?" I learned about situations in which narrow-mindedness and high IQS make for endless, dismal committee meetings in which the ploy "What is mine is mine, what is yours is open to discussion" is often used. The lesson for me was to avoid such situations.

The DOT was getting started, and high expectations marked the recruitment of staff and programs under development. I was caught up in the swirl and served on the NAS Research Advisory Committee and on an Urban Mass

Transportation Administration program development committee, and did many other things.

It was heady stuff, but the dynamic driven by new programs and people began to stagnate, and political expediency rather than considerations that interested me became more and more the rule. The heavy involvement, excitement, and feelings of accomplishment lasted about six years, then I gradually withdrew from involvement. My last gasp was serving as a consultant on the congressional National Transportation Policy Study. Later, I had more limited but similar experiences when the Environmental Protection Agency's social and policy research program got underway. Lots of fun early on, but stasis sets in fairly quickly.

By 1970, I had developed great interest in innovation processes, an interest beginning with my contacts with Torsten Hägerstrand in the 1950s and with Aaron Gellman in the 1960s. This interest was further stimulated during the committee work I had done with Dan Alpert and was augmented by my conversations with Ben Gottas, as well as by service on committees with Cy Herwald, Harvey Brooks, Bill Harris, Jack Fearnsides, and others. Interest expanded the more I read and had contact with people in industry and government agencies. I began to understand that transportation and communications serve as great enablers; they enable innovation in other production and consumption sectors.

By 1970, I was also about halfway along my life's professional path. Although learning in other areas continued, I had pretty much learned who I was and what I could do and liked to do. The path from then on included attempting new venues, making variations on old lessons, and playing with ideas about innovations, including trying some of those ideas.

Zooming along that path, I moved to the University of Pittsburgh at this time to establish an environmental systems engineering program. Dean Hal Hoelscher, a chemical engineer, and I had mutual interests and complementary strengths. I emphasized the *systems* in our program title, and with support from the Environmental Protection Administration we began to do some analyses of the interfaces between built and natural environments. A Sloan Foundation grant provided support for students and gave us visibility on the campus and in the nation. The grant also helped us recruit Kan Chen, an electrical engineer, and sociologist Norm Hummon. Augmenting the program with faculty already in residence, we had a group of about seven intensely interested people, although we suffered a loss when Hal Hoelscher became president of the American University at Beirut.

Striving to influence off-campus clients, we took on a large role in a NSF-funded American Society of Civil Engineers study of the environmental impacts of civil engineering works. In the background was the notion that civil works enhance the environment, so documenting impacts would help tell the story of enhancement. Working mainly with Michael Rowe, a student in the public health program, I traced the history and impact of the implementation of the Allegheny County Sanitary District Plan.

The results aided my thinking about the enhancing issue and more. They said that political considerations led to a large, expensive regional system that concentrated the outfall of the treatment plant. Smaller systems would have been less expensive to build and operate, and would have had a less-damaging diffused outfall among rivers. The experience underscored previous lessons about the relations among political decision making, media promotion, and system designs. I learned to ask about the motivations of those who promote big plans, as well as to question the veracity of big plans.

Zooming further along, I moved in 1973 to the Department of Civil Engineering and the Institute of Transportation Studies at the University of California, Berkeley. There, I tackled academic and research program-building tasks, and although I had striven to reduce my extramural activities, I was still flying back and forth to Washington and elsewhere in committee and professional activity roles. I continued striving to understand better large-system technological development and implementation processes.

Students made a difference at Berkeley. Although I had worked with excellent students at Illinois and Pittsburgh, they were fewer in number than those I had worked with at Northwestern and Washington. It was program size that mattered. Berkeley had large graduate programs, and I mainly worked with students from engineering and planning, and occasionally with students from economics, public health, and geography. Having many students makes a difference.

By 1985, I had pretty much withdrawn from administrative and committee work in favor of teaching and research. Retiring in 1991, I reduced teaching to one course per year and moved into a small, sequestered study. I continue researching, reviewing papers for journals, and doing some committee work, albeit at a reduced level. I do have a good bit of contact with students in residence, as well as with folk with whom I have worked over the years. Playing with ideas continues to be fun, and I learn something new every year.

Looking back over what I've written, I seem to have stressed easy-to-discuss lessons where I learned something clear-cut. What about things that were and are not so clear-cut?

There are ethical questions. Not too many years ago the philosopher Wes Churchman asked me and other of his friends to coauthor a book on ethical world government. In Churchman style, that collaboration involved his writing, our reviewing his work and offering text, and his praising but ignoring what we offered. I'm sorry that the effort lost its energy. I might have learned something because I struggled with ethical questions all along my path, and I'm not certain I handled them well.

The rules are simple enough. In the spirit of Kant, I have tried not to harm others, and I think I managed this and that without doing real harm. The most bothersome cases came about when I served on advisory committees to presidential or senatorial aspirants and made recommendations about programs for those just elected. In these sorts of contexts, "Say this—Say that" recommendations get made that have an unfortunate ends-justify-the-means character. To a lesser degree, such situations also arise when teaching, when reviewing articles for journals, and when serving on scientific and programmatic advisory committees.

Should one avoid such situations? That's one strategy, and it is one a few of my colleagues use. But I've found myself being a little unethical when promoting means and justifying that promotion by highly valuing ends. I'm uncomfortable with that.

There are also situations in which the means determine not so desirable ends. My Allegheny County Sanitary District study sensitized me to such situations, and I see lots of them. That wasn't just a "damn the politicians" situation. Professionals see the world in certain ways, and they may impose their values and ways of thinking on others in situations where such values and ways are inappropriate. Geography seems more benign on this score than are some of the fields in which I have worked. I'm thinking especially of prescriptions by planners, engineers, and economists.

A less difficult problem for me has been the balance between altruistic and self-serving behavior. I've been inclined to lean to the altruistic. If anybody kept score, they would probably tell me that I have spent much too much time traveling here and there, writing gratis committee reports, and reviewing this or that. I'd say that's not true, for these activities have given me great pleasure, and I think they have been socially useful. The great scorekeeper in the sky might say, however, that papers not written and other things not done support the criticism. By giving value to such things, he flatters me.

Again, an unanswered question: How much for myself, how much for society?

During my last few years at the University of Washington, I had a table in the outer office where I'd sit with students for lunch, conversations, seminars, and card games. Not many years later I did away with a desk and worked from a table in what I liked to call my study. Who cares about furniture? The point is that I had a place to sit on an equal footing with faculty, colleagues, students, and visitors. I learned the virtues of that kind of open, many-person, equal-footing interactive situation for learning and playing with ideas. It was a lesson that served well.

How could there be a question about that lesson?

There is because when one works closely with students, one has to ask the question when they should move on and cut the apron strings. Surely most students are happy to move on when the time comes. No matter how much the teacher tries not to impose ways of thinking, he does, and many are glad to be rid of him.

Nowadays, I fear that I have not read such situations well. Have I been too quick to keep apron strings cut by not staying in touch, by being slow to offer advice, and by demurring when asked about this decision or that? I regret that I don't have a good feel for the answer to that question. Costly to me, I may have withdrawn too quickly from personal relations, especially with my many past students.

# You Don't Have to Have Sight to Have Vision

## Reginald G. Golledge

Courtesy of Reginald G. Golledge

*Reginald Golledge (b. 1937) received his B.A. in 1959 and M.A. in 1961 from the University of New England, Australia, and his Ph.D. in 1966 from the University of Iowa. He held positions at the University of Canterbury (New Zealand), the University of British Columbia, and Ohio State before joining the faculty of the University of California (Santa Barbara) in 1977, where he is currently a professor of geography. He is the author or coauthor of fifteen books, a contributor of chapters to seventy others, and author of more than 120 professional articles. As early as 1981, he received an Honors award from the Association of American Geographers, and in 1990 he became a fellow of the American Association for the Advancement of Science. He is an honorary lifetime member of the Institute of Australian Geographers and a recipient of its Gold Medal in 1998. Awarded a Guggenheim Fellowship in 1987 to further his research in helping visually disabled people navigate effectively in complex urban environments, he has also remained deeply committed to bringing geographical perspectives into the larger public realm and is particularly interested in expanding the role of effective geographic education in the schools. In 1999, he was elected president of the Association of American Geographers.*

## The Early Years

I think I've always been a geographer. At least, I've always been interested in place-to-place differences. As I was growing up in Australia, my family moved frequently, largely from one small town to another in the interior of the state of New South Wales. Exotic-sounding little places such as Dungog (where I was born), Katoomba, Lithgow, Tocumwal, Cootamundra, and Yarra became my home at various times. They varied from about five thousand to fifty in population size. Not exactly places that one looks for on a map of Australia. The local environments varied from the rich, rolling hill country of the Hunter River Valley to the rugged, dissected plateau and valley country of the Blue Mountains, to the flat, dusty expanses of the plains of southwestern New South Wales through which the Murray River meandered, to the rich wheat and sheep country of the Australian Riverina and the Eucalypt-covered hills of the Southern Tablelands. The small-town environment and the surrounding countryside favored the development of a state of mind that constantly asked, "What's over the next hill? How far is it to the river? Where are the wild berries and fruits located? Which fields have dangerous cattle in them? If I go north out of town, will it look the same as when I go south out of town? Why is it that we could catch cod in the river near Tocumwal but only perch in the river near Dungog?"

## Scale and Detail

I don't think I was unduly curious about where England was or where the United States or Japan was located, but growing up during World War II, I was faced daily with newspaper coverage (with maps) of how the war was faring in Europe and North Africa. Daily we all examined these maps to see if the combat line had moved. But headlines that screamed "Allies Advance 20 miles on a Broad Front" could not be supported by the mapped lines of combat. These lines looked the same day after day—my first experience with scale!

Acknowledgment: I want to acknowledge the assistance of my wife, Allison, and two of my children, Bryan and Brittany, who refreshed my memory on many aspects of my past life while simultaneously refusing to take any responsibility for any of it.

How different in the latter years of the war, when our interest shifted to the Japanese advance through the western Pacific islands, when it seemed that the daily combat line leapt down the island chains. Little information was known about either the Japanese people or the western Pacific countries. Strange names such as Guam, Bougainville, and Kokoda took the place of France, Italy, Egypt, and Germany. But where were these places? What were they like? Why were our troops now fighting there? If nothing else, the various theaters of war brought world geography into every home, replacing the U.K.-centric knowledge structures that dominated thinking and teaching environments prior to these events with a perspective tied to the United States and the western Pacific Rim.

## Knowing One's Environment

When you spent a part of each day wandering around the Australian bush, it was very useful, and indeed necessary, to have a reasonable knowledge of what might be expected in the environment. Knowing where water could be found, which places most likely harbored poisonous snakes, and how to figure out where you were in a dense eucalypt forest, where every tree looked the same for twenty miles in all directions and there was no moss on the trees to help decide orientation on a cloudy day, were essentials for survival.

I can recall two events from World War II that affected us directly. When we lived in a little Murray River town called Tocumwal (population two thousand), we discovered that there was a Japanese prisoner-of-war camp a few miles away. One night there was a mass escape. Gasoline storage tanks about one mile away were blown up. We were all advised to arm ourselves and stay indoors. Within a day, nearly all the escapees had been recaptured. Over the next few weeks, the rest of them straggled back to prison voluntarily. They did not know how to deal with the harsh sunburnt countryside. They were dehydrated and starving; several had died of snakebites. Many of us kids couldn't understand why this had happened, for we all knew how to find water and food and how to protect ourselves in this same area.

The other incident occurred at the port and steel city of Newcastle, New South Wales. Our family was sitting on my aunt's veranda in the suburb of Stockton, across the Hunter River from the shipyards and BHP steelworks.

Suddenly, about one hundred yards into the river, a small vessel emerged—later identified as a two-man submarine. The occupants began firing a small cannon at the steelworks. We gaped in amazement and horror at the minuscule touch of war on our doorstep. After firing a few rounds, the vessel submerged. It was caught in steel nets drawn across the river at the first sign of trouble. This incident certainly made us more aware of global conflict, and even though we had been complacent because of our distance from the European and North African venues of war, we realized that geographic isolation meant less as the war shifted to the western Pacific.

## A Little Zoogeography on the Side

Snakes have always terrified me. Most Australian snakes are highly poisonous—about the only ones not poisonous are the pythons. I clearly remember many snake incidents to this day. On school holidays, I usually worked for a local farmer as an agricultural laborer or (sometimes with my brothers) ran a professional rabbit-trapping line. Rabbits were at plague proportions in the 1950s in Australia, but the sale of rabbit products paid for my clothes and the extra schoolbooks that helped me get to college. One summer, another teenager and I took a job on a remote farm to help build a small dam for watering household stock. One Saturday we were given a day off; otherwise, because we were very isolated, we worked seven days a week. We borrowed a rifle and shotgun from the "cocky" or "grazier" (i.e., rancher) who employed us (we were both "ex-cadets" who had been given basic military training in high school and thus knew how to handle firearms) and went hunting.

We found a mob of kangaroos. They were almost in plague proportions; it was a drought year, and there was open season on them. I elected to find my way to a small hilltop, and my partner intended to drive the roos to me. To reach the hilltop unobserved, I had to climb a cliff about sixty feet high above a small stream. Shotgun in one hand, I slowly pulled myself up the cliff face, aiming for a small ledge about six feet from the top. I reached the ledge, and, grabbing it with one hand, pulled myself up head high, simultaneously swinging the gun up with the intent of laying it on the ledge. As my eyes reached the level of the ledge, they looked directly into the beady eyes of a six-foot red-bellied black snake coiled there. Reflexes took hold. The arm lifting the gun pointed it blindly; somehow my finger instantaneously found the trigger. The blast blew the snake off the ledge, taking my eyebrows and some skin from my forehead with it. Now, a red-bellied black snake's bite is said to be potentially fatal only about 50 percent of the time if untreated, but a bite on the face may have changed those odds. It took me nearly fifteen minutes to climb the last six feet. And as I stood shaking on the top, I heard my schoolmate screaming at me for frightening off the mob of roos before he could get a shot.

The real point of this anecdote, however, was that our employer thought that we were "city kids," and after we'd been gone for three hours he began to think that we were lost. He panicked. A series of phone calls brought about a dozen neighbors, horse mounted, to begin a search for us. He didn't know that we were fully aware of where we went (even though the area was unfamiliar) and could—in the tradition of millennia of explorers—point directly to home and take a shortcut there. I've learned that this tactic is called "path integration" or "homing-vector" knowledge and that it is typical of navigation by nonhuman species and (some) human travelers. The memory of this episode often returns as today I conduct research on human wayfinding practices with and without the aid of vision.

In our farmhouse at Yama (a hamlet of twenty-five people, a one-room school, a gas pump, post office, and railway station, about six miles from the town of Goulburn and about fifty miles from Canberra), poisonous snakes were a fact of life. I shot them on our front doorstep, in the garden, and on the old run-down tennis court. Once when I came home to what I thought was an empty house, I walked past the open bathroom door and saw my mother pressed against the far wall—stark horror disfiguring her face. About five feet in front of her lay the rippling coils of a very deadly brown snake. I crept past the door, grabbed the shotgun, and, carefully aiming so as not to hit her, blew the snake away. I also blew a leg off the bathtub, a five-inch hole in it, and a fourteen-inch hole in the back wall. Ever try taking a bath in a three-legged tub?

We also had rats, a favorite food of snakes, under the house and in the ceiling. Above my bed, the ceiling boards had warped and gaped open. One Sunday morning I was reading the paper when I heard a scratching, a rustle and skitter, and bingo! A snake dropped through the hole in the ceiling onto my stomach. I believe my scream of fright was heard twenty miles away. I flung myself upward and backward through the glass window above my bed and hit the ground running. I don't know what happened to that snake. I thanked my rugby training for the strength and reflexes to get out of the way and slept on the verandah for the next three months until cold weather forced the snakes into hibernation and me to go inside.

Growing up in the bush made one "notice things." One quickly learned to step *on* logs not over them in case a "Joe Blake" (snake) was basking on a sheltered side; to notice that a bunch of tussocks was really a dug-in spiny anteater *(Echidna);* and to differentiate between edible and poisonous mushrooms— all this was part of growing up. But "noticing things" is an essential part of being a geographer, too!

### Impacts of Formal Schooling

My interest in the different local environments translated into a more general interest in geography in the schoolroom. In elementary school, I systematically read every book in the local library on Australia. The rich descriptions of life in the hostile environment of early settlement captured in books such as Marcus Clarke's *For the Term of His Natural Life* and Rolf Bolderwood's *Robbery under Arms* captivated my senses. Later, John Gunther's "Inside _____" series became another favorite. One of the earliest assignments I can remember doing enthusiastically was a paper on the Fang tribes of Central West Africa. Descriptions of the different culture and different environment fascinated me enough to cause me to go to a local library and even to miss a few cricket games—both previously unheard of activities for me.

From a very early age, I was at the head of the class for geography. Strangely enough, up until the middle years of high school, I was also among the best in mathematics. I'm not sure what changed my attitude to math in my third year in high school, but for the rest of my internment there, math in all its forms (except geometry) was an anathema to me. Geometry, on the other hand, I found fascinating. I learned and understood all of Euclid's theorems and everything else that was thrown at me in that genre. But despite repeated tries in high school, college, graduate school, and numerous postdoctoral excursions into math departments, I have never really been able to get a good grasp of calculus. Markov chains and stochastic processes, matrix and vector algebra, all made sense: they seemed reasonable things for understanding the world or for examining spatial events. But calculus?

Although the family spent most of my early years at interior towns in Australia, each year for two weeks we took annual holidays on the coast. My father was the only one I knew who elected to take holidays in the middle of an Australian winter. He did so primarily because during those weeks in August the ocean fishing was best. I cannot remember how many nights we spent on riverbanks and on sea walls at the mouths of eastern rivers—crouched in the dark, sometimes with sacking (no plastic garbage bags available then) over our heads to protect us from pouring rain, all for the purpose of catching fish.

But I can also remember the long train trips from the interior towns to the coast. Generally we occupied what were called "dog boxes"—really small compartments with direct access to the outside, but no connecting corridor—which inevitably were freezing cold and uncomfortable.

But those trips were exciting: pulling into strange stations at all hours of the day and night, seeing the bustle of railroad station activity at the time when railroad usage was at its peak, wondering about the quality of the food that would be brought back to us from each "refreshment room" (usually meat pies and sausage rolls), and then staring blankly at the lights of city streets or the darkness of surrounding countryside as we chugged our way to our final destinations. And then the long trip home, seeing things from a different perspective, looking at a side of the railroad tracks other than the ones seen on the initial trip, and all the time looking, observing, noticing things, and wondering why.

### Events at High School

Our high school curriculum and classes were rigidly hierarchic in nature, based largely on the Scottish school system. Grade levels were differentiated from A to G in the lower grades, and A to C in the higher ones. Most of the kids in classes D through G graduated after only three years at high school (i.e., at age fifteen), after they had taken an "Intermediate Certificate" exam. In first year, the A-level classes took two languages (usually Latin and French); the B-level classes took French and geography; other classes took no languages and concentrated more on "shop" activities or applied subjects. Latin was the only subject I failed in my school career: a farewell to a medical career. The next year I took geography and listened to our teacher talk about world cities, where humans "lived in caves of steel in cliffs of man-made masonry." That caused raucous laughter, and I related immediately to a subject that could indulge in such flowery description and still take itself seriously.

Except in geography, and perhaps history, I was by no means the smartest in my class. While my classmates were all talking about scholarships and which major university they would go to, my dreams did not extend beyond obtaining a two-year college degree, which would enable me to teach elementary school. It was a tremendous shock, therefore, when the statewide school "Leaving Certificate" results were published and I found that I had grades good enough to qualify me for a variety of federal, state, and Department of Education scholarships to a full university. I was a little scared to face the problem of living by myself in a large cosmopolitan city such as Sydney, so I followed my best friend's actions and enrolled at a very small rural university, the University of New England (UNE), located in a town of about ten thousand people on the northern tablelands of New South Wales.

## The Armidale Experience

The University of New England was *rural* in the true sense of the word. Its campus was about three miles from the edge of town. Two beautiful old buildings, remnants of Australia's "squatocracy" (large rural landholders called "squatters," with significant local power—somewhat like the lords of the manor in England), were the anchors of the campus. When I arrived, most students lived in housing the university owned or rented in the city—large houses in which twenty to thirty students resided. One night, while slinking home along darkened streets at about 3:30 A.M., after successfully "raiding" other residences, I was amazed to see two kangaroos hopping down the main street of the town. Yes, we had really come to a rural university. Strangely enough, twenty years later, when I was in Chicago for an Association of American Geographers committee meeting and on the way back to my hotel at about 2:00 A.M., I was stopped by a police car whose driver rather sheepishly asked if I had seen four kangaroos hopping about! In Chicago? And I thought *I* was a bit looped! Evidently the beasts had escaped from the zoo and headed south. The next day they were seen in Joliet. I wonder how far "south" they got?

On another dark night some years later, tension filled the air again. This time my friend and I were worried about clouds, not raiders or kangaroos. At 11:00 P.M., we were due to view the first satellite. We sat on the hillside, staring at the star-filled sky. Finally a small glowing speck appeared, lost at times in the glorious starry background of a clear Southern Hemisphere sky. The space age had arrived. Our level of excitement was suddenly heightened to the point of excruciation. In the dark, we had unknowingly placed our blanket on a bull ants' nest (fierce little creatures about two inches long). It had taken them some time to find a way out from under the blanket, but after a scouting nip to my toe, they attacked in force. As we fled in pain and panic to the nearest shower, it was brought home to me that even if our attention was to be focused on the stars in the future, we had to be concerned with the present environment also—and that environment sometimes took precedence.

I can't, in all good faith, give details of what a lousy student I was during the first year at college, but I was able to turn things around at exam time. Suffice to say I was a good rugby footballer, a mediocre chess player, and good enough at poker and solo to earn my entertainment expenses. During the day, I spent a great deal of time on those activities and must have seemed like a real flake! I guess my image was of a dedicated jock with little academic potential. But I did try to study at night—which of course was not obvious to my teachers or my peers, who were always surprised at how well I did on exams and papers. I learned that the secret was to work—even if no one knew about it!

In 1955, UNE had about 240 students and was completely residential. For a number of reasons, I count myself fortunate in going to UNE. First, I was welcomed to geography by Ellis Thorpe, then founding head of the department and a truly wonderful person. He knew all the students, their strengths and frailties, their outside interests, their work habits. He was a mentor even more than he was a teacher, and as a teacher he was absolutely first class. From my first year, I developed a respect that ripened into a friendship that lasted until Ellis died a few years ago.

The Department of Geography at UNE at that time was quite different from any other department in Australia. It included Eric Womington, a British expatriate and a maverick if ever there was one (demography and political geography); Jim Rose, a New Zealander (social and population geography); Ted Chapman, fresh from a sabbatical at Ohio State University (historical geography of both Australia and the United States, transportation, economic geography, and locational analysis); Harold Brookfield (cultural geography); Herb King, an Australian Ph.D. from Australian National University (urban geography, including Christaller's central place theory); John Holmes, an Australian (small towns and rural Australia); Gene Fitzpatrick, an American (climatology and meteorology, some basic statistics, and how to fly-fish for trout); Brian Plummer (biogeography and soils); Iain Davis, a Brit via Canada (made geomorphology come alive); and, of course, Ellis Thorpe, an Australian (cartography, map interpretation, mapmaking, surveying, and fieldwork, including fossicking for opals, garnets, and sapphires). Early on I learned that a useful fieldwork tactic was to schedule such work near towns with seafood cooperatives and small breweries. By the time I had graduated (B.A. with honors) at the end of 1958, we had explored Walter Isard's *Location and Space Economy;* examined the demographic transition and Walter Rostow's *Stages of Economic Growth;* immersed ourselves in subsistence cultures; explored location theory and Hoover's transportation modeling; and had been molded by Ellis Thorpe into students who saw a fascinating future opening for geographers. Not incidentally, my first classes were demonstrated (i.e., TA'd) by Robert H. T. Smith before he went to Northwestern University and returned to bring the quantitative revolution to Australia. Incidentally, Bob Smith later (after a distinguished geographic career in the United States and Canada) became the vice chancellor of the University of Western Australia and then of the University of New England.

## The Pitfalls of Fieldwork

I took an honors degree in geography in 1958 and learned many valuable lessons that year. My honors thesis focused on Sydney and compared land-use change in inner-city, suburban, and rural-urban fringe (local-government) areas. I knew little about the city and arbitrarily chose three case-study areas, not knowing what I was in for. Although I was able to work from air photos taken about ten years earlier, I was forced to spend much time in the field updating land-use maps. I was rather surprised to find that the inner-city area that I had chosen as an indicator of high-density urban land use was the main "red light" district of Sydney (King's Cross). Walking those narrow inner-city streets from early morning to early evening, I began to find out things about cities that I had not anticipated. It was a whole new experience and in part served to focus my interest for years to come on cities rather than on rural areas.

My master's degree thesis was on rail freight traffic. Under the supervision of Ted Chapman and with the willing help of Bob Smith, I struggled with the transportation literature. I tried my hand at linear programming, audited some microeconomics and statistics classes, and launched into what seemed like an endless reading of the economics literature on market area analysis. When I became discouraged, I traveled to Canberra to spend a day or two with Bob Smith. He introduced me to Hollerith Cards (the precursors of computers), to the gravity model, and to the "Economic Law of Market Areas," among many, many other things. He assured me that this was the reading that students in the United States were currently doing. No one else in Australia at that time knew much about them. When I punched things on cards and hand sorted them into classes and groups, Eric Womington took me to the next stage and taught me correlation and regression analysis. I soon realized the power of quantitative analysis but accepted it more as a complement to the skills and techniques that I had already learned.

I learned one other important lesson during my honors and master's years at New England. After each trip to Sydney, I would store the collected data in my desk in the physics building. One morning I awoke to the sound of sirens and the banging of doors in the dormitory in which I was a live-in supervisor or warden. I soon found out that the physics department was on fire. We all raced over there, my feet flying a little faster than the others because of the year's work located in an upstairs desk in that building. But we were too late; the fire had taken hold. We got some physics equipment out of the ground floor labs, but mainly I sat and watched it burn to the ground. All my equipment, my data, my notes, my books, and my correspondence were ashes. I had kept no duplicates.

Despite that loss, I was able to reconstruct many of the land-use features in my three chosen areas from aerial photos and a clear memory of what I had observed. I wrote an acceptable thesis and substituted much of von Thünen's agricultural location theory concepts for the hard data that I had lost. The following year I wrote my first professional paper, "Sydney's Rural-Urban Fringe," and had it accepted by the *Australian Geographer*. It had a mix of von Thünen ideas and some correlations between land values and distances that I had salvaged from an earlier trip to Sydney. I believe it was one of the first somewhat quantitative articles to be published by an Australian human geographer.

I heard about my first job offer in a most unusual way. Herb King, our urban geographer, had asked me to come on a land-use mapping trip with him through central coastal and interior Queensland. Herb weighed about three hundred pounds and didn't do much walking. My job was to walk every street of every town and record its land use. When we reached a small town in central Queensland, Herb collapsed with a heart attack. At age twenty-one, I had never driven a car and had no license. Somehow I got Herb out of the motel and into the car, and drove to a local hospital. While he went to intensive care, I went back to the motel and checked out—I had no money to pay for accommodations. I had the equivalent of about a dollar in change in my pocket and that was it. Cautiously, I drove back to the hospital and left his suitcases. I then drove about five miles out of town, pulled off on a back road, and wondered what to do. For three days, I lived on pineapples purchased from a nearby farm at three cents each. They also gave me water. I was getting pretty desperate when a local police officer stopped by and escorted me to the police station. When I got there, after determining what I'd been doing for the last two or three days, they sent out for a hot meal. I showered and changed on their premises. Then the local sergeant explained that they had received a telegram addressed to "Reginald Golledge, Traveling with Dr. Herb King in Central Queensland." The police in about fifty towns had been tracking us down. The telegram said, "You are offered a position as lecturer in geography at the University of Canterbury. Please respond immediately." Well, this was wonderful news; at that time, I couldn't really remember applying for any job at any University of Canterbury. In fact, I thought it was a university in England. I asked the police to telephone my professor and reply that I gratefully accepted their offer. I won't bore you with the trials and tribulations of hitchhiking back to Armidale, but I was immensely surprised when I got back and found that the job offer had come from New Zealand, not the United Kingdom.

## The New Zealand Years

The years from 1961 to 1963 at the University of Canterbury, Christchurch, New Zealand, were tremendously influential in the rest of my life. My master's thesis had been externally reviewed by Harold Mayer (University of Chicago), who was spending a sabbatical at Auckland University at that time and was visiting Canterbury. There he talked to me about going to the United States. The following year, Harold McCarty (chairman of the geography department at the University of Iowa) spent a sabbatical at Canterbury. McCarty and I co-taught courses in economic geography, and he also encouraged me to further my studies in the United States. A year later, he offered me financial support, which I accepted.

While still at Canterbury, though, I was given the task of organizing a field trip to do land-use mapping, soil mapping, and basic surveying. The second-year Canterbury students and I arrived at the little fishing village of Akaroa, nesting inside a caldera on the coast south of Christchurch. Previous field sites had been in the alpine areas of the South Island. I preferred proximity to the ocean and chose Akaroa as our field site. Just how many people have done fieldwork inside an extinct volcano anyway? At 10:00 P.M., playing darts in an after-hours bar, my partner, an Australian Fijian, sent his dart into the double twenty. We had beaten the third team of Maori fishermen who were quietly relaxing with us—at that time pubs in New Zealand closed at 6:00 P.M. Our victories at darts had won a series of prizes: we had bet cases of beer against cases of local crayfish.

Our student group was sleeping in the local elementary school. At 4:00 A.M., I was awakened by crashing noises on the porch, and, staggering out, I found six cases of fresh, lively crayfish stacked there. I hadn't really thought this would happen! But for that day sixty students had cold crayfish tail sandwiches for lunch and steamed crayfish for dinner. I knew the ocean site would prove more attractive than the mountain site!

And in that first year that I was at Canterbury, IBM came to New Zealand! The University of Canterbury invested in an IBM 1620, with thirty-two kilobytes of memory. Geography faculty members (Les King and John Rayner) encouraged me to go to the IBM training classes, where I found to my dismay that all the little exercises consisted of programming mathematical formulae—

none of which I knew. The embarrassment was enough to make me eventually leave the classes, but I did obtain a working knowledge of machine language and FORTRAN programming.

Realizing my quantitative deficiencies, I sat in on King's quantitative methods class for a year and got a good training in spatial statistics. By this time, I'd embarked on a major research project that involved examining one million punch cards (representing freight waybills) per month, provided by the New Zealand Government Railways. I wasn't quite sure what I could do with all these cards. Rayner wrote a summarizing FORTRAN program for me. Thereafter, at the end of each month, I would reserve the computer room for an entire weekend and spend from Friday night to Monday morning running punch cards through Rayner's program to obtain both disaggregate and aggregate tables. Many times Rayner shared these weekends with me, although we fully realized we were not supposed to be bringing meat pies, fish and chips, and bottles of beer into the computer room. Unfortunately, before I had completed the entire year's data summary, I left for the University of Iowa in the summer of 1963. Although I sent funds back to have the remaining data summarized for me by a departmental technician, I never did see that data again.

While I was teaching and reeducating myself, my first wife was completing a master's in educational psychology. Her area of interest was developmental theory, with particular emphasis on Piaget's work. I was curious about the books she read—*The Child's Conception of Geometry, The Child's Conception of Space*—and began reading them myself. I got interested enough to help her carry out her experiments and to run her data through some simple statistics programs using the fabulous IBM computer.

A great deal of activity and learning was compressed into those three years at the University of Canterbury. Through Mayer and McCarty, I had made contact with leading professors in geography from the United States. Through King and Rayner, I had been introduced to computers and had obtained a background in spatial statistics and mathematical modeling. Through educational psychology, I had found Piaget and a new area of interest.

## A Different Place, a Different Clime: The Iowa Years

My wife and I flew to the United States via Fiji, Tahiti, and Hawaii, spending four or five days in each place. We arrived in Denver, where the annual meetings of the Association of American Geographers were being held. There McCarty and King introduced me to virtually everybody who was anybody in North American geography. The gods that I had worshiped from down under turned out to be people in suits and ties. I still can't recall whether or not I went to any papers at that meeting, but I certainly met many people. Afterward, we drove from Denver to Iowa City with the McCartys and the Kings. After a day or so of driving across Colorado and Nebraska, I turned with a puzzled look and said, "Where are all the birds?" Something I had grown up with in rural Australia were flocks and flocks of birds, mostly parrots, with brilliant colors and strange sounds. Crossing the high plains and descending into the Midwest farming area, I couldn't but help notice the absence of birds and beasts. There was no small game or large game in sight or as roadkill; there were no flocks of birds blotting out the sun; there was open range and cattle and fields of ripening corn, but it was static. The movement in the environment within which I had grown up in Australia was missing.

We arrived in Iowa City when the temperature was about a hundred degrees, the humidity about 85 percent, and tornado warnings were being broadcast. Three months later it was twenty-one degrees below zero on the Fahrenheit scale. The shock to my system was immense. The "depths of winter" in Australia produced temperatures around freezing point, but now I was having to deal with temperatures below zero! I would get up in the morning, wander out the front door in my shirtsleeves, and see the sun shining brilliantly; and then my ears would fall off, completely frozen in about fifteen seconds.

Iowa really put the finishing touches to my training as a geographer and allowed me the freedom to develop those newly emerging interests on the cusp of geography and psychology.

At Iowa, I learned the rough-and-tumble and give-and-take of academic interchange, but it was constructive debate and criticism, not the destructive type that is so common of much postmodern and "critical thinking" in geography today. I took courses in philosophy from Gustav Bergman (a member of the Vienna school and a logical positivist), who taught me a new and different way of looking at many problems. But Bergman emphasized the difference between what he called pure science and social science, and he was the first to admit that using logical positivist procedures in social science situations might prove, at the least, extremely unproductive. Nevertheless we all tried it.

## Entering the Job Market Again

I suppose I could leave off here. For those wanting to know how I became a geographer and in particular how I became the sort of geographer I claim to be today, what I've written so far should provide a pretty good background. But the story's not finished yet. Like many academics, I became committed to constant retraining and exploring new ideas. New problems, new ways of stating problems, new ways of solving problems—all drove me to undertake these searches.

After Iowa, I went to Ohio State University, where my cooperative work with John Rayner was, I think, a good example of how physical and human geographers can work together. Remember that Rayner had helped me out with computer programming at Canterbury, and we had jointly taught surveying and mapping together. At that time, different members of the Ohio State geography department taught a seminar open only to other members of the faculty. I learned spatial analysis (King), differential equations (Emilio Casetti), spatial diffusion (Lawrence Brown), and social networks (Kevin Cox). I taught mathematical learning models, and Rayner taught spectral analysis. We began thinking of how we could use the methodology of spectral analysis to deal with problems of urban and economic geography. Stimulated by Tobler's paper "The Spectrum of US-40," we wrote a short paper using two-dimensional spatial spectral analysis, "The Spectrum of US-40 Re-examined." This paper was followed by an examination of the spectral properties of settlements in parts of Oregon, South Dakota, and Pennsylvania. We mapped the town patterns in each case and undertook cross-spectral analysis between those patterns and patterns of slope, drainage, transportation systems, and economic activity. Several years later Rayner and I joined up once again to analyze the spatial competence of selected populations, this time using spectral analysis to compare the cognitive maps and behavior patterns of deinstitutionalized retarded citizens with those of low-income, multiethnic groups from the same neighborhood. In much of the work we did, I think we proved that individuals from physical and human geography could work productively together if they used a common language and were willing to be educated in alternative viewpoints. I also think that this cooperative work was possible because both of us were exposed to solid programs in physical and human geography as well as in geographic techniques during our undergraduate years.

## The Simonett Connection

Christmas 1976. My wife, Allison, and I had completed a six-month sabbatical/honeymoon at the University of Auckland, New Zealand, and were stopping off in Australia before returning to the States. It was hot. It was the kind of hot you can get only in Australia on a midsummer day. We were frying on the beach at a north shore suburb in Sydney when a shadow blotted out the sun. I opened my eyes, turned my head, and saw a pair of skinny freckled legs with pants rolled up to the knee. My gaze drifted upward, and against the sun I saw the blur of a face—a face I knew immediately. "How are ya' mate? Have you decided to come to Santa Barbara?" Dave Simonett, about six months earlier, had offered me a job at the new department he was building at the University of California, Santa Barbara (UCSB). He knew I would be either in New Zealand or Australia around Christmas and physically tracked me down. To this day, I don't know how he did it. Finding me on one of the dozens of harbor and oceanside beaches in Sydney, with tens of thousands packing each beach, was a Sherlockian effort.

In 1970, I had taken a leave of absence from Ohio State and had spent six months at the University of Sydney, where Simonett was chair. I gave a public lecture on the use of multidimensional scaling in geographical analysis—a pretty harmless and not too exciting topic I thought. To jazz it up, I opened the discussion with a series of overheads showing how you could extract a latent spatial configuration from a set of nonspatial data. The data I chose were the basic statistics of Playboy bunnies for the preceding twelve months. The solution came in two easily identified dimensions. Then I launched into the more formal part of the discussion. When I finished and turned off the overhead, the auditorium lights went on. I looked at the audience and gulped. The first two entire rows were filled with nuns.

## The Santa Barbara Experience

With Simonett's guidance, I learned about university politics. Within the University of California system, the academic senate is very powerful. Simonett identified a number of key committees and made sure that our department was represented on each of them. Very quickly I learned the power structure of the campus and began finding where all the skeletons were buried. This knowledge was to stand in good stead when I later became department chair.

We knew that a geography department interfaced between the natural and the social sciences. Yet the theories and methods were so different that it was often difficult for people in those areas to communicate. It was agreed that the best way to encourage communication was to establish a common language. The languages we chose were mathematics, statistics, and computer literacy. The new department therefore evolved with the structure that included physical science, human science, and techniques. Mathematics, statistics, and computer competence were supplemented by cartography and computer graphics. Later, geographic information systems (GIS) analysis was added and has become the most active and significant technical area in the department.

Although I had interacted extensively with psychologists at Ohio State University, it was really after I moved to Santa Barbara that my associations ripened into cooperative work. At UCSB, I met psychologist Larry Hubert. We met for a beer at a local pub called "the English Department," where on Fridays we played darts. The geographers soon proved to be the darts champions (I guess we were there most frequently) and maintained a superiority for many years until one day we were beaten by the math department. Members of that department had convinced a scorekeeper, who was already deep in his cups, to subtract geography scores and add mathematic scores during a simple game of cricket. What perfidy! That single event made me realize that mathematicians were human, too—and were equally capable as geographers of exhibiting base behaviors.

Larry Hubert and I offered two papers to the 1980 International Geographical Union (IGU) meetings in Tokyo, and both papers were put in the same session. There were five presenters, but at presentation time only Akin Mabogunje (the session chair) was there. I presented the two papers to Akin, then we went off for a drink.

After Larry Hubert left UCSB, I began working with another educational psychologist, James Pellegrino. Initially, we were part of a joint team that began examining the use of computational processes and artificial intelligence models in geography. Pellegrino and I had begun working on the empirical basis of children's wayfinding decision processes and carried on a productive program for almost a decade.

## My Work on Disability

White still at Ohio State, I began working with a social psychologist and social worker to study special populations. Our research was a mix of ethnographic, phenomenological, scientific, and analytical approaches. I still become extremely annoyed when I remember reviews and criticisms that used political correctness to criticize us for this work—criticism largely based on our use of terms such as *retarded* rather than on the quality and contribution of the research. I still become annoyed and irritated by those—usually postmodernist thinkers and social theorists—who are content with labeling our work *positivism* and *scientism* rather than understanding the human-responsive and careful way in which we designed the experiments and undertook the research. (Here I can see a bunch of our "antiscientific " social theorists shaking their heads at the mere idea of running "experiments" on humans, themselves neither understanding what experiments are nor knowing that they can be designed in such a way as to protect the interests of the participants while simultaneously providing a rich source of information for the advancement of knowledge about those participants.)

The sudden occurrence of an ischemic optic neuropathy in 1984 left me severely vision impaired and legally blind. The sudden onset of vision loss was devastating. I don't believe that many of our current writers on disability, many of whom are able-bodied, can appreciate what a dramatic impact such an event can have on one's quality of life and on the lives of people around them. And let me say up front, it makes no matter what type of social system you are in or what type of empowerment, political clout, representation, or amount of welfare is given to such groups (e.g., those with vision impairments)—if you can't see, you can't see. And if you can't see, one of the most critical problems is to find some ways to survive, not to navel gaze and meditate about empowerment.

After returning to work in the fall of 1984 and trying desperately to find out how I could continue an academic life—and at first failing miserably to do so—I was visited by two UCSB psychologists, Roberta Klatzky and Jack Loomis. They wondered what they could do to help. Ignoring my rudeness, they sat and talked to me. Bobby Klatzky, whom I had met briefly when attending seminars in the psychology department, knew of some of my work. Eventually she asked: "Haven't you spent the last fifteen years working on cognitive maps and spatial abilities?" I said, "Yes." She then asked: "And haven't you worked with special populations of disabled people before?" Again I answered, "Yes." She then asked, "Why don't you use the expertise that you have developed over the last fifteen years to help yourself?" I was flabbergasted. The thought had never entered my mind. That simple question changed my life. It also began what is up to now a fourteen-year period of collaboration with Bobby and Jack on theoretical, empirical, and methodological questions relating to the spatial abilities of blind or vision-impaired people.

This cooperation started when, after that initial visit, Bobby and Jack met with me every Friday afternoon for two hours. Each week they selected a set of relevant articles concerning spatial behavior, cognitive mapping, tactual mapping, instrumentation to assist people with vision loss, and so on. They effectively gave me a new focus in which I could continue building geographic knowledge.

After listening to a marvelously innovative concept paper Jack had written, our group decided to focus on the problem of independent navigation. In 1985, Jack suggested that we try to use a global positioning system (GPS) and satellite tracking to guide a person's travels through an environment. He imagined the environment to be represented as a digitized base map, with analytical functions similar to the early GIS work. And to crown it all, he suggested a user interface consisting of an auditory virtual display. The National Eye Institute funded our work in the late 1980s, and this collaboration has occupied a good part of my research activities since that time.

I remember going to a departmental party several months after I suffered vision loss. I was still quite depressed and almost unable to cope in public with my condition. A friend, a dentist by trade, was casually talking to me and suddenly said, "Reg, you don't have to have sight to have vision." Fourteen years later, my brain is still reeling from the consequences of that simple statement. I even volunteered to go next week and have my teeth cleaned by the guy! Not knowing my general aversion to dentists, persons other than my family may not fully understand the significance of this particular sacrifice.

Now where does geography fit into all this? Obviously, our studies with the special population of retarded individuals were a mix of social work and geography. Cognitive maps, spatial abilities, geographic understanding of complex environments, mobility and navigation, daily activities and time budgets, navigation and transportation, awareness of social and political and legal systems, and the importance of information flows and information comprehension—all came from my geography background. Through my dealings with vision-impaired and blind individuals, it has come home to me very forcibly that geography is highly dependent on vision. Its findings are represented in tables, graphs, and maps—most of which are inaccessible to those without sight. Yet should we ignore the vision-deficit population? Should not those with even minimal understanding of their environment be encouraged to learn more geography? Do we ignore those people who can't use computer-based GIS or see the multicolored exotic representational forms of maps, atlases, and on-screen displays? There is no doubt that in the geographic domain vision is the most powerful sense. But there are other senses that at times are all that can be used and that geographers almost universally ignore. For some years now, my mission has been to try to find ways that geographic information can be accessible to everyone—in particular to those who are blind or vision impaired. It has been an exciting journey.

I remember how in 1988, when I was working with Don Parkes as he developed for me the tactile auditory information system (later called NOMAD and marketed worldwide), an eight-year-old congenitally blind girl took the seat in front of NOMAD and began tactile exploration of a world map that was being used as the basic display. As she explored the path of raised lines representing the boundary of Australia, a synthesized voice told her that this was Australia and gave a simple description of the different areas of the country that she explored tactually. I will never forget the pure pleasure in her voice as she said, "So *that's* what Australia looks like!" Other blind people at that same demonstration were amazed at how a simple ten-minute exploration of the auditory and tactile information at their fingertips showed them what a representation of the world was like and how it gave them an idea of the isolation of Australia from the other continents and the distances to places that they constantly heard about—such as the United States, England, Japan, and the Soviet Union. Providing these insights to those who have not been able to experience them before has been a part of Don Parkes's life now for more than a decade; he is truly making geography come alive for those who can't see its wonders. I think my efforts have been channeled along similar lines.

While Parkes was developing NOMAD, Jack Loomis, Bobby Klatzky, and I were laying the foundations for what we later called a "personal guidance system for blind or vision-impaired travelers." True to Jack's original concept, we had designed and built a GPS-based navigation system that used GIS to represent and explore a spatial database consisting of the UCSB environment. Although we are still developing and refining this device, our attempts have been given worldwide exposure. And at this point I will try and anticipate my critics by repeating what I have often said in public presentations of our research: that we do not imply that disabled people lack a rich and varied life or are inferior to other people.[1] But based on my own experience and my dealings with large numbers of others who have become disabled after being able-bodied, I believe that there is no question that one's quality of life degrades from that previously experienced. But our research can turn the tide and enhance life.

Some years ago I was asked to write a paper on my experiences as a blind person. The paper was quite difficult to write. I wanted readers to understand the nature of the personal struggle that is required to deal with a disability, but I also wanted the story to be objective and to focus on the devices and techniques that I had found helpful and that others could use. My first draft was returned with comments from the editor and reviewers that helped me to re-

1. See R. Imrie's "Ableist Geographies, Disablist Spaces: Towards a Reconstruction of Golledge's 'Geography and the Disabled,' " *Transactions of the Institute of British Geographers* 21 (1996): 397–403.

think parts of the paper to make it even more personal, but also with some annoying comments, for there was an attempt to liken dealing with a disabling condition to the struggle by feminists and minorities to gain social and political power. This comparison confuses a personal struggle with a social movement—two very different animals. After trying to lay bare my soul (at least that's what I thought I was doing) concerning the struggles, disappointments, and triumphs that I had experienced in dealing with my disability, I was asked to try and make the story "more interesting"! Enough said.

At this time, I am both appreciative and disappointed with the growing work on disability in geography. I am appreciative because of the growing number of people interested in this problem; this work can only end up being beneficial to different disabled groups because I believe that geographers can offer solutions to many problems faced by blind or otherwise disadvantaged people. I am disappointed to a very large degree because of what (at least to me) is an excessive concentration on the fuzzy social-theoretic concerns about disability. There is more being written about the political correctness involved in using the term *disabled* than there is about solving disabled people's problems or even understanding those problems. There is much written about the need to empower such groups, but without in-depth analysis of what such empowerment can do to alleviate the daily problems they face. There is more of a demand for everyone to think with the same social-theoretic tunnel vision about disability than there is for encouraging a multitude of approaches toward examining disability problems and (as Ruth Butler aptly put it)[2] for working *with* disabled groups, not writing *about* them. Despite the constant carping from my social-theoretic critics about the way I conduct my research and the inhumanity with which they claim I deal with participants in that research,[3] I plan to continue my current research practices and to try to find ways to use geography to help different disabled groups in whatever ways I can. And I would challenge my critics to do something useful apart from intellectual navel gazing.

2. See R. E. Butler's "Geography and Vision-Impaired and Blind Populations," *Transactions of the Institute of British Geographers* 19 (1994): 366–68.

3. See Imrie's "Ableist Geographies" and B. J. Gleeson's "A Geography for Disabled People?" *Transactions of the Institute of British Geographers* 21 (1996): 387–96.

Today, I'm still a geographer. Sometimes I'm called a psychologist. Some geographers do not like the type of work I do. I have constant arguments with members of the discipline about the scale at which I often work—the laboratory or rooms or small patches of grass on a campus. I'm frequently told that this is not geography, but no one seems to be able to tell me what *is* geography or *why* my work is not! They are not able to say whether geography starts at the tabletop or at one hundred meters or at one mile or whether it starts beyond the horizon. But many are firmly convinced that what I do is not geography, even if I use the same concepts, processes, and theories that they do.

I believe I am still growing as a geographer. My life as a geographer has been one of change. One might be surprised at this; for decades I have been tarred with the same brush that paints me a logical positivist or a quantitative analyst. It's obvious that people who use those words have read nothing that I have written since the 1970s. But in my constant evolution as a geographer, I have not been seduced by the byways of Marxism, by postmodernism, by social theory, by critical thinking, or by any of the other dozen or more *isms* that have ebbed and flowed in the discipline. I think that by the time I left Iowa, I knew what I wanted to do for the rest of my life. At Ohio State, I laid out for myself a set of long-term research guidelines. I've been pursuing these goals now for approximately thirty years. I have, like the discipline itself, constantly evolved.

Could I or would I have done these things had I been trained in any other area? I don't think so. I think the peculiar ways of looking at the world that are part of being a geographer provide certain insights that one wouldn't get from being involved in another discipline. I have usually worked in a multidisciplinary context, and I believe that more geographers should do so. Although we are still developing a strong discipline of our own, there is still much that we can glean from many others.

The things I remember most about becoming a geographer? Not everyone has had the opportunity to pull an internationally renowned cartographer from the raging surf on Copacabana Beach. Not everyone has had the privilege of seeing a sober demonstration of the Highland fling in the light of a brilliant summer full moon on a spice-drenched grassy knoll at Bellagio in northern Italy, or seen that followed by an exhilarating Irish jig. Not everyone has picked blueberries in the Uppsala Forest while exchanging views on birds in eggs and society and space. Few people have drunk champagne and eaten cold crayfish on the windswept and flotsam-covered beaches of New Zealand's South Island with friends from Sweden, Greece, New Zealand, and Australia. And probably very few geographers have been charged $120 for a shot of Japanese whiskey while they soaked up nightlife after an IGU meeting. Do these events and many others make me any less or more of a geographer? I don't know, but I probably would not have experienced any of them had I not pursued a professional career in geography. I've always thought that my being a geographer is at least owing in part to the way I have lived my life.

I think I've always been a geographer.

# Memories and Desires

## David Harvey

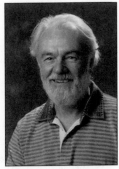

*Courtesy of David Harvey*

*David Harvey (b. 1935) received all three degrees from Cambridge University (Ph.D., 1962) and today is the world's most-cited geographer. His teaching, research, and political engagement reflect a deep concern for social justice, as exemplified by one of his most influential books,* Social Justice and the City *(1973), a work that had a profound effect on an entire generation. Most of his teaching and research since 1973 have been conducted at Johns Hopkins University, with a break of six years as the Halford Mackinder Professor of Geography at Oxford and in the acceptance of twenty-four visiting or short-term scholarly appointments, nineteen plenary addresses, and lectures at ninety-four universities around the world. In 1993, he returned to Johns Hopkins, where he has written eight highly influential books, including* The Limits to Capital *(1982),* Consciousness and the Urban Experience *and* The Urbanization of Capital *(1983),* The Condition of Postmodernity *(1989),* Justice, Nature, and the Geography of Difference *(1996), and* Spaces of Hope *(2000). He has received many honors, including the Retzius Gold Medal from Sweden, the Patron's Medal of the Royal Geographical Society, and the Vautrin Lud International Prize. He has also been awarded honorary degrees from the University of Buenos Aires, Argentina, and Roskilde University, Denmark. In 1988, he was elected a fellow of the British Academy.*

> In a universe divested of illusions and lights, man feels an alien, a stranger. His exile is without remedy since he is deprived of the memory of a lost home or the hope of a promised land.
>
> —Albert Camus, *The Myth of Sisyphus*

149

When I was around twelve years old, I had my first lesson on the geography of North America. We drew a map of the eastern seaboard and marked on it something called the "Fall Line." It stretched from New England to Georgia and recorded where the rolling foothills of the Appalachians abutted onto the flat alluvial coastal plane. Its name derived from the numerous waterfalls to be found there. These waterfalls had social significance because they provided water power for innumerable mills that spawned towns and villages and eventually large cities. Today, the Fall Line is roughly marked by Interstate 95, which connects a whole string of cities up and down the eastern seaboard of the United States.

As I studied that map in the dark and gloomy days of postwar Britain, I dreamed that one day I might visit North America and explore its wonders. The idea seemed hopeless then, even though I had relatives in the United States (they sent us food parcels in the Second World War). We were too poor, and it was all too far away. Little did I then imagine that I would spend more than half my life living on that Fall Line. The deep ravine I see from my study window marks it. I walk two blocks and find myself in the old mill village built in the mid-nineteenth century by the owners of the cotton mills (closed many years ago and now converted into spaces for artists and artisans). These mills lie in the valley where the Jones Falls River sluices across the rocks through channels modified to turn the wheels that brought the industry that made the town that grew into the city where I live now. That city is Baltimore. Let me look backward at how I got here.

I am a well-known human geographer with a substantial (and somewhat controversial) international reputation both within the discipline of geography and beyond. I teach at an elite private university (Johns Hopkins) and have written many books. Because these books are important markers of my career trajectory, I list them in order: *Explanation in Geography* (1969); *Social Justice and the City* (1973); *The Limits to Capital* (1982); *The Urbanization of Capital* (1985); *Consciousness and the Urban Experience* (1985); *The Condition of Postmodernity* (1989); *Justice, Nature, and the Geography of Difference* (1996), and *Spaces of Hope* (2000). I have received many awards and much recognition (including honorary degrees). I work within a framework known as "critical geography," which in my case rests on Marxian theory buttressed by political commitments that are "of the left." My substantive work rests primarily on studies of the political economy of urbanization, most particularly in North America and Europe, coupled with wide-ranging interests in the geographical themes of space, place, and environment in relation to social, economic, cultural, and political life. I have sought to encourage the development of a perspective called "historical-geographical materialism" within geography and tried to take insights gained from such a perspective into a wider world of academic thought (both Marxist and non-Marxist) as well as into cultural and political affairs. I am perpetually trying to merge the geographical and historical imaginations into something that is excitingly present. I want to light up the sky with that knowledge and through that illumination help create a much better world in which we can live, not without conflict, but with reasonably equal life chances coupled with intense respect for our differences (geographical as well as social) and deep understanding of our commonalities. Wherever I go, I try to speak to these themes.

I get pressing invitations from all over the world to lecture and talk. The invitations come from all manner of different institutions (governmental, cultural, and political as well as academic) and disciplines (everything from law and humanities to sociology, international relations, and economics). There are, it seems, many people interested in what I have to say. I accept some of these invitations and travel the world often—Singapore, London, Paris, Barcelona, Melbourne, Milan, Los Angeles, São Paulo, Seattle . . .

## Early Years

If you had told that twelve-year-old boy living in the despairing days of immediate postwar Britain that this was the future in store for him, every muscle in his body would have been aquiver with excitement. He longed to travel the world, studied atlases, assiduously copied maps of strange and faraway places, and pored over the one or two books he found at home with titles such as *Peoples of All Nations* (written, I think, in the 1910s at the high point of British imperial experience of the world). That boy had to rummage in his imagination (TV was yet to come, and he got to the cinema as a treat only once every six months) to construct flights of fancy that landed him in Rio, Rangoon, San Francisco, or Benares. And more than once he decided to run away from

home and explore the world only to find that if it was sunny in the morning, it
was raining in the afternoon (an elementary fact of British meteorology), and
that sharing a hollowed out tree in the rain with an assortment of insects was
not anywhere near as comfortable as bathing in the maternal warmth of home.
And so it was that his interest in what I now call the dialectics of space and
place (the way experiences in place always mesh with broader spatial relations)
began.

But that young boy also came to consciousness in the midst of a bitter and
disruptive war that raged across the world with names such as Malta, Singapore, Tobruk, El Alamein, Stalingrad, Burma, Guadalcanal, imprinted on the
memory along with innumerable maps in the newspapers of tank thrusts here
and naval strategies fought out there (in the Atlantic, the Mediterranean, the
Pacific). He learned to interpret and follow those maps (Rommel's defeat in
North Africa and General Patton's incredible breakthrough into Germany).
The boy lived in Dover shortly after the British forces were ignominiously
ousted in the Dunkerque evacuation, and then he moved to the safety of
Gloucester when the shelling of Dover began. When he was later returned to
his hometown of Gillingham in Kent, he had a view from a bedroom window
across the River Medway (noting its distinctive tidal rhythms) and watched
battered ships limp at high tide into Chatham Dockyard, where his father
worked (often day and night) to get them back into battle. From there, too,
the boy watched antiaircraft fire bursting around German aircraft on some
nights and, most scary of all, the autopropelled V1 rockets bumbling like vast
insects overhead (we learned after a few days to keep one ear cocked for the
sudden silence that betokened the rocket was about to crash to earth).

It is difficult to assess the effects of being raised in the midst of that. There
were strong social solidarities in evidence (even across the class divide—a fact
that may have had something to do with the victory of the Labour Party in the
1945 election and my own subsequent embrace of socialism). But there were
shortages (no candy treats, a restricted diet—I saw my first grapefruit when I
was ten and mistook it for a large lemon) and a sense of insecurity that even
the most devoted parents could not assuage. My early sense of geography intertwined with military strategies, patriotic fervor, and an intense admiration
for the Royal Navy. I longed to join that navy (my grandfather, who died
around the time when I was born, had been a lieutenant commander, the

highest rank a nonaristocratic officer could then achieve in Britain's class-bound system). I thought of applying to the Naval College at Dartmouth. It seemed a good way to live out the fantasy of escape, live dangerously, rise above insecurities, explore the world, and be someone important and admired! Why exactly I didn't go through with it I cannot now recall.

Intense patriotism was everywhere in those early days. I stood in the crowd on Edinburgh's Prince's Street to celebrate the victory over Japan in 1945 (after the traumas of nuclear devastation of Hiroshima and Nagasaki that were relayed to me at age ten as a triumph of science and righteous conduct against a yellow and barbaric infidel). It all felt very different when I went to Hiroshima a few years ago, but by then I had long decided (with a little help from Renoir's film *Hiroshima Mon Amour*) that the use of nuclear weapons was one of the most heinous and inexcusable acts in human history. My first really serious political demonstration came when, as a young college lecturer at Bristol, I joined people in the streets to protest nuclear threats in the Cuban missile crisis (I remember the jeers of the onlookers, in particular those who had risked their lives in the Second World War "for the likes of you" as they put it).

## An Empire's Decline

The patriotism connected to all the symbols of Empire surrounding me (was that not what the Royal Navy existed to protect?). The fantasy of escape and exploration was nourished by the knowledge that the world was open to be explored. Those maps with so many parts of the world colored red as somehow "belonging" to Britain indicated wide-ranging choices of territories available for inspection. The postage stamps I collected from India, Kenya, Jamaica, Sarawak, Rhodesia, Nigeria, and South Africa all proved it—the head of a British monarch was on every one. But British power was declining. The empire was crumbling at an alarming rate. Britain had ceded global power to the United States and was having a tough time coming to terms with its second-rate status (those "awful Yanks" was a frequent topic of comment). It was difficult not to feel the effects of that. The map of the world began to change color. The atlases that had seemed so fixed were being redesigned with new colors and new names. The geography of the world was in motion, and

Britain's role in it was visibly shrinking. Yet those comfortable books on the imperial experience still lay around the house, and Edwardian nostalgia for "the good old days" of a civilized Pax Britannica was everywhere in evidence.

Many of my teachers at Cambridge had experience in the military or colonial service. They seemed to regret the loss of Empire, while accepting that it should evolve, though only in "sensible" ways. It is easy in retrospect to criticize their imperial vision, their paternalism, their colonial thinking, and even their nostalgia for the good old British Raj. But what stays with me more positively is the incredible love they evinced for the countries and the peoples they worked among and studied. My geography tutor, Benny Farmer, was not only progressive in many ways but passionately interested in what was then called Ceylon (Sri Lanka). After three years, I felt I, too, had an intimate knowledge of the country, its ethnic variations, its history of lost civilizations, its physiographic and climatic regions, its different systems of agriculture (dry and wet zone). And through innumerable anecdotes, I learned of the curious conflicts among the colonizers and the colonized, as well as the more obvious symbiotic conflict between the two over what to do and where to do it. The everyday struggles around the use of the land, over power and social relations, over resources and meanings came alive in those anecdotes and remain fundamental to my geographic education.

The long decline of Empire and of everything that went with it formed a curious backdrop to geographical learning. Not only was the decline ignominious, but it also shed much light on the ignoble and self-serving side of British imperialism. It was possible to justify for a while the fight against the insurgency in Malaya (I had a cousin who was there) as a fight against the communist infidels and to write off the dreadful consequences of the partition of India (the press photos still hover in my mind) as a morality tale of what happens when "sensible" British rule gets replaced by irrational natives. But the attempts to suppress independence for Cyprus and the long, drawn out fight against the supposedly barbaric forces of tribalism in the Mau Mau emergency in Kenya raised serious doubts. The catalytic moment for me occurred when Britain, France, and Israel colluded to try and take back the Suez Canal from Egyptian control. Even my father, who never expressed any overt political opinion but gave off the aura of a respectable working-class patriot who accepted that the aristocracy had been born to rule benevolently over the nation and the empire, expressed disgust. Corresponding as it did with the Soviet suppression of the Hungarian uprising, this act made the "the angry young man" generation in Britain rise up in wrath at the perfidious policies of a nakedly self-interested and rapidly fading imperialism that looked no better than the nasty communists that we had been told were the prime enemy. A watermark year for me was 1956. I was just twenty-one and in my final year as an undergraduate. I abandoned my studies for a whole term to argue vehemently about politics and turned resolutely anti-imperialist thereafter.

Many years later, shortly after arriving in Baltimore in 1969, I participated in a popular local radio call-in show on Flag Day. This day has particular significance in Baltimore because it was at Fort McHenry that the British were defeated in the war of 1812. Francis Scott Key wrote the "Star-Spangled Banner" there. I was asked how I felt as someone from Britain about that event. I said I was thoroughly ashamed at the British attempt to suppress the ambitions of a newly independent nation, but by the same token I thought every American should be ashamed of what their country was then doing in Vietnam. The response was electric and virulent—caller after caller told me to go back where I belonged (Britain or Hades as the case may be). The experience of Suez had carried over into my automatic and instantaneous support for the antiwar movement in the United States. I am appalled every time the United States takes the big stick to any country that gets out of its line and place (Guatemala, Cuba, Chile, Grenada . . . ).

## From Global to Local

But what kind of geography was possible given the sense of disillusionment toward the imperial fantasies I had earlier entertained? A geography complicit with imperialism and war was not acceptable. Interestingly (and it is only now that I realize it), I have steadfastly avoided setting foot in most of those territories colored red on that old map of Empire (with the exception of the white settler colonies). There is an inverse correlation between my international travels and the countries that featured so prominently in my stamp collection. Fortunately, I had another option. "The eyes of the fool are on the ends of the earth," proclaimed S. W. Wooldridge, an influential professor of geography at London University sometime in the early 1950s. This set the stage for a retraction of global ambitions among many geographers in postwar Britain and a turn to deeper consideration of national and local questions. My tutor scoffed at the thought, but I was not averse to concentration on the local. Wooldridge wrote a wonderful book on the Weald, a fascinating area of south-east England close to where I lived and of which I had intimate experience. I spent many happy hours traversing that landscape in my teens. I biked and hiked everywhere without constraint, up and down lanes now dominated by murderous cars and thundering trucks, across fields and through woods. I did some of that during my active days as a Boy Scout, too (we did lots of camping and excursions, the most instructive of which involved being taken blind-folded in a car to some unknown place in the countryside and being left to navigate home alone with only the help of an ordnance survey map). I learned to navigate and to read maps, but I also learned about the geology, noted the different vegetation and agricultural land-use patterns, and before long was lurking in the local archives looking at old maps of buildings that were hop kilns or granges on eighteenth-century maps.

There was (and still is) something deeply restful about that land. Traversing it so often I got a sense of the "palimpsest effect"—what happens when layer after layer of settlement prehistory (with its megaliths), Roman (with its villas), Anglo-Saxon (with its place names ending in -ton and -ham), Norman (with its churches and castles), late medieval (with its abbeys, moated houses, Tudor and vernacular town and village architecture), early modern (with its landed estates and agricultural improvements), nineteenth century (with its markets and railways), and present-day efforts all welded together in a common frame. I became deeply immersed in that world. I did my undergraduate thesis on fruit cultivation in the nineteenth century in my local area (I picked fruit to earn extra money—a pittance—every summer from age fourteen onward) and continued such studies through to the end of my doctoral dissertation, "Aspects of Agricultural and Rural Change in Kent, 1815–1900." Exploring and excavating the deep roots of my own landscape, my own locality, became a central obsession. Intimate sensual contact with the land went hand in hand with study of it.

As part of my dissertation research I read the local newspapers from 1815 to 1880. It took me a whole summer to do it, and it was an incredibly rewarding experience. Anecdote upon anecdote all adding up to an intricate picture, much as Benny Farmer had constructed for me of Ceylon, that saw personal lives articulated with abstract social forces making for those glacial changes that in the end have incredibly deep consequences for the landscape and social life. But there was something else at work within that history, and that "something else" jumped out again and again at me as I pored over issue after issue of the local newspapers of Mid-Kent that summer.

To begin with, the newspapers changed their format and their social content as the century wore on. In 1815, they were little more than news reprinted from the *London Times* (such as Napoleon's escape from Elba) two days late, with advertisements for sales of property and goods and a few local notes about this or that (human dramas mostly, with a special issue dedicated to the last public execution on Pennenden Heath near Maidstone in the early 1830s). By midcentury, they were vibrant media for conveying national information into a local context, coupled with mobilizing expressions of regional consciousness. By 1880, the local press had lost out to the London tabloid press as a means to communicate information of national importance. The sense of regional consciousness or debate was much diminished. The newspapers were back to real estate concerns and local crime and scandals. When I sat back and reflected, I recognized that I had witnessed the rise and fall of a certain kind of regional consciousness (even the production and decay of a viable region), an upheaval that depended on changing means of transport and communications, as well as on more general economic, technological, and social changes. The speed and spatial range of communications was clearly a dynamic shaping force in historical geography.

### From Local to Global

Yet there was another theme writ large in that experience. For example, one focus of local agitation in the 1840s was the question of tariffs on sugar. This issue connected sugar interests in the Caribbean with those of fruit growers in Kent, who wanted cheap sugar to ensure their market in jams and conserves (sold primarily to the working classes of London, the Midlands, and the North). When, much later, I read Mintz's book *Sweetness and Power*, which is about the long history of the inner connection between the Caribbean sugar plantations (slavery) and working-class sustenance in Britain, I understood perfectly well what he meant. But the general point was that life in Mid-Kent was inexorably connected to conditions of that wider world (of commerce, technology, politics) in which it was embedded. And when I looked at the data on the hop industry, the cycles in plantings, output, and spatial spread and contraction correlated almost exactly with business cycles in the British economy. Agricultural distress or affluence in Mid-Kent was a function of changing discount rates in the London financial markets, which depended on trade conditions more generally. The map of hop cultivation spread outward onto marginal lands in good times and contracted backward when interest rates rose. Finance capital and geographical forms were, as I now would put it, intimately and dynamically connected.

These issues even have a peculiar connection to my daily life. I often ask students the all-important geographical question: Where does your breakfast come from? As they try to unravel answers, they find themselves in a very intricate world of geographical relations and ecological dependencies (sugar from the Caribbean, tea from Sri Lanka, coffee from Costa Rica or Brazil . . .). Well, a key ingredient of my breakfast is bitter marmalade (I love it). But how did this item became so important in British diet (it is not very popular anywhere else)? And how come these very bitter oranges from Seville in Spain got into the nineteenth-century British jam industry (so important, as Mintz shows, to working-class sustenance)? A brief answer is that producers needed to keep their fixed capital fully employed throughout the year. When British fruit was not available in the winter, oranges from Seville could be used because they could be shipped in bulk and were available from January through March. My dietary taste derives from a peculiar historical-geographical moment in which an industry needing to secure the full employment of its fixed capital created a demand for a product that would enhance the nutrition of a strongly urbanized proletariat! I often think of this at breakfast and how I uncovered the connections in that long summer of reading the local press.

I record all this because I now think of that summer as one of the most formative experiences of my intellectual life. The reading occurred, of course, against the dual background of the local intimacy that I had long cultivated and the fantasy of escape to a wider world that I had long harbored. But the experience gave me insights and resources that I have drawn upon ever since;

it plainly underlies much of what I myself write about the circulation of capital and the spatial and temporal dynamics of global and local relations. The local and the global, as we would now put it, are two faces of the same coin, and they even intersect on the breakfast table.

## A Love of Place

Yet there is, I acknowledge, something seductive and securing about attachments to particular places. It is easy to live daily life where one is as if nothing else matters and to erase the imperial connection. Whenever I go back to Kent (to visit with my older brother, who has never lived more than a few miles from where we were born), I feel the tug and pull of the depth of its history and the complexity of its meanings. It was (and I guess still is) easy to slide into the comforts of a "little Englander" antiquarianism in that world. I can empathize more generally with those who write about place (or region or locality) as if it is the one concept that really matters. I, too, feel the temptation. I have strong loyalties to places and to the people who live there. I have lived in Baltimore, on the Fall Line, for thirty years now (with a six-year break in Oxford). I feel it is my hometown. Well-known academics usually move about much more than that!

But, it could be said, I indulge my taste for movement, escape, and exploration by all the travel I undertake. I have, it seems, a perfectly happy resolution to my childhood dreams and desires: my search for a meaningful tension between exploring space and securing place ought to be satisfied. But there is something seriously lacking. And it is not difficult to say what. The manner of the travel has become too formulaic. "Above all," wrote the great geographer Carl Sauer in a memorable phrase, "locomotion should be slow." And he was right, for it takes time and concentration to understand places and peoples. Whizzing in and out of airports (that all look the same) into lecture and seminar rooms (that exhibit very little variation) and, worst of all, piling into one of those dreadful international hotels (which make a fetish of serial monotony) where conferences so typically occur do not do the trick. I have become a privileged member of a globalized faculty club. I end up seeing the same faces and trading the same ideas in Singapore, Barcelona, and Berlin. I have, it seems, become a viable commodity in the circulation of academic ideas. I have been globalized along with ethnic foods, Coca-Cola, Levi jeans, rap music, and the omnipresent McDonalds. This was not at all what I had in mind.

I try, of course, to vary the schedule and to spend a bit of time off the beaten track, and those local people (often geographers) who know me are often very generous in showing me around and getting me to places that I would not otherwise see (imparting much of their own intimate local knowledge in the process). But I still prefer to strike out on my own, just wander the streets and travel the buses in unknown cities and to unknown places, let accidental encounters and experiences occur as they will. In my early years, I did a lot of hiking and rambling in the summers (walking the Yorkshire Dales and the Lake district twice), and I spent two long vacations as a student hitchhiking at random all over Europe (though with a heavy emphasis on Norway, where I joined a geographical expedition to the Jotunheim to study the movement of a glacier, actually living on the glacier for three weeks or so). On the trips around Europe, I picked up an occasional job washing dishes here or planting trees there, stayed in youth hostels and basically lived off a diet of bread and salami unless some generous motorist stood me a meal. To hitchhike in those days was a special experience. On one memorable occasion in Norway, only three cars passed by in one day, forcing me to return to the youth hostel (a primitive cabin on a mountainside) for a second night. I didn't mind. The place was so beautiful, and I had happily spent the day just inspecting everything and taking in the lights and shadows, the shapes and forms of the land, the sounds, smells, and feeling of the pine needles and the grass to the touch. Long after, when reading Marx, I knew instantaneously what he meant when he dwelt on the importance of sensual interaction with the material world.

So my preference is to go places and just hang out. I did that on many trips to Sweden in the 1960s (usually to visit with Gunnar Olsson, who invariably had to lend me money to get me back to England) and in Paris for several summers in the 1970s, when I was collecting materials and writing up on the historical geography of the Second Empire. Long Greyhound bus trips in the United States in the 1960s, with stopovers to visit with people such as Bill Bunge in Detroit (a very formative influence), or down into the depths of Mexico (where I designed the format of *Explanation in Geography*) were typical of my wanderings. The international conference circuit hardly satisfies that craving. Furthermore, the world has become so full of tourists and tourist traps that aimless wandering becomes less and less easy. Or maybe it is that my toleration for living in fleabag hotels without showers has radically declined.

## The Arrogance of Class

The formal world of academic life too often seems removed from such tactile experiences of the world. I am known for writing in a theoretical and abstract way. And many may feel there is some acute disjunction between the experiential world that I have described as underlying my geographical sense of the world and the books and articles I have produced. The disjunction is present, of course, but not exactly in the way many might presume. There is always a connection between what I write and what I feel, and what I feel depends on where I place myself and how I react to people and situations. To walk the streets of Baltimore or to talk with workers in a Burger King, for example, is to experience outrage at the waste of lives and opportunities, the patent injustices and stupid inefficiencies, the gross neglect that demands rectification. Experiences like that impel me to write (*Spaces of Hope* records the connection directly). They fuel my academic rage. I was therefore happy when the *Singapore Straits Times* reported, regarding a speech I gave in that city to the 1999 World Congress on Model Cities, that I was walking proof of the adage "that rage increases with age" among "honest intellectuals." But why the rage?

I recall a visit to Colombia in the early 1970s, when my university was setting up joint projects on public health with schools of medicine there. I went ten days early and roamed the country by bus, getting into remote places. I arrived at the official meetings in Cali somewhat disheveled, with a backpack, rather longish hair and dusty beard. It had been a wonderful and eye-opening experience, and I felt I had much to say because of it. But no one would listen. I was not to be taken seriously. Some of the younger Colombians (doubtless on their way to become important officials in some international healthocracy) openly mocked me. (One of their older colleagues apologized for their stupidity, telling me how much he appreciated that I would take the time and risk some dangers to get to know his country with the kind of intimacy it deserved).

I had encountered that kind of arrogance before. It was exactly the kind of dismissal that greeted me at Cambridge when, as a naïve student from a state school, I encountered the public school students with their superior ways (Raymond Williams gives a parallel account). The experience of class difference and segregation at Cambridge was both dramatic and formative. Not a single one of my friends was from a public school background. The more obnoxious examples of the latter had cars (which in the 1950s was really something), spent weekends in London, swilled beer in the college bar (I could afford only a half pint per night), and on other weekends walked round with beautiful well-dressed debutantes on their arms (Cambridge was very monastic for the rest of us). I fought class war there with the only weapon I had—intelligence—and vowed I would best them come the examinations. I did, only to find that it mattered not a jot—they all went off to become lawyers, stockbrokers, and judges, leaving me to my doctoral studies and a penurious position as an assistant lecturer four years later.

Call it class envy, prejudice, or war, but Cambridge taught me about class in a way that I had not earlier experienced, for I had been raised in a fairly humdrum town where the main class distinctions were between professionals (mainly military) and respectable and dissolute working classes. My own family aspired to (and eventually succeeded in joining) the first of these groups, while beginning in the second and openly disparaging the third (I was strongly advised not to associate with "those kids from the estates"). But my class conditioning was a little more complex than that. My father's mother was the last of an aristocratic line that had fallen on hard times. She disgraced herself by marrying a professional naval man, but the aristocratic heritage was there. One of her ancestors had been chancellor of the exchequer under Queen Elizabeth and owned Sissinghurst Castle in Kent. The ruined estate was later turned into a beautiful house and garden by some of the famous Bloomsbury set. I occasionally visit it as if it were still part of my own lost heritage. Another in that line was a rear admiral who died in the Battle of the White Nile and has a plaque in Westminster Abbey. Like my father, I was prepared to accept some notions of aristocratic privilege until I got to Cambridge and felt its abusive qualities firsthand.

My mother's father, on the other hand, was of "the aristocracy of labor." Scottish by origin, he was a skilled worker in the Amalgamated Engineering

Union (we found his Freemason's paraphernalia much later). He said little and was quiet and dutiful—which was just as well because his wife was extremely determined, strong, and opinionated. She was the ambitious and pushy daughter of an agricultural laborer and had been "in service" to the aristocracy as a cook (she made the most wonderful pies). She was also a strongly outspoken socialist who would shop only at the co-op. I recall her standing, in the middle of the Second World War, in the co-op denouncing Winston Churchill (our much-revered war leader) as a "rotten bugger" who cared nothing for the working classes. She responded to the surprised looks by admitting that Hitler was an even "rottener bugger" and admitting that it probably "took one rotten bugger to get rid of another rotten bugger." So I was not surprised when Churchill failed to get elected in 1945. And because my grandmother was one of the very few people who could ever face down my warm but somewhat oppressive mother, I entered into a strong alliance with her on all things, including, of course, her socialist politics and some of her ability (or was it an urge?) to challenge any authority anywhere no matter what. I evidently inherited some of her political rage. If you look at my class heritage as a whole, of course, only one significant class is missing: that of capitalist. I sometimes think I inherited anticapitalism in my DNA. But at Cambridge I found a more immediate target for political anger: that ruling class of aristocratic and rich ruling privilege must go. The only question was how and by what means.

My mother believed deeply in education as the route to personal security. I was born in the depths of that great capitalist nightmare called "the Great Depression," and with my father unemployed and no money in sight it must have been desperate days for my parents facing the need to support a second child. "Get yourself an education" and avoid the nightmare of unemployment was the message drummed into my brother and me from the earliest age. Education was viewed as the privileged means of personal empowerment, but I also saw it as a weapon in class war. Of course, my parents wanted me to get an education in something "useful" such as electrical engineering or chartered accountancy. On that point, I demurred. My passion for geography was already set, and because I loved history and literature, those were my chosen preferences.

It should be clear by now that I don't think of geography primarily as a formal field of study but as a fluid and dynamic state of mind derived from a whole host of complex experiences of and fantasies about the world. The formal field of study could give expression to some of those experiences and fantasies, and it could also deepen and broaden the meanings attached to those experiences. To learn how glaciers or rivers erode and shape distinctive landscapes, how mountains form and beaches shift, how monsoons and trade winds develop, how agricultural practices change soil profiles (for good and bad), how soil erosion can threaten sustenance, how cities and towns build into networks of commercial interchange, how cities assume distinctive spatial patterns and moral orders, how regions form and cohere, how cultural values gain distinctive forms of expression on the land—learning all of this through my school years with two talented and interested teachers provided some sort of basis for a collective sense that there was something called "geographical knowledge" that was distinctive and important. But if that formal knowledge was not adequate to express and encompass geographical experience, then a struggle ensued in which the formal structure, with all its accumulated rigidities, was pressured to embrace new experiences and circumstances. I never conceived of geography, therefore, as a fixed field of study that one merely learned, but as a dynamic form of knowledge that not only could but should be changed according to individual and collective needs, wants, and desires.

### Graduate Years and Changing Gears

My transition from undergraduate to graduate student status at Cambridge was crucial. The general atmosphere among the graduate students was to insist that authority must earn respect rather than be automatically accorded it. Mine was the generation that spawned the *Footlights Review,* which became *That Was the Week That Was*—a television show that mercilessly ridiculed the ruling classes as well as almost everything else that might be regarded as "traditional" in British life. Cambridge was populated by an intellectual elite, and if something was seriously wrong in the state of Britain (and many thought there was), then this elite was surely positioned to do something about it. The modernization of Britain was firmly on the agenda, and a new structure of knowledge and power was needed to accomplish that task. For me, and for many others who came from a similar class background, socialist moderniza-

tion (automatically eroding the power of the upper classes) backed by technological efficiency seemed to hold the key to a better future. The path was not
revolutionary, but gradualist and progressive.

The quest for some new knowledge system occurred in an environment in
which that peculiarity of Cambridge education known as "the college system"
actually worked. Graduate students in the college came from all disciplines. I
shared digs with an archaeologist (Warwick Bray), had fascinating discussions
over dinner with students in comparative literature (such as Michael Wood,
now at Princeton), while perpetually being involved in discussions with fellow
geographers (such as David Grigg, whose brilliance was not in question). We
discussed Camus, Iris Murdoch, Lawrence Durrell, Osborne's *Look Back in
Anger,* and other products of the "Angry Young Man" generation, and wal-
lowed in the ideals of Colin Wilson's influential tract *The Outsider.* I learned to
take ideas from all over the place and bring them close to geography if need
be. Knowledge was not bounded by disciplinary affiliation. Add to that general brew the advent of Richard Chorley and Peter Haggett as well as Tony
Wrigley as young faculty in the Department of Geography, and the impetus
for change became overwhelming.

For those of us involved in geography, rational planning (national, regional, environmental, and urban) backed by "scientific" methods of enquiry
seemed to be the path to take. Tony Wrigley's approach was deeply philosophical, and I soon found myself reading about August Comte and positivism, about "covering laws" in history. Chorley and Haggett (incredibly
opposite in personality but somehow forever bonded in their life's work thereafter) were exploring scientific methods and statistics as a means to rationalize
and systematize geographical enquiries. I was immediately enmeshed in that
approach, incorporating all sorts of ideas from them into my dissertation
(which received enthusiastic support from Michael Wise, the external exam-
iner, but skeptical horror from Jean Mitchell, a very traditional historical geographer). I remain intensely grateful to that trio of Chorley, Haggett, and
Wrigley for the liveliness they imparted to geographical work in those years
and for many years thereafter (culminating in the famous Madingley symposia
of the late 1960s).

But what was I doing working on nineteenth-century hop cultivation in
Kent when the paramount issues were modernization, rational planning, sci-

entific method, and the like? I needed to change gears, but doing that is not always so easy. I had acquired a Leverhulme scholarship to study in Sweden for a year, mainly on the grounds that the Swedes had better population data going back into the eighteenth century. But a demographic project would hardly help. Fortunately, when I arrived in Uppsala in 1960, I was unceremoniously dumped into a room alongside some strange bear of a figure named Gunnar Olsson, an event that both of us freely acknowledge was one of those fortuitous accidents that have longstanding consequences. Gunnar's former office mate had, I later gathered, smelled rather badly, so Gunnar sniffed the air suspiciously for an hour or so. I apparently passed the test, and we thereafter became friends for life, in spite of the enormous philosophical differences that subsequently came to separate us.

But at that time both of us felt that some sort of modernization or even revolution in geographical methodology was imperative, and we spent several years arguing about exactly what that might mean. Along with many others (Chorley, Haggett, Ullman, Garrison, Brian Berry, Morrill, and Hägerstrand in leading roles), we helped to bend the structure of formal geography, against considerable opposition, to our collective will. The immediate effect was that I dropped the research project on Swedish demographics (Leverhulme miraculously didn't seem to mind) and just hung out all over Sweden, learning a tremendous amount about what it meant to live in a strange and foreign land, while retooling myself with all sorts of ideas and prospects for undertaking new kinds of research armed with different philosophical foundations and methods. This was the project that was to preoccupy me throughout my subsequent years at Bristol University, with a talented faculty that included people such as Michael Chisholm (in his sensible years), Barry Garner (a wonderful drinking companion), Peter Haggett, Allan Frey, and Mike Morgan. This project (aided and abetted by a year teaching in Penn State with Peter Gould as mentor) was to culminate in the writing of my first major book, *Explanation in Geography,* in which I sought to explore the rational and scientific basis for geographical knowledge by way of the philosophy of science.

But I confess to being inwardly torn throughout much of this period. On the one hand, the political, intellectual, and hence professional project pointed toward the unity of all forms of knowledge under the umbrella of positivism and toward the rational application of such knowledge to the general task of social betterment (via "social engineering," as Gunnar would call it). On the other hand, I still had that lust to wander and diverge, to challenge authority, to get off the beaten path of knowledge into something different, to explore the wild recesses of the imagination as well as of the world. In the same year that I met Gunnar, I also met Michael Moorcock in Uppsala, now a novelist of considerable accomplishment. We immediately struck up a friendship because I had been greatly impressed by the novels of Mervyn Peake (a person Mike knew well). We spent the summer rambling (sometimes dangerously and other times hilariously) over the whole of the north of Sweden in my Austin Mini car. He brought his guitar along and sang for our suppers in several small hamlets and towns. We slept in barns and alongside lakes, got bitten by mosquitoes, had a rough time climbing a mountain (I shudder now at how reckless we were). This sort of wanderlust did not sit easily with rational planning. I had the creative urge to write and even wrote a short story or two (one of which came out in *New Age Science Fiction*). When I was offered the job of assistant lecturer in geography at Bristol, I hesitated and for at least the first year felt the strong urge to chuck it all in, write fiction, teach English in Sweden or wherever. Such atmospherics were fed as the '60s wore on, with the Beatles, the Doors, Jimi Hendrix, Janis Joplin, the Rolling Stones, Pink Floyd, and the Grateful Dead, to say nothing of my growing appreciation of serious jazz in the form of Coltrane, Miles Davis, Thelonious Monk, et al., loud open concerts in Hyde Park, and the general emotional and often anti-intellectual turmoil that the youth movement of those years created.

But one thing I had solidly learned from my father, is that projects can be completed if you put your head down and ignore everything else (he lived almost his whole life like that). And that is, in essence, what I did while writing *Explanation*. I turned in the manuscript in the summer of 1968 with near revolutions going on in Paris, Berlin, Mexico City, Bangkok, Chicago, and San Francisco. I had hardly noticed what was happening (except for being profoundly upset at the assassination of Martin Luther King Jr. and being appalled at the TV images of cities in flames across the United States). I felt sort of idiotic. On the one hand, I had undertaken a major enterprise and disciplined myself to write a long and hard-honed book. For me, that entailed and continues to entail a great deal of single-minded concentration, a willingness to lay everything else aside, to retreat from all manner of immediate interactions with the world. I had a lot of energy, ego, and intellectual capital invested in that book. I was proud of it and wanted it to do well (which it did). On the other hand, it seemed absurd to be writing when the world was collapsing in chaos around me and cities were going up in flames. It was like Emile Zola getting mad at the revolutionaries in the Paris Commune of 1871 because they delayed the publication of his first novel! What on earth did I think I was doing, pretending to be progressive (in some undefinable long run that never comes), but just sitting on the sidelines when action actually hits the streets? I was even more mortified when Richard Kirby, a student in my tutorial group at Bristol who had been active in a student sit-in, was railroaded and disciplined by liberal professors (who equated Nazis of the right with Nazis of the left). Richard subsequently went off to China and wrote on that world and now teaches at Liverpool.

The balance between active engagement and academic work is always tough to negotiate, and the whole issue still bothers me immensely. I now recognize that doing academic and intellectual work has its own demands and rules of engagement that cannot be evaded (any more than anyone else in any kind of job can pretend it does not put specific demands on time, energy, and concentration while circumscribing actions in often quite draconian ways). I try to retain an activist connection by attachment to social movements and some level of participation where I can. Such participation always remains an important source of inspiration, and I hope I can translate some of that inspiration into the world of academia. That is the best I can do, but it always feels inadequate. I admire Bob Colenutt's choice. He studied as a graduate student with me at Bristol, got a good position in geography at Syracuse, and dropped it all to become a community activist in South London, where he has been fighting battles around there ever since. It is not a question of who was right or wrong. We both did what we felt was best given our temperaments, capacities, and powers. And we stay in contact.

## Hopkins and Baltimore

In any case, I felt a crying need to retool myself again, to take up those moral and ethical questions that I had left open in *Explanation* and try to bring them closer to the ground of everyday political life. In 1969, I was hired at Johns Hopkins in Baltimore to work in an interdisciplinary program dealing with geography and environmental engineering. I think I was hired because of my interest in the unity of science under the positivist umbrella. But by the time I arrived, I was set to do something entirely different (much as had occurred in Sweden a decade earlier).

The internal attraction of Johns Hopkins was its interdisciplinarity. I thrived on it, much as I had as a graduate student. I interacted over the years with many talented individuals from different disciplines. Figures such as Vicente Navarro, Rick Pfeffer, Nancy Hartsock, Donna Haraway, Emily Martin, Katherine Verdery, Ashraf Ghani, Alejandro Portes, and Neil Hertz (to name just a few) became part of my intellectual firmament; none of them were geographers. I had hardly any contact with geographers on a daily basis. Even Erica Schoenberger, the only other human geographer in the department since the late 1970s, was trained as a planner. Not only did I feel free to explore ideas wherever they came from and wherever they went (by, for example, participating in seminars all over the university), but I could do so in companionship with anyone who cared to listen. I could explore geographical knowledge independently of the formal discipline of geography.

The other attraction was my location in Baltimore, a city deeply troubled by social unrest and impoverishment, one of those cities that had gone up in flames the year before I arrived. I wanted to put my skills to work to try and deal with urban issues and to do so in a directly reformist and engaged way. My new department already had some research under way on inner-city housing. I immediately became engaged in that work and, together with my first graduate student at Hopkins, Lata Chatterjee, did some very detailed studies on housing finance and government policy in the city. This formed the empirical background to my thinking about urban geography in a new way. I gained confidence in the new framework when I used, without attribution, the ideas of Engels to describe the dynamics of the housing market in reports to the city and found that even the landlords thought it "a pretty damn good" representation. I learned then what has proven the case many times since: there is no better framework for understanding capitalist dynamics on the ground than that provided by Marx and Engels. More broadly, however, these housing studies introduced me to the city. The travails of Baltimore have formed the backdrop to my theorizing ever since (they have a central place in my latest book, *Spaces of Hope*).

How was I to get intimate knowledge of a city—a city that was not only strange but in many respects forbidding with its class distinctions and racial divisions, its ethnic exclusions and convoluted political practices, and, above all, its overwhelming sense of imminent violence and disintegration? The task was very different from cycling around Kent or wandering around Sweden. I am somewhat shy, not easily gregarious, and political or community organizing is not one of my strong points. Bill Bunge (on that memorable first visit to Detroit in 1964) correctly insisted (as he set up his "geographical expeditions" to the inner city) that theoretical work and embeddedness in the world of inner-city suffering must go hand in hand. I was not sure his particular vision was possible, but I knew that some sort of connection was necessary. The best I could do was periodically to position myself (in political practice as well as in research) in ways that were somehow outside of my respectable home as a professor at Hopkins. And if my attempts in this regard were not much more successful than my attempts to run away from home when a child, they were nonetheless equally instructive.

In this project, I had a fantastic source of help. The chaplain at Johns Hopkins, Chester Wickwire, was well known in the civil rights movement, a politically engaged white clergyman with credibility among black leaders in the city. He entrained me into all sorts of activities. When a national conspiracy arose to kill off the Black Panthers (with the killing of Fred Hampton in Chicago—an act later proven to be gratuitous police violence), Chet organized white students and faculty from Hopkins to sleep on the pavement for three weeks outside the Panthers' headquarters in Baltimore. I spent several nights there and learned much about life in the inner city (as well as how to behave if a gunfight erupted). This sort of experience provided the emotional heft behind a seminal transitional essay I wrote called "Revolutionary and Counterrevolutionary Theory and the Problem of Ghetto Formation." Chet remained thereafter a vital source of connection to the city and its grassroots politics. Though long retired, he continues to play a role thirty years later (we now work on the campaign for a "living wage" in Baltimore and in Hopkins).

Chet was not, of course, very popular with the university authorities. His strong support of African Americans in a very white traditional university set in a predominantly black city placed him at a crossroads of conflict. His strong support of the antiwar movement put him against the university hierarchy. I was in much the same boat. How we both survived in such a conservative and reactionary institution as Hopkins is an interesting story. All members of the faculty who actively opposed the war were somehow gotten rid of—except me. In fact, almost anyone who had progressive ideas in those days was forced out (Donna Haraway, for one). Those of us in opposition to the university developed some strong bonds, but few of my close friends in those early years lasted. Somehow I did. I had two things going for me. I had a strong academic reputation (largely based on *Explanation in Geography*) and by 1972 became a tenured full professor. The institution of tenure, although not foolproof, provides a key protection without which I could not have done what I did. I cringe at its contemporary weakening.

But I also had the backing of Reds Wolman, a powerful and highly respected figure in the university who was for many years the chair of the Department of Geography and Environmental Engineering. Reds is a wonderful person. Not only is he intellectually engaging, an exciting interdisciplinarian who makes my own efforts in this regard look pale (he bridges engineering, science, and the social sciences with incredible ease), but he is also a deep believer in freedom of academic enquiry. He often disagreed with me but always insisted (against, I later learned, considerable carping from the university power structure) that I had the right to do and say whatever seemed right to me. He saw no point, he once told me, in making the United States a mirror image of the Soviet Union in its denial of freedom of thought. He had become head of the department back in 1954 in an attempt to heal a split resulting from the red-baiting of Owen Lattimore (a famous China scholar and a wonderful geographer and historian whom I later met and wrote briefly about) largely on the basis of information provided to Senator Joseph McCarthy by George Carter (another talented but very conservative geographer who chaired the department prior to Reds). So Reds was no foreigner to political turmoil and intrigue. I suspect his support was more crucial to my survival than I ever knew or will likely ever know. Over the years, he also helped temper my ways, teaching me that good things lurk in unlikely places, that the political judgments so necessary to firm academic work do not have to be turned into personal vendettas, that it is often possible to reach across political and intellectual positions and to learn gratefully from opponents as well as from friends. I regard him (along with Peter Haggett) as worthy mentors of how to be generous in relation to fundamental diversities in university settings.

All of this set the stage for the publication of *Social Justice and the City*, a book that contrasts what I called "liberal" with "socialist" formulations of urban issues (with the essay on ghetto formation supplying the divide). I began to read Marx seriously around 1970. Some very talented faculty and students (such as Dick Walker, Gene Mumy, and Jorn Barnbrock) and I set up reading groups. I have taught Marx's *Capital* (volume 1) every year bar one ever since. The journal *Antipode* had been launched a bit earlier, and socialist, Marxist, anti-imperialist, and anarchist thought had an organized focus within the Association of American Geographers (AAG). Ben Wisner, Jim Blaut, David Stea, and Dick Peet at Clark University, along with many of their graduate students, formed the nexus, with the awesome but difficult figure of Bill Bunge always lurking in the background. They remained stalwart figures in the struggle to reform geography thereafter. We formed a socialist caucus within the AAG.

These were heady days of discovery for me, both in Baltimore and beyond. For once in my life it seemed that professional, personal, and political life merged into one turbulent stream of continuous innovation backed by revolutionary fervor and cultural power (I went nearly every week to the Left Bank Jazz Club, one of the few places where the racial divide seemed to melt away in Baltimore under the power of music). There were powerful barriers to confront and not many of us available to do the work. Quite a few, less fortunate or secure than I in their academic position, did not survive the repressions. We were accused of "emotionalism" and "politicizing the academy." Our research was dismissed as unreliable and irrelevant. The publisher's readers of *Social Justice and the City* called it incoherent and unreliable, and recommended its rejection. It took persistence at Hopkins, the support of John Davey, who was then at Edward Arnold in London, and the intervention of a youthful Ira Katznelson to turn the rejection around. This scenario was not unusual. A key essay I did, "Population, Resources, and the Ideology of Science," was rejected out of hand by the *Annals of the Association of American*

*Geographers* on the grounds that it had nothing to do with geography (it came out in *Economic Geography* instead). An essay called "Class Structure and the Theory of Residential Differentiation" for the Ron Peel festchrift at Bristol was rejected by the editor, Michael Chisholm, as totally unsuitable (the manuscript came back to me with learned comments such as "Ho Ho" scribbled in the margins). Some of the other Bristol contributors protested, and Peter Haggett intervened to put the essay back in. A couple of years later Chisholm attacked my arguments at an Institute of British Geography conference on the grounds that class categories such as landlords, capitalists, and workers were gross oversimplifications that could have no substantive meaning. I replied that his book *Economics for Geographers* itself appealed to the categories of land, labor, and capital, but the only difference was that he evidently preferred to deal with dead things rather than with the living dynamics of class relations and social struggle.

By then the pattern was set. I had a relatively secure position in the field. I could use my intellectual resources and powers to a political end and was fiercely determined, along with many others (I think of Doreen Massey, Dick Peet, Jim Blaut, Michael Eliot Hurst, and a host of younger scholars among whom some of my own graduate students, such as Dick Walker and Neil Smith, figured large), to do so to the hilt, much as I had as an undergraduate when confronted by aristocratic privilege. By the time the battle was over, the opposition conceded (though usually with multiple caveats) that the field of Marxist geography might be intellectually coherent, even empirically relevant, but refused to engage with it for reasons of politics (the very principle we were initially accused of illegally importing into the discipline!). Like the battle with the aristocrats at Cambridge, the intellectual victory at Hopkins was Pyrrhic. I should have learned the lesson then: political battles are never won by intellectual prowess alone! Nevertheless, for a while in the late 1970s, the innovations coming out of the Marxist, anarchist, ecological, and socialist strains of geography were powerful enough to entrain many younger scholars onto new terrains of study with long-lasting but, from my standpoint, increasingly unintended consequences.

## Limits and Paris

For myself, there was a more pressing analytical problem within the larger political issue. I wanted to understand the political economy of urbanization in general (and of Baltimore in particular). My studies of Marx and Marxism had made me aware that there was a long tradition of theorizing about social processes in general (the work of Paul Baran and Paul Sweezy and of Maurice Dobb being crucial). But this tradition rarely encompassed contemporary urbanization. The group that formed around the *International Journal of Urban and Regional Research* set out to remedy that gap. Ray Pahl, Enzo Mingione, Chris Pickvance, Edmond Preteceille, and Christian Topalov were central, but above all it was the figure of Manuel Castells who was the focus of interest. I had discovered earlier that Henri Lefebvre had written on the urban question (I cited him briefly in *Social Justice*), and it seemed that the urban question was taken seriously in France. Manuel had published *La question urbaine* in French in 1972. I met and listened to him with great interest several times in the mid-1970s. He encouraged me to come to France and created an attachment in Paris. I got a Guggenheim to go there in 1976–77. Manuel, as it turned out, was often not there (though we continued to interact at a distance), but Topalov, Preteceille, and others were welcoming. My French was weak, and I struggled to understand matters as best I could. But I also discovered something alarming about many Parisian intellectuals, including the Marxists. They were often as arrogant and dismissive as any other elite (although Manuel, Lefebvre, and the urban group were not like that). My reactions were visceral and negative, mirroring my response to upper-class arrogance in Cambridge.

It is one of those sad commentaries on life that, to adapt my grandmother's opinion of Winston Churchill versus Hitler, it often takes arrogance to confront arrogance. Cambridge had taught me how to be intellectually arrogant if necessary, and I have on occasion used that capacity as a political and personal weapon. I am not proud of it, but I sometimes tell myself, by way of weak justification, that it takes a little arrogance (or conviction if you will) to write anything. The difficulty is not to let it spill over and dominate everything else (including personal interactions). For whatever reason, I came out of the French experience wanting to convert my attempt to construct a better urban political economy into nothing less than a project to overhaul all of Marxian theory in order to encompass historical and geographical questions more competently. The grounds for so doing were strong. The theory of urban political economy demanded proper consideration of the use of the land as a resource, finance capital, community formation, the local state, investment in the built environment, and, above all, the production of space. These elements were either ignored or poorly understood in Marxian theory, so I conceived and wrote *Limits to Capital* to remedy the problem.

If I began with heady arrogance, the writing of that book was terribly humbling. I first needed to gain a deep command of all of Marx's work, which proved difficult to do (it took me almost a decade of hard slogging with a lot of help from students and even then was not complete). It demanded intimate engagement with the thought of someone of astounding intellectual powers but who had failed to complete his work, particularly in the directions I had in mind. I felt overwhelmed and at times unable to function. Crucial support came from Dick Walker (by then at Berkeley) and from Neil Smith (whose key thesis work, which later became the book *Uneven Development,* intertwined in all sorts of ways with my own project) and Beatriz Nofal (both then graduate students), who shared some of the agony. Once again, I did what my father taught me. I put my head down and battled it out, largely in isolation (though Neil was then, as he has been since, a tremendous comrade in arms). I became like a monad trying to internalize the universe of Marx's thought. The more I worked at it the more complicated it became. It seemed an endless and hopeless project. I tried simultaneously to simplify and to make the theory comprehensible (though many may doubt this when encountering the finished product). But publishers' readers intervened again and complained I had not taken this or that bit of arcane Marxist scholarship into account (Marxian scholasticism is a particularly obnoxious variety of the species). So the project became more and more complicated. I was desperate. I dropped everything and took off for Oxford one summer, where John Davey, of Basil Blackwell— the prospective publisher—put me up. John had published both *Explanation* and *Social Justice* while at Edward Arnold, and we had become close friends. With his help and encouragement, plus a rigorous schedule of work and good cooking, I managed to complete the text (albeit with many rough edges and short cuts) over that one summer in "a sometimes sunny corner of his kitchen" (as I put it in my preface).

*Limits* tested my own limits. It humbled me to write it. I knew then how much I could never know. Yet it was also a serious achievement. It gave me an intimacy with a body of thought (as opposed to a landscape or a city) that I had never had before. It gave me ways of thinking and understanding, of working through questions, of dissecting and reassembling problems, that were to be the foundation of everything else that I was ever to write. For me, it is the most important work I have ever written and will always remain so. It forms one limit beyond which I will never go, but it provides an extraordinary foundational resource for exploration of the world in other directions. Like all resources, it can be applied and reshaped for many different purposes in multiple ways. It does not rule with iron hand over my thought. It instead provides an intellectual home from which to explore the world.

To my surprise and disappointment, *Limits* was neither widely read nor, as far as I could tell, influential with anyone very much apart from those specifically interested in geographical and urban questions. I then discovered the limits that disciplinary tribalism placed on free exchange. Economists (even of the Marxist variety) would not take geographers seriously, and the sociologists (some of whom did react well to it) had their world system theory, and so on. Like most geographers, I read widely in other disciplines, and so it always seems odd to find that the habit is not reciprocated. It was, it turned out, far easier to bring Marxism into geography than to take geographical perspectives back into Marxism. By the time *Limits* was published in 1982, Reagan and Thatcher were in power, and the bloom was quickly fading off the fashionable interest in Marxian political economy. All sorts of schisms and divergences, many of them of long gestation, were occurring in left thought as feminism, ecology, concerns about racism, and, above all, those overwhelming movements that became known as "the cultural turn" and "postmodernism" began to grab hold. Nevertheless, like any author who has invested nearly ten years of hard slog in the production of what he thinks is a vital text, I felt seriously unappreciated. Not a very noble or even justifiable feeling, but all too human.

Writing a book like that takes its toll. Fortunately, friends and colleagues dragged me into political activities. Neil Smith was particularly insistent in those years, frequently pointing out the pitfalls of lapsing into theory with no political praxis (I lost count of the number of picket lines he had me walking on)! I also became involved with solidarity work in Central America and spent time with a former colleague and his wife, Chuck Schnell and Flor Torres (she later became a special assistant to Daniel Ortega), doing support work for the Sandinistas (including occasional journalism with my then partner, Barbara Koeppel), operating out of Costa Rica. I witnessed there the devastating effects of U.S. imperialism firsthand. Why it is that the United States finds it so necessary to crush almost any kind of alternative, however experimental, other than that which is consistent with its own narrow interests continues to bemuse me, as does the general support that such policies acquire from the U.S. populace at large. I sometimes think that a proper education in geography might be an antidote, for almost all those institutions and people who had any knowledge of the real material conditions in Central America on the ground (the church groups such as the Jesuits, the Maryknoll nuns, the Protestant churches, and organizations such as OXFAM) recognized the necessity for social and economic revolution in the region. Ignorance of real geographical information (as opposed to competency in the techniques of geographical information systems) increasingly appears to me as a deliberate means for the prosecution of narrow and self-serving U.S. imperial interests. No wonder geography is so marginalized and so badly taught in the U.S. educational system! It allows a privileged elite to have its way with the world without any serious protest.

The reason I wrote *Limits* was to better ground my understanding of urbanization not imperialism (though *Limits* did have much to say about the latter). During my Guggenheim year in Paris, I drifted away from the project I had initially formulated (much as I had done in Sweden many years before) and became fascinated with the Paris Commune of 1871, that awful Basilica of Sacre Coeur that could be seen from everywhere, and the history of the Second Empire. So I tried to set up a dialogue (or dialectic, if you will) between the theoretical abstractions of *Limits* and the historical geography of urbanization, concentrating in particular on Paris between 1850 and 1870. The long essay on Paris in *Consciousness and the Urban Experience* was the result. It was a terrific exercise. I emphasize the idea of "dialectic" and "dialogue" because the point is not to try and "read off" from theory into the rich complexities of historical geographical change, but to put them in motion together in order to see the gaps in theory while identifying hidden meanings in the historical geography. But to do that well required an intimacy with both the theory (I had that with *Limits*) and the historical geography of the city. The latter took me time to acquire, but it was a joyous thing to do. Summers away from the heat of Baltimore in Paris libraries and cafés was hardly punishment. I regard the Paris essay that resulted as a pretty good example of how to bring theory and materialist historical geography into contact with each other in fruitful ways.

The two books published on urbanization in 1985 would seem to mark a high point in my career. In fact, this turned out to be one of the lowest periods of my life. *Limits* and the urbanization books—a joint project that had absorbed me for a dozen years—were dead and gone. I was exhausted intellectually and emotionally and felt I would never write again. The books, although respected and respectable, were hardly howling successes, and their general message seemed to be falling on stony ground. Geographers and others did not care for the Marxism, and Marxists did not care for the geography. Reagan and Thatcher were pushing reactionary pro-market policies and zapping it to labor everywhere. Progressive institutions and forces were under assault and pretty much everywhere in retreat. The left was increasingly fragmented; political economy was increasingly discredited; and many, including an assortment of leftists, seemed hell-bent on attacking Marxism (sometimes for good but more often for gratuitously opportunistic reasons). The visions of the early 1970s, however problematic, were either totally defeated or rechanelled in ways I did not appreciate. Baltimore seemed in terminal decline as a liveable or civilized city. My university was turning more and more into a money-spinning corporation with less and less interest in critical studies or even basic learning. My personal life was a mess, my intellectual project a failure, and I was exactly fifty years old.

My sole support came from an incredibly talented group of graduate students—Kevin Archer, Patrick Bond, Michael Johns, Phil Schmandt, and Erik Swyngedouw. Together with Erica Schoenberger, we would assemble for seminar and dinner every week and go over and over all sorts of intellectual intricacies of Marxian thought, reading Gramsci, Hilferding, or whoever. But the group was also politically active, most particularly in the campaign to get Hopkins to divest from firms that did business in South Africa. Kevin was seriously burned across his back when some fraternity students decided "for a joke" to firebomb the shanty erected as part of the campus protest against apartheid. The support I had from this quarter was not enough, however. I felt I was spinning my wheels, gradually winding down, and I knew that when, as always happens, this talented group would disperse and go on its way, there would be nothing left.

## Oxford and Postmodernism

Fortunately, I was then offered—to my considerable surprise—the Halford Mackinder Professorship of Geography at Oxford. For all sorts of odd reasons, I went there in January 1987, even though I was reluctant to leave Baltimore (by then my adopted hometown) and my salary was cut by half. It is difficult to assess the effect of the move. Certainly I felt less of a failure. I had some power and influence by virtue of the professorial position (Oxford was still very hierarchical), and I confess to a certain smug satisfaction at gaining one of the commanding heights of the British class educational system. I was, for the first time since 1969, also working in a large geography department, which had a strong attraction. Being in the company of Terry Eagleton, G. A. Cohen, Andrew Glyn, and many others certainly helped. But, above all, I think Oxford reminded me of my origins. The question of England and its relation to what had been Empire was still all over the place. Furthermore, the college faculty at Oxford treated me as if I had just arrived from Cambridge. The twenty-six intervening years were viewed as some unwanted exile overseas as I waited to be called back to that center of the universe, where the brilliance of intellects made up for the fact that the sun failed to shine most of the winter (an American friend commented that he finally understood the term *dark ages* after wintering in Oxford).

Depression in Baltimore got converted into annoyance, even anger, in Oxford. I was forcefully reminded of what I had long ago rebelled against. The smugness of much of Oxford repelled me, and I could not understand how the left in Britain, within geography and without, was taking such wishy-washy positions in relation to Thatcherism. I wrote a few tendentious articles to that effect and got everyone annoyed. And then I launched in and wrote *The Condition of Postmodernity.*

*Condition* was the easiest book I ever wrote. It just poured out lickety-split, without a moment of angst or hesitation (perhaps that is why it is so readable). I sometimes characterize it as my least serious book, but I think that is the wrong way to put it. I was irritated and annoyed at what was happening and wanted to say something immediately political about the situation in academia in general. I used every intellectual resource I had accumulated since a young student and put everything to work in that one book. I used my knowledge of Second Empire Paris (one of the acknowledged centers of the birth of the modern) and my knowledge of the travails of Baltimore; I used the analytical frame of *Limits* to understand the dynamics of capital accumulation; and, above all, I used my understanding of the experience of space and time, of history and geography, to get beneath the cultural shifts that were preoccupying so many academics and intellectuals in the late 1980s. The writing came so easily because the stance I took was simply to say whatever it was I could say given my accumulated knowledge about this somewhat undefinable entity of thinking called "the postmodern."

I wanted to prove that Marxism was not as dead as some proclaimed and that it could offer some very cogent explanations of the dynamics then occurring (much as it had in the case of the Baltimore housing market in my earlier work). The argument worked well enough to provoke considerable and occasionally irate opposition among many postmodernists (particularly of a strongly feminist persuasion). Because I had been deliberately provocative, what should I expect? As time went on, it became clear that *Condition* was a forceful account to be reckoned with and that it had helped many people put a perspective on events that had hitherto been lacking. It became a best-seller (widely translated into foreign languages). It established my global reputation. Furthermore, it marked a moment in which contributions by geographers—Ed Soja's *Postmodern Geographies* and Derek Gregory's *Geographical Imaginations* did similar work—could break out from their disciplinary ghetto and spill over into different disciplines.

But there was an interesting sidekick in all this because the kind of Marxism that was working here was quite different from that which had dominated in the early 1970s. Marxists as a whole had never taken very seriously questions of urbanization, of geography, of spatiotemporality, of place and culture, of environment and ecological change, and of uneven geographical development. There were obvious exceptions (such as Raymond Williams or Manuel Castells), but they were treated as marginal figures in the 1970s. The *New Left Review*, a premier journal of Marxist thought, had only one article on urbanization (and that on its role in the transition from feudalism to capitalism) prior to 1984, when it finally came into the picture in Fred Jameson's article "Postmodernism as the Cultural Logic of Late Capitalism." This neglect, in a historical period of massive urbanization across the globe, had always seemed inexcusable to me. *Condition* worked precisely because it took these geographical matters as central rather than peripheral to Marxian thinking. Conventional Marxists thought *Condition* made far too many concessions to postmodernist sentiments and ways, when in fact it sought to show how much of what passed for postmodernism was better understood in terms of the urban, spatiotemporal, and geopolitical dynamics of capitalism. Nevertheless, in Marxist circles I found myself often referred to as a postmodernist.

The peculiarly ambivalent position that *Condition* assumed in the opposition between postmodernism and conventional Marxism is, it turns out, not so unusual with respect to geographical knowledge in general. This is an important point on which everyone in geography needs to reflect. Geographical knowledge tends to get shoved aside, marginalized, or even dismissed because it poses very awkward and even disruptive problems for conventional forms of disciplinary wisdom. The fundamentals of economic theory, for example, are specified as if economic transactions occur in a spaceless nondifferentiated geographical world. Insert space and, as Koopmans and Beckmann showed in a key article back in 1957, most of economic theory can't work because equilibrium prices cannot be defined. I had long argued that the insertion of space into any social theory (including that generally dubbed as Marxist) is disruptive of its central propositions. Most disciplines have preferred not to confront the problem and have swept it under the rug, usually with a few disparaging remarks about the uninteresting or trivial qualities of geographical knowledge.

But postmodernism actively sought to disrupt general theory and thereby to liberate thought from the constraints of any overarching "metanarrative" or "grand social theory" (including that of Marx). It found the idea of space—sometimes real material space but more often space as a metaphor—a convenient means to accomplish its goals. Postmodernism invoked geography, cartography, spatiality, and all manner of other geographical themes as part of its intellectual armory in ways that general social theory (including Marxism in its standard form) could not. Postmodernism even appeared to be about geography! And some geographers, Ed Soja being by far the most successful and the most prominent, reveled in the connection. But it plainly was not about my kind of geography and therein lay the difficulty and the challenge.

I had learned from Marx how important it is to know one's enemy. Marx studied classical political economy voraciously. I thought it equally necessary to study the poststructuralist and postmodernist thinkers then very much in vogue. This goal became imperative because writers such as Foucault made frequent reference to questions of space and spatiality, sometimes as disruptive of received meanings (as in *The Order of Things*) or sometimes as manifestations of power (as in the image of the panopticon in *Discipline and Punish*). I had long enjoyed Foucault's writings and not seen them, at least in their early manifestations, as antagonistic to Marx (I noted all sorts of parallels between the arguments in his *Discipline and Punish* and the materials that Marx assembled in *Capital* to explain how the worker became disciplined to the work regime of wage labor). I was surprised, therefore, to find Foucault viewed as deeply antagonistic to Marx and wanted to know why. In these enquiries, I had much graduate student help from Clive Barnett, Adrian Passmore, Argyro Loukaki, and the more skeptical Andrew Merrifield and Mike Samers, backed by Erik Swyngedouw, who had just been appointed to the School of Geography as a lecturer. This research provided a forum for intense and lively debates over Foucault, Derrida, Althusser, and the like. Sitting in a smoky pub in Jericho, an interesting neighborhood of Oxford, with a bar billiard table to divert attention, sometimes joined by John Davey, we had all sorts of arguments. An evaluation or direct response to all these strains of thought, beyond their somewhat polemical treatment in *Condition*, seemed necessary.

It was difficult to concentrate at that time. I had become a father and was really enjoying parenting no end. Delfina, one of the best things that ever happened to me, gave me much to think about as well as plenty to do. I wonder what kind of geography she will acquire. She knows the difference between grapefruits and lemons and flits the world regularly between North and South America and Europe with the Discovery Channel easy to hand. But she cannot roam the city and the country on her own the way I did when I was a kid. It is a different experiential world, and different fantasies enter into her constructions.

## The Challenge of Geography

I knew, however, that the grounded geographical angle had contributed to the success of *Condition*. By way of the numerous invitations that flowed my way after its publication, I began piecemeal to explore the geographical angle more systematically, in part as a defense against criticisms from both Marxists (who mistook the geography for postmodernism) and postmodernists (who equated the Marxism with dogmatism). I wrote several essays on the significance of place, space, and the dialectics of environmental change, and it more and more occurred to me that some kind of synthetic statement, joining these all together, might be helpful to reflect on the nature of geographical knowledge more generally. The idea of a synthetic statement had been gestating for many years, most notably in an article in *The Professional Geographer* in 1984, "On the History and Present Condition of Geography: An Historical Materialist Manifesto," where I first raised the question of the disruptive effect of spatial arguments within general theory.

My embeddedness in a large and somewhat "traditional" geography department also helped. It was good to listen in to seminars on the nature of the peasantry and debates on historical geographical transformations or political organization and struggles over water in the Middle East. It was good to be in on the debates surrounding the formation of a unit to study environmental change. I wanted to say something more directly to geographers about the basic conceptual apparatus of the discipline—about concepts of space, place, time, and environment being fundamental—but I wanted to do it via general theory. This last desire stemmed from my long-standing belief (indeed, what had my whole career as an academic geographer been about?) that geographical knowledges are not outside of theory and that the usual dichotomy between universality and general theory, on the one hand, and geographical particularity and incomparable specificity, on the other, is a false distinction. The fact that a version of this distinction had been raised to the status of a virtue within postmodernism made such an effort doubly important.

And yet I was not going to back off on the politics either. The threats to close the Rover Company's Cowley works in East Oxford led to a protest movement in which I participated tangentially by way of a collective effort, led by Teresa Hayter, a well-known activist with Trotskyist leanings, to do research and present evidence to various enquiries. A book, edited with Teresa, on the story of the Cowley autoworkers eventually emerged as a side product of that struggle. At the editing stage, I found myself at loggerheads with Teresa on a number of important issues. My account of these differences eventually formed the first chapter of *Justice, Nature, and the Geography of Difference*. In several respects, this chapter defines my differences with that version of the Marxist (and in this case Trotskyist) tradition that focuses on the workplace to the exclusion of much else. The fact that I also built the chapter around an analysis of Raymond Williams's novelistic treatment of space, place, and environment (noting that one of his novels was set in and around the divide between the university and the car works in Oxford) was also significant because it highlighted a particular tradition of Marxism that was far more sensitive to the issues that concerned me.

At the other end of the spectrum, I needed to consider the criticisms of the postmodernists, in particular the feminist wing of it. I had been somewhat taken by surprise by the feminist attacks on *Condition*. Not all of these attacks could be characterized as "postmodern," and they varied a great deal in tone from matronly moralizing and psychoanalytic meanderings to vigorous denunciations of my "cultural" sexism. Not all feminists thought this way, however, and there is and was a considerable divide between feminists who identify with postmodernism and those who do not. It was difficult to sort out my own lapses from major differences over politics and method, but the often personalized and insulting tone of the criticisms seemed unnecessary and gratuitously hurtful (unfortunately, this form of personal attack has become very much par for the course in this arena of work in recent years). It had not generally been my practice (though there were a few notable and quite famous exceptions such as my contretemps with Brian Berry in the early 1970s!) to respond to individual criticisms or even to engage too much with critical evaluations because it had always seemed to me that the general stance of opposition to capitalism was best served by making creative contributions rather than engaging in destructive slanging matches (of the sort that so often accompany sectarianism). However, in this instance, I felt I had very little option. The editors of *Society and Space* did not offer me space to reply to the critical articles they published, so I sent my response to *Antipode,* as an article I called "Postmodern Morality Plays." Nobody took much notice.

But there was more to the problem than that. I had not made as much use as I should have of the very good feminist work available, and I needed to reevaluate somewhat the role of writings by Emily Martin, Donna Haraway, Nancy Fraser, Iris Marion Young, and several others who had made invaluable contributions in a number of areas. This I set out to do in a variety of contexts. On the other hand, I was resolutely opposed to the kind of feminism that either totally merged into postmodernism or became so stuck in a narrow definition of identity politics as to ignore problems of class, the workplace, and what was happening through the dynamics of an increasingly neoliberal process of capital accumulation. I used the example of a horrendous workplace fire in a chicken-processing plant in Hamlet, North Carolina, in 1991 to reflect on relations between class, race, and gender in the quest for a more just social order. This work also formed a key essay in *Justice.* Given the disproportionate attention this one essay subsequently received at the hands of reviewers (largely protesting against my arguments, sometimes vehemently), I guess I hit a raw political nerve.

### Back to Baltimore

By 1993, I was back on the Fall Line in Baltimore. It was not an easy move. I went from a position of power in a large geography department to being a minor and marginalized figure in a department dominated now by engineers in a very corporatist Engineering School that cared only about grants and sponsored research. Indeed, the dean of engineering had initially refused to reappoint me, and it took a strong protest from colleagues in Arts and Sciences to get that decision reversed. It meant I returned to Hopkins under unfavorable conditions, though a position of sorts was found for Haydee, my wife (an oceanographer with strong training in fluid dynamics), and I had support from many individuals in other departments. But putting the essays together for *Justice* did not prove easy. These days, the older and more prominent you become, the more you become a target for criticism (justified or not). You also become acutely aware of that status. No quiet retirement into the role of ancient and venerated scholar! And there were other barriers to completing this project. Though I did not know it, my own permanence was seriously threatened by lack of blood flow to the heart. I sometimes think that my articulation of dialectics as a relation between flows and permanences subconsciously reflected that condition! The five heart by-passes came just after I had done the index to *Justice*.

For all of its lapses, I regard *Justice* as one of my most profound geographical works. Unfortunately, the politics seems, at least in the short run, to have masked appreciation of that fact. The chapters on nature and environment, space-time, and place, and the use of dialectics to formulate a perspective called "historical-geographical materialism" are for me the key to a certain understanding of how geographical knowledge works. The book does for me in a geographical way what *Limits* did for me from the standpoint of political economy. It grounds a conceptual apparatus, but it does so in an open and flexible way that acknowledges how spatiality and geography are disruptive of other theoretical formulations. It tries to find a narrow path between the atheoretical particularity to which geographical thought is always prone and the dismissive treatments of matters geographical in conventional forms of social theory (including Marxism). Geographers may have been adepts at navigating the world in the past, but they have never proven very good at navigating the continents of knowledge.

A project of this kind is not easy to pursue. A similarly noble vision had animated the embrace of positivism as a universalizing method in the 1960s (a movement I then endorsed). However, the limited practical success of positivism and its reduction of the rich complexity of geography as lived experience (as materiality, as representational forms, and as the living out of human wants, needs, and desires) to geometrical forms and statistical representations indistinguishable from any others hardly amounted to a realization of that noble vision. For the record, let me state that I am not antagonistic to statistical enquiries or geometrical representations, but they must be embedded in some larger sense of what geography is and can be about. So why, then, did I hope that *Justice* could point to a more fruitful way to secure geographical theory in the pantheon of academic knowledges?

"Hope," says Balzac, "is memory that desires." I have (we all have) a rich store of geographical memories, both in an intellectual and experiential sense. My desire to construct a different kind of world—ecologically more sensitive and socially more just—is not diminished. Trying to keep the connective tissue of experience, of thought, of writing, and of just being in the world, all together is what life is all about for me. So if I have just written a book called *Spaces of Hope*, which focuses on possibilities and conversations about alternatives (the missing chapters of *Justice*), and if the book draws inspiration in part from the "living-wage campaign" in Baltimore and Hopkins (with some of my students heavily involved), and if it uses Baltimore as both foreground and backdrop, and if it draws upon thirty years of teaching Marx's *Capital*, and if it harks back to a sense of dynamics acquired by a graduate student in a long summer reading local press reports, and if it turns a certain imperial gaze acquired in the waning days of the British Empire upon the continents of knowledge . . . well, that is just how books get written! I hope, because I have both the memory and the desire to change the world into a far, far better place than that which now exists. Life without hope is the death of desire.

And so I offer a conclusion. At the heart of my geographical theorizing lies the concept of the dialectic. It sounds like a difficult and mysterious term, but it is simple. It rests on the idea of a dynamic relation between processes and things. I live my life, as you do, as a continuous process. That process often gets channeled into specific projects—gaining a qualification, raising a child, becoming rich and famous, learning to swim, writing a book, building a home. Some of these projects are freely chosen, and others are forced on us by fate and circumstance. But as we live our lives, we also produce things for which we are more or less accountable, and we are usually judged in the public realm by the things we produce rather than by the processes that led to their production. I am judged academically by the books and articles I have written. Students know this feeling well enough. I sometimes watch a student learning immensely through the process of study only to produce a terrible examination or paper at the end. Should I grade on the basis of the process or the thing produced? In situations of reasonable educational intimacy (largely impossible in the factory conditions of modern universities and absolutely impossible with distance learning), I can make allowances. But in the world at large, it is the things that count. The processes so fundamental to their production fade into nothing. We can hope, of course, that the things capture something about the joys and frustrations, the irritations and sublime moments, entailed in their production. But I still regard my books as essentially dead things crystallized out of a continuous lived process of learning and exploration.

One of the crucial insights that drew me to Marx was his powerful commentary on how we so often let the things we produce return to dominate us. This is what he meant by alienation. The worker produces the machinery to which the worker must then submit. More generally, we produce a world of things (such as highways, automobiles, and shopping centers) around which we have to arrange our living processes in submissive ways (like so many shopping projects). It is very easy for academics to let the books they produce return to dominate them. That is something I have always steadfastly resisted. My books and articles are what they are, and I cannot let them dictate my life as process and let them rule over me as an alien force. I signaled this long ago when responding to Stephen Gale's review of *Explanation:* I noted he had a singular advantage over me because he had read the book and I had not. It sounded a flip comment, but it really was profound. If I have, as I believe is the case, a record of continuous innovation and exploration in my writing, it is precisely because I understand (at first intuitively and later theoretically) how the dialectic works. I have tried in this essay, which now begins to assume the form of a dead thing, to convey something about the processes that have animated my geographical works. I have learned much in so doing, but now I must let the text go to stand in the world as a fixed, static, and unchangeable document. But the dialectic of living does not stop here, neither for you nor for me. But you now know where you can find me. I am on the Fall Line. Dreams can come true, if only for that small segment of space-time in which we are able and willing to sustain them. But hurry:

> If I tell you that the city toward which
> my journey tends is discontinuous in space
> and time, now scattered, now more condensed,
> you must not believe the search for it can stop.
> —Italo Calvino, *Invisible Cities*

# The Life of Learning

## Donald Meinig _____

*Donald Meinig (b. 1924) served in the Corps of Engineers during the Second
World War before receiving his B.S. from Georgetown University in 1948, and
the M.A. in 1950 and Ph.D. in 1953 from the University of Washington. In
1994, Syracuse University awarded him the Doctorate of Humane Letters,* honoris causa. *He began his teaching career at the University of Utah but has been at
Syracuse since 1959, where he is now Maxwell Research Professor of Geography.
The recipient of many professional awards, medals, and invitations to distinguished lectureships, he was the first native-born American to be elected, in 1991,
as a fellow of the British Academy. His books on the historical geography of Australia and Anglo North America,* On the Margins of the Good Earth, The
Great Columbia Plain, Imperial Texas, *and* Southwest, *culminated in his
magnum opus, the eventual four-volume work* The Shaping of America. *The
first three volumes were published in 1986, 1993, and 1998, with volume four
anticipated in the near future. The successive volumes have been hailed not only
for their fine scholarship but for the breadth of vision they display in illuminating
Anglo-America in its geographical setting.*

Had the idea of such an invitation ever crossed my mind, I would have
thought the chances of being asked to give the Haskins Lecture as a
good deal less likely than being struck by lightning. I found it a stunning ex-

Professor Meinig gave the Charles Homer Haskins Lecture at the Hyatt-Regency hotel,
Chicago, on April 30, 1992, and is here reprinted with kind permission of John H. D'Arms, president of the American Council of Learned Societies.

189

perience, and I cannot be sure that I have recovered sufficiently to deliver a coherent response.

I can only assume that I was selected because I am one of a rare species in the United States—an historical humanistic geographer—and someone must have suggested it might be of interest to have a look at such a creature, see how he might describe himself, and hear how he got into such an obscure profession. Geographers are an endangered species in America, as, alas, attested by their status on this very campus [the University of Chicago], where one of the oldest and greatest graduate departments, founded ninety years ago, has been reduced to some sort of committee, and the few remaining geographers live out their lives without hope of local reproduction. I shall have more to say about this general situation, for while I have never personally felt endangered, no American geographer can work unaware of the losses of positions we suffered over many years and of the latent dangers of sudden raids from preying administrators who see us as awkward and vulnerable misfits who can be culled from the expensive herds of academics they try to manage.

I have always been a geographer, but it took me a while to learn that one could make a living at it. My career began when I first looked out upon a wider world from a farmhouse on a hill overlooking a small town on the eastern edge of Washington State. My arrival on this earth at this particular place was the result of the convergence (this is a geographer's explanation of such an event) of two quite common strands of American migration history. My paternal grandparents emigrated from a village in Saxony to Iowa in 1880, following the path of some kin. My grandfather was a cobbler and worked at that a bit, then got a laboring job on a railroad, and before long had purchased a farm. He had three sons (my father being the youngest and the only one born in America), and as they were reaching adulthood, he heard that good farm land in Washington State could be had for a third of the price in Iowa, and so in 1903 he moved there and settled his family on a fine four-hundred-acre place. My mother's parents were born in upstate New York and what is now West Virginia, met in Minnesota, where she was born, and about the same year migrated to the same town in eastern Washington, where my grandfather dealt in insurance and real estate. My forebears were not pioneers but moved to places that were developing with some prosperity a generation or two after initial colonization. That prosperity eluded almost all of them, and, in time, most of my aunts and uncles and cousins joined in the next common stage in this national pattern and moved on to Seattle, Tacoma, and western Oregon.

The view from that farmstead was one of smooth steeply rolling hills to the south and west and of local buttes and a line of more distant forested mountains in Idaho on the north and east. This was the eastern edge of the Palouse country, a regionally famous grain-growing area. Physically it is a unique terrain, famous among geomorphologists for its form and texture, and it can be beautiful in the right season, especially just before harvest. To me it was interesting in all seasons. As far back as I can remember, I was fascinated by that panorama. I wanted to know the names of all those features; I wondered what lay beyond; I explored on foot for miles around, climbed all the nearby summits to gain a broader view. Two branch-line railroads ran along the edge of our farm, readily visible from the barnyard. And so I also became fascinated with trains, watching them every day, counting the cars, learning to recognize the different engines, deciphering all those mystical letters, emblems, and names of the railroad companies, poring over timetables and maps obtained from indulgent station agents. From an early age, I was collecting road maps as well and avidly reading about places, mostly faraway places. Geography was, of course, a favorite school subject; but more than that, it fired my imagination. There was something about maps, and names of places, and the way they were arranged in space; about rivers and railroads and highways and the connections between places that enthralled me—and they still do. Whereas other children might have imaginary playmates and adventures, and write about them, I had imaginary geographies: I made up railroad systems with names and emblems, engines and schedules, and put them on maps, with mountains and rivers and ports, and all the places named and their relative population sizes shown by symbols.

But what does one do, really, with such interests? I remember announcing at some point in my boyhood enthusiasm that I was going to be a "geographic statistician." That came after some hours of poring over my big Rand McNally atlas and memorizing the 1930 census population of every city, town, and hamlet in the state of Nevada. But of course there soon arose in my own little mind the deflating question of what possible use would such a person be? If anyone wanted to know such information, wouldn't they simply look it up in an atlas, as I had, instead of hiring me to tell them? So there I was, even at so young an age, a skilled person with a bleak future, a living data bank no one wanted.

However, larger horizons were being created by the hammer of world events and emblazoned in the headlines of the Spokane newspaper, announced in the clipped tones of H. V. Kaltenborn, featured on the cover of *Time* magazine:

- the Italian invasion of Ethiopia
- Japanese attacks on China
- the bewildering chaos of the Spanish Civil War
- Nazi pressures in the Rhineland, Sudetenland, Danzig—World War II

I tried to follow it all closely in my atlas, and when I graduated from high school a few months after Pearl Harbor, I was ready to go across the mountains to our big university and begin to train for a career in the U.S. Foreign Service. That seemed a logical combination of geography and history, of places and events, with exciting prospects for actually seeing a lot of the world.

When I look back upon my preparation for this undertaking, I am rather appalled at how thin it was in all formal respects. Neither of my parents had more than an eighth-grade education. My father read the Spokane newspaper every day, but I can never remember him reading a book. My mother read a good deal, but other than her Bible we had almost no real literature in the house. Nevertheless, they assumed that my older sister and I would make our way as far as we might want to go and did everything they knew to encourage us. Even though I soon could look back and see that the 1930s were very stringent times, I never felt touched by the Great Depression. We might not have electricity or running water (in that we were somewhat behind the times even locally, chiefly because my father was so fearful of debt), but I always had new books and pencils and tablets and new clothes for school. But I cannot recall in any detail just what I was being taught, what kind of academic groundwork was being laid, what books I was reading. I remember lots of English drill on grammar but only a few excerpts of great literature. As for classes in history, the only one that comes clearly to mind was the joke of the curriculum: while I was in high school, the state of Washington suddenly decreed that every student must have a course in Washington history and government. We had no textbook, and to be led through the State of Washington Constitution by the music teacher was far from inspiring.

Such small-town schools were not bad schools. I was given a foundation in basic subjects, but never pressed very hard to excel. I had conscientious teachers, but no really inspiring ones. The most extraordinary person was a talented young drama teacher fresh from Seattle (hired mainly, of course, to teach typing), who generated such interest and discipline that our little school won the state one-act play contest two years in a row, and the town was so thrilled that they raised enough money (something like three hundred dollars) to send us across the country by car to the national contest at Indiana University. I had a bit part, and the long journey to the Midwest and back was an important stage in my geographical education. Our teacher characteristically insisted we make the most of it, and not only plotted a route by way of such features as the Mormon Tabernacle, Royal Gorge, Mt. Rushmore, and Yellowstone, she took us to Chicago and arranged for us to stay a night at Hull House. Now, sixteen-year-olds from the country would have much preferred a modern hotel on Lakeshore Drive, and it was hard to grasp just what a "settlement house" was, but a walk through the immigrant ghetto and incredibly congested Maxwell Street market left a powerful new impression of American life.

The deficiency of my schooling I was first to feel was the lack of foreign languages. These were not required for college entrance and apparently were taught only when there was enough interest or a teacher available. I know that French and German had been offered, but not, as I recall, to my class, and I later regarded these as a burden in my university work. The broader limitations of such a place only became apparent later and have never been a cause of great regret. Those of us who enjoyed school and all of its activities never thought of ourselves as country bumpkins. We were well aware of a larger world, in part because of our geographical situation. Washington State College and the University of Idaho were less than twenty miles away, and those campuses were familiar ground. Although it was common for students to drop out of high school, most graduated, a few each year went on to college, and I never doubted that I would.

When I now think about those formative years, I conclude that the weakness of my formal training was in some degree offset—especially in view of my later work—by the experience of how lively small-town life could be. For hundreds—probably thousands—of towns like Palouse, Washington, one has to go back at least to 1941 to find that vitality, for things changed with the war and changed rapidly—drastically—after the war. And it may seem a contradiction, or at least a paradox, that the 1930s—the Great Depression—was a period of great activity in such places, at least in that part of the country, for crops were good even if the prices were low, and there was an influx of people from drought-ridden Montana and Dakota. If almost no one was making much money, a great many were trying hard to scratch out a living. In that town of 1,100 people, there were fifty shops and businesses, several doctors, dentists, and lawyers, a weekly newspaper, half a dozen churches, busy farm suppliers, ten passenger trains a day, a usually packed movie theater, occasional traveling shows, evangelists, and lots of sports. Saturday night in harvest time, when all the stores stayed open, was so packed you had to come early to get a parking place. I am glad to have experienced all that. I think it has given me some real understanding and feel for what a large segment of American life was like in many regions over a considerable span of our history.

I went off to the University of Washington in 1942 because I was just seventeen, but knew that I would soon be in military service. There was, of course, much talk and plotting among all male students as to how we could get into some branch that might be exciting or at least interesting. Unlike many of my friends, I had no interest in going to sea or flying. That left the army, and the ominous possibility of being arbitrarily assigned to cooks-and-bakers school or something equally awful. Concluding that the only thing I knew much about was maps, I spied a course in cartography in the winter term offerings and went to the geography department to enroll. It turned out to be an upper-division course full of Naval ROTC students, but after a conference with the chairman, he agreed to let me take cartography and a prerequisite course simultaneously. And it worked. At the end of that term, I enlisted and was assigned to the Corps of Engineers as a topographic draftsman—and as soon as I completed basic training, they saw that I could type, and I was put in a dull office job and never had a drafting pen in my hand.

I'll not give an account of my illustrious wartime career. I never got out of the U.S.A. The only pertinent thing is that three years in the army provided a much-needed maturing and did nothing to dampen my interest in the foreign service. The GI Bill opened up heady new prospects, and I remember that with unsullied naïveté I sat in my boring army office and sent off for bulletins from Harvard, Stanford, and Georgetown to decide which might offer the best training. After careful study, I chose Georgetown because it had the most specific curriculum and because it was in Washington. I had glimpsed some of the attractions of Washington, D.C., while in officer's school at nearby Ft. Belvoir.

Naïve as I was about universities, I have never regretted my choice. The School of Foreign Service was certainly an uneven place, but I had a few first-rate professors, and my interest and enthusiasm never flagged. As everyone knows who was a part of it, it was a wonderful time to be at any university. Even though classes were packed, staff was short, and we went day and night, the year round, there was a maturity and seriousness about it that was quite unprecedented. One's classmates varied in age from twenty to forty, from all walks of life, and with a great diversity of experiences. I never had a small class, but some of the lecture halls crackled with excitement: as with Carroll Quigley on "development of civilizations" and Shakespeare with John Waldron. After my first term, I returned West to Colorado to be married, and needing extra income, I got a part-time job as assistant to a remarkable academic character, Ernst H. Feilchenfeld, a Jewish refugee, doctorate from Berlin, who had taught at Oxford and Harvard before happily settling in, as he put it, "under the benevolent despotism of Jesuit Georgetown" as professor of international law and organization. He ran an Institute of World Polity, more or less out of a file cabinet, and my job was not only to take care of his correspondence with a distinguished board of consultants scattered about the world, but to sit and listen to him talk. He was a garrulous and lonely man, and after two years with him I was tempted to think that about 50 percent of my education at Georgetown was from Feilchenfeld and 50 percent from all the rest.

So it was an immensely stimulating time to be at that unusual school in the capital of the new superpower. Many of us participated in small networks of contacts with the lower levels of various government departments and agencies. But there were many dark clouds as well, and they rapidly thickened. Senator Joseph McCarthy and many little McCarthys were running amok. Foreign Service officers were being pilloried as traitors, the State Department increasingly demoralized, and the whole prospect of having one's life work bound to and constrained by such a government created a vocational crisis for me—and for many of my classmates. There were other factors, as well. One of the virtues of the School of Foreign Service was the practical segment in its curriculum: one studied accounting, business law, and a consular practice as well as history, government, and literature. Even a glimpse of the actual chores of consular work, the endless forms and regulations, responding to imploring citizens and would-be citizens, began to tarnish the glamor of my adolescent view of overseas service.

But where to turn? I floundered for a few months. I tried to think about what I most enjoyed. Railroads? I got an introduction to some railroad officials in Washington, but all they could describe for me was to become a salesman and solicit freight. Geography? Read and learn about the world? But how to make a living out of it? I have no explanation for why I was so stupid as not to see what was so obvious; it finally did dawn on me that that is what professors do: read and study and talk at great length about that which most interests them—they have a great deal of freedom to do it in their own way, and they have captive audiences forced to listen to them. Once I had that belated breakthrough, I had no doubt about what I wanted to be: an historical geographer. I knew of a book or two by that name, but neither I nor anyone else I talked to knew if there really was such a field. But I had spent many hours, usually fascinating hours, in history classes and had read rather widely, and I already knew enough geography that I was always visualizing a map and often thinking how much more effective the teacher or writer might be if the narrations and explanations had been informed with maps.

I had no advice whatever as to where to go to graduate school, but I knew there was a big geography department in Seattle, where I had taken two courses as a freshman and had actually talked to the chairman; and besides that, I think we were a little homesick for the West. It was not a very good department. Shortly after my time there, it was revolutionized under a new chairman and mostly new faculty, and became one of the most influential centers of a "new" geography in all the Euro-American world, but it was distinctly mediocre in 1948. Within a short while I realized that I should have gone to Berkeley, but practical reasons impelled me to persevere in Seattle. The not-very-taxing geography courses provided a sound foundation, and I read widely and roamed the campus in search of interesting lectures and courses. Among the most memorable was the packed hall—standing-room only—of Giovanni Costigan's lectures on English history; what I should have sought was a solid semester in historiography.

However, I happily acknowledge my debt to one professor who took a real interest in me and was helpful then and thereafter. Graham Lawton was an Australian, a Rhodes scholar who had taught briefly at Berkeley. He sought me out when he learned that I, having seen an announcement on a bulletin board shortly after my arrival, had applied for a Rhodes scholarship. He did his best to help shape my rather exotic statement of interests (as I recall, I declared a research focus on Northwest Africa—mainly because I hadn't found much to read on that corner of the world and was curious about it). As was not uncommon in our region, some bright fellow from Reed College won the Rhodes, but I had gained a very supportive advisor.

I had arrived from Georgetown with a head full of Quigley and Toynbee and Mackinder and other sweeping worldviews, and it took a while for my geography mentors to bring me down to earth, to get my feet firmly on the ground and eventually on my native ground, in the prosaic little Palouse country. Graham Lawton guided me into British and American historical geography—not a large literature—and I soon tried my hand at it.

What started as little more than an exercise, a convenient thesis topic, soon developed into a much larger and self-conscious work. I wanted to put my home area into history, to see how it fitted in as part of American development. To do that, one had to create a rather different version of history, one that was focused on the land and places rather than on politics and persons. I wanted to find out what the early explorers actually said about all the various localities, just where the earliest farmers and townsmen settled, spread into other districts, and domesticated and developed the whole region with the way of life I had known in boyhood. I avidly reconnoitered the countryside, visited every locality, studied old maps and documents, read hundreds of country newspapers, plotted data from public and private records. I had a lot to learn about my native ground, but I already knew about some important matters. I knew a lot about farming and livestock raising because I had done them. Our farm was small by Palouse standards but nonetheless real—indeed, more real for my purposes than others, for my father was the last farmer in that area to use horses rather than tractors. He loved those big workhorses as much as he hated all the high-powered machinery that was already essential to successful farming. And so I grew up with them, learning at an early age how to take care of them, harness them, and work in the field with them—and thereby I was in contact with an older—indeed, ancient—world of farming.

I found great satisfaction in that research, and I wanted to share it with others. I wanted to write a book that could be read with pleasure and enlightenment by local residents who had some serious interest in their homeland. I overestimated that potential, but a sprinkling of letters over the years assures me that *The Great Columbia Plain* has helped a few.

At the same time, I wanted to write a book that would command attention in professional circles. I wanted to help create a literature that would at once exemplify something of the character and value of the geographical approach to history and the historical approach to regional study. I was convinced that professional geography in America badly needed that kind of literature. Human geography and regional geography were too largely textbook in form, stereotyped descriptions of a set of standard topics with rarely any historical or interpretive dimension at all. Certainly no geography book told me what I most wanted to know about my country. I thought my approach was a valuable way of looking at a region. It answered most of the questions I had at the time, and I hoped it might encourage others to do something similar on other regions—though, in this, too, I seem to have overestimated that prospect.

For a while I had in mind more such studies myself, and I did, in fact, write another book (before I completed this first attempt) from an opportunity provided by a Fulbright to Adelaide—where Graham Lawton was now head of the geography department. A surge of settlers into the dry country north of Adelaide had created Australia's premier wheat region. Emerging at the same time, working with the same general technology, and competing for the same Liverpool market, this South Australian episode offered illuminating comparisons with the Pacific Northwest. Regional geographers are often accused of being too focused on particularities and diversities, but any geographer's global training should provide analogues and generalizations as well.

But I did not proceed with more historical studies of agricultural regions. Two experiences of residence in "foreign lands" brought about a shift of focus, a change in emphasis. One of these was that year in Australia, where another branch of English-speaking pioneers had created a nation on a continental scale. "The most American" of lands beyond our shores was a likeness many Australians were ready to assert and most Americans seemed happy to accept. There were, of course, grounds for such a characterization, but I was struck more by the differences, and they helped me to see my own country in a clearer light. The thing that most impressed me from my reading, research, field studies, and general observation was the difference in the general composition of the population: the homogeneity of the Australians as compared with the kaleidoscopic diversity of the Americans. And one was more alert to the comparison because the Australian population was just beginning to change toward the American type by the unprecedented postwar influx of emigrants from Continental Europe: Germans, Dutch, Poles, Italians, Greeks, Maltese. Their number was not really large, but they were clearly injecting a new variety and vitality into Australian life. Australian commentators, novelists, dramatists were giving attention to the many individual, familial, and social challenges of immigration, acculturation, assimilation—themes that were century-old clichés in America, and I returned with a heightened appreciation of the stimulus, the energy, the creativity, and the special problems generated by the marvelous ethnic and religious complexities of American society.

The other so-called foreign experience was congruent with that. I began my professional career at the University of Utah. I knew, of course, that Salt Lake City was the seat and symbol of the Mormons. We all knew of the Tabernacle Choir and something vaguely about their peculiar history—polygamy, Brigham Young, and the Great Trek to the desert West. But I didn't realize just what we were moving into when my wife and infant daughter and I settled into the Salt Lake Valley. We found ourselves classified in a way we had never thought of: we were "Gentiles." We had unwittingly moved into a dual society wherein everyone was either a Mormon or a Gentile (giving rise, of course, to the local cliché that "Utah is the only place where a Jew is a Gentile"). This binary character was a subtle but pervasive reality: two peoples, interlocked in much of daily life, not at all visibly distinct to the casual observer, without any overt antagonism between them, each subdivided into two complex varieties within—yet ever conscious of being two distinct peoples. That Mormon-Gentile dichotomy seemed to permeate everything, and it gave a special interest, flavor, and edge to life in Utah. One also came to see that the local landscape, rural and urban, was different from adjacent areas. The farm villages, the ward chapels, tabernacles, and temples, the rigid squares and the scale of those big city blocks, stamped a visible Mormon imprint on the area.

And so one came to realize that the Mormon Church was not just another of the many denominations in the remarkable diversity of American religion, but was the creator and vehicle of a distinctive people, of a highly self-conscious, coherent society that had set out to create a large region for itself in the desert West and had essentially done so, for Gentiles were a minority and generally regarded as "others," "outsiders," even at times "intruders." Nine years in Utah taught me something new about America, heightened my consciousness of such social groups, made me feel that the historical geographer would do well to focus on the kinds of communities that were characteristic of various regions. Despite powerful pressures toward standardization and conformity, the American West was far from even an incipient uniform or united area.

And so with a heightened sense of life and locality, I began to examine the West as a set of social regions. I wrote an extensive essay on the creation and dynamic character of the Mormon culture region, then a small book on Texas, followed by another on New Mexico and Arizona. Each of these gave considerable attention to ecology and spatial strategies, as in my earlier books, but the main focus was on the various peoples shaping discrete regional societies. In this kind of human geography, one was not describing simple regional patterns, fixed in form and place, but continuous geographical change. That is, changes in limits and relationships; in internal character as a result of migrations, diffusions, demographies; in economies, transportation, and other technologies; in regional attitudes and perceptions.

These more interpretive writings were well received outside geography. A number of historians seemed to find in them a fresh perspective on a general topic still dominated (twenty-five years ago) by the Turner frontier thesis. And I must also tell you that they were the means of snaring my favorite student. He is a fictional character in James Michener's vast volume on Texas. I got to him in the middle of it, on page 504, and changed his life. He was already a football hero at the university, but, in Michener's words: "he read a book that was so strikingly different from anything he had ever read before that it expanded his horizons. *Imperial Texas* . . . by D. W. Meinig, a cultural geographer from Syracuse University, . . . was so ingenious in its observations and provocative in its generalizations that from the moment Jim put it down, he knew he wanted to be such a geographer."

Michener sends him off to Clark University instead of to me, and I lost track of him in the further depths of that book. I've never heard from him, but I take satisfaction in the fact that whatever one may think of Michener's fictions, it is generally agreed that he gets his facts right.

I had in mind to do a large book on these American Wests—I had done considerable work on California and Colorado as well—but then came another sojourn overseas and from it another shift in scale, if not in perspective. In the fall of 1973, I had a very pleasant visiting position at the University of St. Andrews in Scotland. I was expected to give one lecture a week, ten in all, on the United States; the rest of the time I could do as I pleased. In the winter, we shifted to Israel, where I repeated that course on America at the Hebrew University of Jerusalem. How to treat the United States in ten lectures made one search for a few major themes and to generalize at a broad scale in time and space. And thinking about such matters in those places, and later on as we settled into a small village in Gloucestershire, forced one to consider things from the beginning: How did Europeans reach out, make connection with, and get all those colonies started in America? Once one began to think seriously in terms of oceanic, intercontinental connections, one was caught up into a vast field of action, and inevitably American Wests became but small pieces within a large system. One had always known that, of course, and it didn't make the West of any less intrinsic importance than before, but it altered the balance and made seeing the West in the fuller context of nation, North America, and, indeed, an Atlantic system the principal goal.

And so I slowly got under way with the rather audacious task of writing a "geographical perspective on five hundred years of American history."

I suppose my whole writing career could be seen as a geographer's version of the search for the self—of who one is, and how that came to be, and what is the meaning of it all. For the geographer, that means close attention to where one is, what the place is like, and what the summation of the localities of life might reveal. Thus, the geographer began his search on his native ground, expanded into the next larger encompassing region, and so on and on through successively larger contexts in a search for an understanding of his whole country, of what the United States of America is like and how it got to be that way.

The Haskins lecturer is asked to reflect upon "the chance determinations" of a life of learning. I have suggested some, but two others come prominently to mind: going to Salt Lake City rather than to London in 1950 and going to Syracuse rather than to Berkeley in 1959. When I was finishing graduate course work, I needed a job; I had a family to support. I had applied for a Fulbright to London many months earlier, but the process in those early days of the program seemed interminable. I was unable to find out anything about my status, and so in early June I accepted a position at the University of Utah and felt I could not ethically back out when the award came through later that summer. After all my talk about foreign service, it was a painful choice, and I have occasionally wondered what might have happened had I gone to Britain on the threshold of my career. The University of Utah proved to be a lively place for a beginner, starved for funds by a niggardly legislature, but home to some excellent faculty, engaged in considerable experimentation under a new dean fresh from the University of Chicago. The teaching load now seems like a killer, but I was young and energetic, involved in many things, including a TV lecture series in 1953.

In 1956, I taught a summer session at Berkeley, and the next year Carl Sauer invited me to join his staff. At the time, that was generally considered the best possible thing that could happen to a young historical geographer. But there were complications, at his end and mine. It turned out that the position was not as yet firmly authorized as permanent, and by that time I was already committed to go to Australia for a year. Mr. Sauer agreed that I must go there, and he would see what could be worked out for the year following. On our return voyage from Australia, a letter awaited me in London from a new chairman, explaining that Sauer had retired and regretting that no position was available. I have always assumed that I was not the new chairman's choice, and, of course, I was greatly disappointed at the time. But we had barely settled back in our mountainside home when the chairman at Syracuse telephoned and invited me to come for an interview. Looking out my window at the sunshine on the snow-capped mountains looming above my backyard, I very nearly said "no thanks." I had never thought of going to Syracuse or that part of the country and had no real interest in doing so. But I did have sense enough to realize that it would cost nothing to go and have a look. In fact, it cost a good deal, for I returned in a serious quandary. I didn't really want to leave the West, for a variety of reasons; I had assumed I would spend my career somewhere in the mountain West or Pacific Coast, but the prospects at Syracuse were so much better professionally and the region so much more attractive than I had realized that, after much agonizing—and a strong nudge from my always more sensible wife—we did decide to go. It was a chance determination of major consequence for us. Syracuse University provided a far better working environment, the geography department was very good and kept getting better; the university was never rich in funds, but it had some riches in talent, and for thirty years its leaders at every level from department chairman to chancellor have given me much help to do whatever I most wanted to do. Equally important, upstate New York was a beautiful region and an excellent location, and we quickly settled in contentedly. Our relatives, all westerners, regard us as living in exile, but those who have visited have had to acknowledge the attractions.

To conclude on "chance determinations," I would add that I was fortunate to meet at the outset of my career (in one case quite by accident) two of the foremost scholars and teachers of historical geography, Clifford Darby of London and Andrew Clark of Wisconsin, and to receive their cordial welcome and respect as if I were already a worthy member of our small guild. That meant a lot to a beginner.

Geographers work at various scales; it is expected that we can move easily and skillfully up and down the general hierarchy. My own published work has been mainly at some sort of regional scale, and my current project retains something of that emphasis, for a central purpose is to assess the United States as, simultaneously, an empire, a nation, a federation, and a varying set of regions. But my life of learning has been strongly influenced by both larger and smaller views of the world.

Geography, like history, provides a strategy for thinking about large and complex topics. Stephen Jones's observation that "the global view is the geographer's intellectual adventure" has always had a ring of truth to me. I began adventuring at that scale through boyhood fascination with a big atlas, learning locations, shapes, and names, and added substance to that framework through reading at progressing levels about places and peoples. It was always

a minor thrill to discover some thick book on an area one knew little about—
McGovern's *History of Central Asia* comes to mind—and a challenge to try to
make historical sense out of some complicated geographic pattern, such as the
world map of languages. One was not simply accumulating facts packaged in
convenient areal compartments; one was seeking concepts that helped one to
make ever greater sense of the complicated natural and cultural patterns of the
world. That sort of study has a very respectable lineage, dating from the mul-
tivolume works of Humboldt, Ritter, and Reclus, but never really got a firm
hold in America. Modern single-volume versions only belatedly appeared
from the writings of the Berkeleyite geographers, Rostlund, Kniffen and Russell, Spencer and Thomas, but these remained marginal, and increasingly anti-
thetical, to the main stream of American geography. Similar comprehensive
works in anthropology, such as Linton's *The Tree of Culture,* and in history,
such as Ralph Turner's two-volume *The Great Cultural Traditions,* and the
polemical interpretations of Lewis Mumford (especially *Technics and Civilization*) also nourished my appetite during my early growth. In time, I would
work out my own ways of presenting the historical geography of the great
world cultures to undergraduates. Helping students to make sense out of their
world in such a manner has been a very satisfying experience, and I have never
understood why such knowledge has been so persistently undervalued in
American universities.

Much the most challenging intellectual adventuring was to be found in
those heavy ambitious works that asserted deeper meanings, especially Spen-
gler, Toynbee, and F. S. C. Northrup. One didn't swallow them whole, for
reading critiques and alternatives was part of the fare. For example, at the
same time I was devouring some of these works, I was being led methodically
through the dissection and analysis of "culture" and "cultures" of Kroeber's
*Anthropology* by the formidable Erna Gunther, a student of Boas. It was not
the audacious claims and portentous conclusions of these metahistorical
works that were so fascinating; it was their sweeping perspectives and attempts
to integrate an immense range of knowledge in order to grasp the wholeness
and the vital springs of the great cultures and civilizations.

A few months ago I mentioned to a musicologist friend of mine a book that I
had read about but had not yet seen. He said, "Well the author tried to synthe-
size a whole society by looking at its art, but," my friend said, "it didn't work, it

can't work; it was grand, but it was a failure." (You may infer that we were talking about Schama's *The Embarrassment of Riches*). I said that I was especially interested in grand failures. I was, in fact, trying to break into the business. I was confident I could be a failure; I dreamed of being a really grand failure.

The American Council of Learned Societies generously refers to the Haskins lecturer as "an eminent humanist." You would do well to regard me as marginal on both counts. Although my kind of geography belongs in the humanities, for much of it seeks to be a form of portraiture, a depiction and interpretation honed into literature, my understanding of humans is not exactly *humanistic* in the most common modern uses of that term. Rather, it is grounded upon the old, rich, and rather severe view of Man and all his Works as expressed in *The Book of Common Prayer.* That book has been a routine part of my life for forty years. It provides a larger scheme of things, however mysterious, that helps put one's own work in perspective. I find nothing therein to keep me from accepting whatever real truth science may offer and a lot therein to help me keep a certain detachment from whatever the latest popularisms of the academy may be. More specifically, in relation to my own specialization, it provides a quiet but insistent warning about some of the characteristic tendencies of American society and culture, as expressed in its exaggerated emphasis upon freedom, individualism, democracy, materialism, science, and progress. By providing wisdom and hope rather than cynicism and despair, it helps to mitigate the anger and alarm one often feels about the drift and disorder of one's own country. Furthermore, that book and its associated rituals offer a code of conduct and a rehearsal of the follies and perversities of mankind that can have a salutary bearing upon daily life. To be reminded year after year that "thou art dust and unto dust thou shalt return" is a specific against the vanities and posturings so endemic in professional circles—and it comes with the insistent warning that none of us is immune from such temptations.

Geography has sometimes been represented as a kind of moral philosophy, primarily in the sense that those who have a deep fascination for the earth needs must have a special concern for the care of the earth. An old definition of geography has been coming back into favor: the study of the earth as the Home of Man—or, as we now say, of Humankind. We have recently become aware that the earth as Home is in alarming condition, and geographers, like

many others, are eager to tackle urgent problems of home repair and of re-
modeling the way we live. I have no practical skills to put to use on such projects. I can only add my small voice to the few urging the need, as well, for a
much longer perspective on such matters, a far better understanding of how
we got to where we are. And that sort of historical investigation must surely
lead to a sobering meditation on the human situation on this earth. There are
mysteries there to haunt the mind. In such matters, I can be no more than a
faint echo of the wisdom of Carl Sauer, the only really philosophical geographer I have known, who, while working quietly over a long lifetime mostly in
remote corners of time and space, spoke and wrote eloquently about these
grand themes, calling for geographers to "admit the whole span of man's existence" to our study and to press for "an ethic and aesthetic under which man
. . . may indeed pass on to posterity a good earth."

For me, meditations on deeper meanings are more likely to be prompted
by a walk in the country than by trying to contemplate the globe. It is this
other end of the scale, that of landscape and locality, that most enlivens my
sense of ethics and aesthetics. *Landscape* has always been an important—and
troublesome—word in geography, referring to something more than a view,
setting, or scenery. What lies before our eyes must be interpreted by what lies
within our heads, and the endless complexities of that have stimulated important work. I have paid particular attention to symbolic landscapes as representations of American values and generally tried to use the landscape as a kind of
archive full of clues about cultural character and historical change that one can
learn to read with even greater understanding. At the same time, landscape is
always more than a set of data; it is itself an integration, a composition, and
one tries to develop an ever keener appreciation of that. It is here that geography makes its most obvious connection with aesthetics, with writers and poets
and painters and all those who try to capture in some way the personality of a
place or the mystery of place in human feelings. If geography's old claim to be
an art as well as a science is as yet backed by relatively little substance, the logic
and the potential are there.

I was rather slow to appreciate these truths, in part at least, because I was
never trained to see them, and there was then little American literature on the
subject. I did have the good fortune to happen upon an obscure new magazine in the 1950s called *Landscape,* published and edited by a J. B. Jackson,

from a post office box address in Santa Fe. A few years later I arranged to meet this modest, refreshingly unacademic man who would eventually be re-garded—even revered—as the principal founder and inspiration of cultural landscape studies in America. By happy coincidence, I also met Peirce Lewis of Penn State on that very same day, and he has served as my principal academic mentor in learning to read the landscape. This dimension of my life was steadily enhanced by personally exploring British landscapes with increasing regularity. I found there a wonderfully rich literature, by scholars and special-ists of many kinds and by those splendid English creatures, the devoted, gifted amateur. I got acquainted with William G. Hoskins, the foremost historian of English localities, who by talent, perseverance, and personality reached out with several sets of books and a splendid BBC television series to bring this kind of historical-geographic appreciation to a broad public. In the 1970s, I devoted much of my time to landscape studies, to a lecture series, seminars, and field trips. I tried to bring together the best of what I had found in Britain and America with the hope of stimulating some fresh work. A few of my stu-dents responded quite creatively, but although I itched to do so, I never pro-duced a substantive study myself. The actuarial tables warned that I dare not delay my larger project, and it is one of my few regrets that I have had to give up doing something on Syracuse and Central New York.

Quite by chance I was able to participate in a really vast outreach to the reading public. In the 1980s, a former student of mine, John B. Garver Jr., served as chief cartographer at the National Geographic Society, and he in-vited me to guide the preparation of a set of maps depicting the historical re-gional development of the United States. Seventeen large sheets, each containing a set of maps, were issued with the magazine over a span of five years. Each distributed to 10,600,000 subscribers around the world, it must have been my most effective teaching even if only a very small percent were ever studied carefully. (When I see these maps in bins at used bookshops for fifty cents a piece, I'm always tempted to buy them, they are such bargains.)

I am a peculiar geographer in that I almost never travel with a camera. This is surely a limitation, even a flaw, but I have tried to compensate. I carry the images of thousands of places in my head, all partial and impressionistic, of course, but obtained with a cultivated "eye for country," to use an old saying. Perhaps I got both the eye and the preference from my father, and my resistance to technology makes me as archaic and crippled in my time as he was in his. My colleagues aptly sum me up as the man with the quill pen in an age of word processors.

Travel is, of course, an important part of a geographer's learning. Though I have traveled fairly extensively, I have not deliberately set out to see as much of the world as possible, as some geographers do, but I find it uncomfortable to write about areas I have not seen, and over the years I have used every opportunity—meetings, guest lectures, vacations—to obtain at least a passing acquaintance with every part of the United States and adjacent Canada. What few research grants I have sought have been used in some degree for such reconnaissance, thereby continuing in modest personal form the famous role of the geographer as explorer.

I mention Canada deliberately because it has come to have an important place in my life of learning. As a geographer, I must regard Canada as an essential part of the context of the United States. And I refer not just to its physical presence on our northern border and the many practical interactions between the two countries, but, as well, to the presence of a companion empire, federation, nation, and set of regions that can provide invaluable comparisons with our own. I regard the common indifference to and ignorance of Canada by Americans as arrogant and stupid. To learn and ponder the fact that the basic foundation of Canadian nationalism is the desire not to be American ought to be an instructive experience for all thoughtful Americans. This is not the place to expand upon this topic. I only wish to declare that I feel much the richer for having gotten acquainted with a good deal of Canadian territory and literature. I have been especially interested in writings on nationalism and regionalism, technology and social philosophy, and I have found the ideas of George Grant and W. L. Morton particularly instructive and congenial.

Although I have pursued my own interests with relatively little attention to what was exciting many of my colleagues, I nevertheless claim to speak for geography in a quite literal sense: for "geo-graphy," "earth writing," "earth drawing," the task of depicting the actual character and qualities of the whole surface of the globe—at various scales and at various levels of abstraction. Such a field does not fit comfortably into modern academic structures and has suffered for it. To the not uncommon question "Is geography a physical or a social science?" almost all geographers would answer "both." That in itself can become an annoyance to tidy administrators (as at the University of Utah, where in my day geography was in the College of Mineral Industries as one of the "earth sciences"). My answer to such a question has always been "both, and more." That is to say, while much of our work is a form of physical or social science, the larger purpose is of a quite different character. I accept the old Kantian concept that geography, like history and unlike the sciences, is not the study of any particular kind of thing, but a particular way of studying almost anything. Geography is a point of view, a way of looking at things. If one focuses on how all kinds of things exist together spatially, in areas, with a special emphasis on context and coherence, one is working as a geographer. The ultimate purpose is more synthetic than analytic. Of course, no one can master all that exists together in any area. Every geographer must be selective, and we follow the usual division and identify ourselves as social geographers, economic geographers, biogeographers, or whatever. The great temptation for administrators is to dissolve geography departments and allocate their residual members to these various disciplines. Such taxonomic logic is not only arbitrary and intellectually suspect, it is deeply destructive. It denies the legitimacy of a venerable field and the coherence vital to its nurture. It implicitly declares to the student that there is nothing there worth devoting one's life to.

If my remarks have taken on a polemical tone, it is because such matters have been an ever-present part of my life and because my life of learning has always extended far beyond my formal life in the university. I hope I have conveyed to you that geography has been more to me than a professional field. We are odd creatures. Geography is my vocation, in an older, deeper sense of that word: vocation as an inner calling—not what I do for a living, but what I do with my life. The born geographer lives geography every day. It is the way one makes sense out of one's world, near and far, and it is the means of appreciating the immediate world—of whatever lies before one's eyes. Every scene, every place—one's daily walk to work as well as one's traverse of unfamiliar ground—can be an inexhaustible source of interest and pleasure—and pain, for there is plenty to deplore in what people have done to their surroundings. It is difficult to convey the intensity and fullness of such a thing. To such a person, geography is not simply a profession, it is a never-ending, life-enriching experience.

I have no idea how widespread this aptitude and hunger for geography are. There are relatively few geographers in total, and a considerable number who call themselves such are of a narrower technical kind who would not really understand what I am talking about—indeed, will be embarrassed by what I have had to say. I have no doubt that there are others who never think of themselves as geographers but who are also responding to the vitalizing attractions of such interests. There are some encouraging signs that the crisis of social science and new confrontations with a complex world may cause the value of professional geography to become more recognized in America. One hopes that thereby not only will the number of persons with the requisite skills for productive work be enlarged, but that the prospect of becoming a geographer will become much more widely apparent so that the young natural-born geographers among us can be nurtured to the full wherever and whenever they may appear.

I was one of the lucky ones. Like most American geographers of my time, I only belatedly discovered that there was such a profession, but I did so just in time to make the most of it. It has been such a richly satisfying thing that when I reflect upon my life, in the way that your kind invitation has encouraged me to do, it seems as if from the moment I first looked out in wonder across the hills of Palouse, I have lived happily ever after.

# Pausing for Breath

## Richard Morrill

Courtesy of Richard Morrill

*Richard Morrill (b. 1934) received his B.A. from Dartmouth in 1955 and his Ph.D. from the University of Washington in 1959, then took a postdoctoral position at the University of Lund, Sweden. He taught at Northwestern University before his appointment at the University of Washington in 1961, from which he retired as professor emeritus in 1997. His research career has been marked by the application of formal analytical methods to real and pressing social, economic, and political problems, including those involving the provision of health care, residential segregation, regional planning, political and school redistricting, local transportation, water management, radiation hazards, and urban growth. In 1972, Seattle's Federal District Court appointed him special master to oversee and recommend redistricting plans for the congressional districts of the state of Washington, and he has served as a consultant to Mississippi and California for similar redistricting options. His service at the national level includes advisory committees of the National Academy of Sciences, the Census, the Association of American Geographers, the American Association for the Advancement of Science, the National Research Council, and the Regional Science Association. He has also served on four editorial boards and has been president of both the Association of American Geographers (1981) and the Western Regional Science Association (1992). His seven books and monographs as well as more than one hundred research articles summarize a geographic life devoted to the analysis and illumination of concrete problems for which the spatial perspective of the geographer has been crucial.*

I just received my Medicare card, an inescapable sign of "senior" status, enough to put me in a contemplative mood about my life, so I appreciate the invitation to share some thoughts about what I've done and why.

In the process of retrospection, I found certain themes or continuities in my life. First, I have an underlying skepticism with a corollary commitment to science as a way of learning and a resistance to any a priori theory. Second, I have a long-standing concern about inequality and for social justice; this concern has impelled much of my professional activity and public service. Third, and this some may disbelieve, I am a conciliatory person who prefers cooperation to competition. Fourth, I am not narrowly focused but eclectic, preferring to do research and writing over a wide range of human geography, economic, social, and political. Fifth, I am often aware of a tension between an intense sense of individual self-realization and an intellectual belief in the collective good.

## Early Years

My recollection of early years is rather sketchy, a mixture of what I really remember and what I have been told about or seen in old photos and films. But one experience remains quite vivid and detailed—when we lived in Honolulu, Hawaii, from the summer of 1941 to the summer of 1942. My father, a civil engineer in Los Angeles, was on leave from the city to work with the Army Corps of Engineers at a dam project in Oahu, and we sailed there in luxury on the Matson liner *Mariposa*. My mother got a job as a map librarian for the navy in Pearl Harbor. This was an exotic life for a seven-year-old. Then came Sunday morning, December 7. My sister and I stood outside our Hickam Field army housing, in the middle of the strafing, and watched the Japanese attack on Pearl Harbor, not comprehending that it was real. A couple of weeks after the attack, we were moved to a cottage on the beach at Kahala, where in the middle of the day the guards permitted us to go under the barbed wire onto the shore. The next months of curfew and barbed wire, blackouts, air-raid drills, and gas masks extended the drama of the experience and perhaps prepared us to take later California disasters such as fires, floods, and earthquakes more calmly than we otherwise might have.

My father was born and raised in New Hampshire, where the "family" has conservatively stayed since 1636. He was the youngest and by far the most radical and rebellious of nine children, abandoning the hearth for Los Angeles in 1925. A civil engineer for the city for thirty-five years, he shared the family's pervasive work ethic and sense of private and public morality and order. Some of my mother's roots were Calvinist, French Swiss, Jura mountain folk; a woodcarving and musical tradition prevailed in the Riffo clan, and at least the music (violin, choir) extended to me. But this artistic, and perhaps reckless, spirit had to contend with a more stolid German strain (Redekers, Pfaffs), which included shopkeepers, small manufacturers, and professionals. Although I know little about the occupations of either family before this century, what is significant is the almost total absence of farmers and the predominance of middle-class, small-business, and professional endeavors. I say this because I can recognize behavior and predilections transmitted through my parents from the wider families.

Each family's work ethic was supported by a commitment to education, with both sons and daughters expected to achieve some college education. Because my father was a broadly educated civil engineer and my mother a librarian and teacher, our home was rich in intellectual encouragement, and I learned to read by the age of two (so my mother has told me). Throughout the 1940s and 1950s, we went to plays, concerts, and lectures. I read vast amounts—I mean several books a week, year after year.

I was born in Los Angeles in 1934 in the depths of the Great Depression and lived there with my parents until I was seventeen, when I headed east to Dartmouth, my father's university. Although there were difficult economic times when my parents had reduced hours and pay, we were basically middle class. In fact, reflecting the prewar social structure, I even had a nanny! Except for brief periods, my mother either worked or returned to school, always because she wanted to. Long before women's liberation, my parents (and my sister and I) shared housework, and I learned to cook reasonably well by the time I was thirteen. Both of my parents and many of their friends were quite liberal, even somewhat radical, and unusually socially tolerant as well. Thus, we were outraged when our Japanese American neighbors were sent away to a desert internment center.

Being a kid in Los Angeles was fun. There were beaches and piers and fun houses, huge parks, farm areas interspersed among residential areas, rivers and giant storm drains to explore, mountains to hike in, interurbans on which to cross the metropolis. It may seem incredible to imagine now, but before 1950 we kids did all this on our own by the time we were eleven or twelve. Sometimes, while traipsing around the west Los Angeles hills, we would meet a real movie star!

My family loved hiking and camping, and year after year we explored the West, as well as visiting family in Montana, where my mother grew up and taught school before heading for the big city in 1926. Like a good "son of the West," I learned to drive early and often. I'm still addicted to hiking and driving and to the wide open spaces of the West, tolerating the East only for short periods. Like my parents, I hated apartments and have always preferred a spacious house and yard. Home ownership was a goal and expectation, and achieved by the time I was thirty-five. I go into this history because it helps explain my impatience with, even fear of, the current wave of new urbanism, the postmodernist thought that views cars as evil, houses as selfish and isolating. It also helps explain why I can be an environmentalist and support the Nature Conservancy and the Trust for Public Lands, but simultaneously defend logging and mining and especially suburbanization—a.k.a. "sprawl." Growing out of the Depression, I found the vast change and technological development of the 1940s and 1950s to be wondrous rather than frightening.

When my mother became a librarian for the Beverly Hills schools, she was able to have me admitted as an "out of district" student. Beverly Hills was the elite public high school in southern California, so of course this change profoundly influenced my life, providing exceptional educational opportunities: I had classes in Greek as well as Latin and economics, and played the violin in a superb orchestra. Although I liked and did well in all subjects, my most influential teachers were in social science ("problems of democracy") and history.

But despite its typical wealth, the Jewish intellectual group with which I associated was quite radical; I even attended an occasional Communist Party youth cell, abandoning it when we were expected to conform to predetermined Stalinist positions. I was more an Edward Bellamy *Looking Backward* idealist.

Los Angeles was an exciting and diverse city, riven by extreme anticommunist hysteria in the late 1940s. I was involved in the Senate campaign of Helen

Gahagan Douglas because I knew her son Peter; his father was actor Melvin Douglas. This experience led to my abiding extreme dislike of Richard Nixon, who defeated Douglas via a famous smear campaign. I loved growing up in the big city of Los Angeles, and although I enjoyed my four years at Dartmouth in rural New Hampshire (*vox clamantis in deserto*), I could never live outside a metropolis.

## Dartmouth Years

My attendance at Dartmouth was "predetermined" because that was where my father and uncle went, and I wanted to gain new experiences in new settings. Because Hanover was a remote place and a college for men, and I was neither aggressive nor affluent, developing relationships with girls really was put on hold, so I put my energy into class work and, of course, into hiking and traveling (hitchhiking) around New England. I couldn't afford a fraternity. Despite a good scholarship and family support, Dartmouth was expensive. I worked both in the library and in the college snack bar in the evening—the extra free food was critical to my survival. I always looked forward to visiting my New England aunts and uncles, enjoying both their company and their cooking.

I started in Air Force ROTC (or else I would have been drafted for Korea), but although I got along well with the ROTC professor (I played violin with him and his wife in a string quartet), we both realized I was incompatible with ROTC when I refused to regurgitate the "correct" answers in a "world political geography" course! So in 1954, to avoid imminent draft, I joined the army reserve. This experience turned out to be rather interesting because I was associated with the Fiftieth General Hospital reserve unit in Seattle from 1955 to 1962. I attended several summer camps as well as years of weekly meetings, but miraculously was never called up for active duty or even for basic training.

My original intent at Dartmouth was to pursue history and secondary education. As in high school, I took a large and diverse curriculum, but little math and no statistics! I quickly became disillusioned with secondary education, so my classes were mainly a variety of social sciences and humanities. In what is a cliché among geographers, I came to geography by accident. I'd had no geography since the seventh grade, but on my roommate's recommendation took Geography 1 with Trevor Lloyd. His was a world regional course, and although the text was pretty awful, he was astoundingly good at explaining the variations in living across the globe. He made the interaction of physical and human phenomena so fascinating that I was totally hooked. So I became a geography major. The geography curriculum was not large; I took most of the courses, physical, human, and regional, as well as related courses in botany, economics, geology, and government. Although geography's exciting topic was northern studies, I instead chose urban and economic geography, taught by Al Carlson, a very applied, business-oriented man. For my senior essay, I decided to write on the role of hydroelectric development on the Columbia River—I cannot recall why. I wrote to the University of Washington to ask for information (does this sound familiar?), and Marion Marts responded with a wealth of material. That helpfulness and Al Carlson's advice that I should go to Washington to study with Ed Ullman convinced me to apply (only) to Washington, where I was accepted for fall 1955.

Looking back at my graduate school experience, I see the power of peers and the serendipitous effect of courses and events. I was baptized at once into a new world of statistics (not just elementary)—models and location theory and regional science, microeconomics, theory and statistics in sociology and anthropology and transportation—through courses and seminars with William Garrison, Edward Ullman, and Marion Marts in geography, Charles Tiebout and Phil Bourque in business and economics, Edgar Horwood in civil engineering, and Calvin Schmid in sociology. I was entranced, excited, and converted; theory and statistics promised to raise the discipline from an "inferior" descriptive service course to a higher level of analysis and understanding. The goal of trying to explain human territorial behavior, to understand how the landscape evolved, and to predict how it might change was (literally) inspiring. The sense of being in at the start of something transformative was a further elixir.

The skepticism and opposition, and sometimes harassment, from the "Establishment" was a goad both to even harder work ("we'll show them") and to tighter solidarity, mutual support, and conviction as the rightness of our cause. Our peer group established a beachhead and, to some degree, prevailed at the highest level, succeeding in gaining positions at major universities, in publishing books and articles, in getting grants, and in diffusing the message across the country and the world. But the victory was a veneer because, I suspect, the majority of geographers and institutions were only moderately convinced and transformed.

Within the realm of what was called the quantitative revolution, which might better be called the scientific and theoretical revolution in geography, I developed a special interest in transportation and movement. In an econometrics class with Arnold Zellner, I wrote a paper on "spatial price equilibrium" models (fancy versions of the "transportation problem"), which became the methodological core of my master's thesis (on wheat flows) and then of my dissertation (concerning the impact of freeway development on health service use), and was useful in many later consulting projects and several publications, especially in redistricting applications.

Many faculty and student peers helped shape me in these years: Bill Garrison foremost, through his blending of theory, method, and encouragement; Ed Ullman, through his ability to think theoretically; Marion Marts, for keeping me grounded by real data and by field work; Charlie Tiebout, for providing skepticism and brilliance, and for showing that economics really did have something to offer; and Ed Horwood, for his superb interdisciplinary perspectives and early immersion into programming and computers. To this group must be added Torsten Hägerstrand, second only to Bill Garrison in influencing my subsequent career. Visiting the University of Washington in 1959, he imparted not only his theories of diffusion, but the idea of testing theory through historical analysis, the significance of the individual decision, and the importance of migration in the evolution of landscape. My student peers were critical to my development as well, through constant discussion and feedback. Among them the most long-lasting influences have been Brian Berry, for his easy blend of theory and method, and Bill Bunge, by the sheer power and vigor of his insights and arguments.

Meanwhile, my life changed in other ways. For my four years at Dartmouth, I had been in the Dartmouth Christian Union's square-dance band, affiliated with the Congregational Church. We played at small-town churches and granges in parts of New Hampshire and Vermont. Besides playing the fiddle, I learned the guitar, mandolin, and banjo, and for my last two years was a square-dance caller! On arriving in Seattle, I surmised that joining Pilgrim

Club, the college-age group at University Congregational Church, would be a good way to begin a broader social life—that is, meet girls (there were still very few women graduate students in geography before the late 1960s). This decision proved fortuitous because at Pilgrim Club I made many close friends and connections to the wider community, and met both my first and my present wife.

Within ten days of coming to Seattle, I went with a new friend on a hike and climb in the Cascades—the first of hundreds. Instead of joining the university's symphony, I opted for a Spelmansgladje—a Scandinavian fiddle team—and became active in folk dancing for the next ten years. For the conservative 1950s, I was politically very radical and fairly unconventional otherwise, too. I was a founding member of the University of Washington's Students for a Democratic Society (SDS) chapter and joined the local Socialist Party U.S.A. A group of us moved into a houseboat, considered very countercultural in those days; indeed, the next ten years included our political struggle against the city and state, who wanted to "abate" the floating slum by banning and burning. We won, and now the houseboats are for upscale yuppies.

When I was twenty-four, in 1958, I married Margaret Peterson. Margaret was a local Seattle girl whose father, Arnold, a steelworker at Bethlehem Steel, had been born in Sweden and grown up in Vancouver, B.C. Her mother, Jean, was from Edinburgh, Scotland, before her family moved to Vancouver, where she met and married Arnold. Margi's parents were extremely supportive and wonderful for as long as they lived.

### From Student to Faculty

At the 1958 Association of American Geographers (AAG) meeting in Los Angeles, I interviewed for jobs at Michigan and Northwestern. I was offered the Northwestern job (an assistant professor at the age of twenty-five!), and Margaret and I arrived in Evanston in September 1959. My year at Northwestern was exciting and productive. Ned Taaffe was a mentor and ally, and immediately involved me in his research projects. Peter Gould, Maurice Yeates, and other early-generation revolutionaries were finishing up and continued the atmosphere of exploration and fervor I'd enjoyed at Washington. I continued radical involvements, helping organize the Student Peace Union march in downtown Chicago.

I might have stayed at Northwestern longer, but because my University of Washington advisor, Bill Garrison, accepted an offer from Northwestern, a spot opened up for me, and I in turn accepted a bid to return to the University of Washington. Meanwhile I applied for a research grant from the Office of Naval Research (ONR), then the primary funding for geography, to study the development of urbanization and migration in Sweden as a kind of post-doctoral experience at the University of Lund with Torsten Hägerstrand. The proposal was transferred to the National Science Foundation (NSF), so that in 1960 I may have been the first geographer to receive an NSF grant.

My research in Sweden was intellectually important because it both established my interest in population and settlement, and also showed the value of historic evidence in corroborating geographic theory. It was also professionally important because it led to my first book and *Annals of the Association of American Geographers* article. Some highlights of the year included accompanying Torsten on field trips with a class (wow, professors could actually get respect!); joining a whole group of Swedish geographers on a two-week excursion to Yugoslavia (lots of physical geography); participating as a "professor" in the very formal Lund University parade and graduation; being part of the 1960 Lund symposium, where I met Walter Christaller and other legendary figures; and participating in the Stockholm International Geographical Union (IGU) Congress.

My wife and I enjoyed experiencing a Swedish Christmas season. With very favorable exchange rates, we were able to buy a car (a used Peugeot), dining— and living-room furniture, dishes and glassware; almost everything but the car is still being used today. We were able to travel to Finland and Denmark, and early in the summer of 1961 drove through parts of eastern Europe—Poland, Czechoslovakia, and east Germany (before the wall!), including an illegal side trip to bombed-out Dresden. In 1961, I was ardently socialist and anxious to visit eastern European socialist experiments: the dismal reality, especially East Germany, was sobering and probably instilled some doubts about centralized socialism in practice. On our way home to Seattle, we stopped in England to visit relatives and to tour northern England and Scotland before taking the *Empress of Britain* from Liverpool to Montreal.

My years as an assistant professor at Washington (1961–64) were crammed with activity—professional and personal. I was astonishingly energetic. Although my dissertation was in the transportation area and I was a Garrison economic geographer, the economic track was more or less covered by Morgan Thomas and transportation by Edward Ullman. I suspect I was really hired to teach the infamous 426 statistics class. I did offer "regional specialization," which I later converted to a location theory course. Many economic geography graduate students were preempted by the senior faculty (academic life was less democratic in those days), so by default I inherited students who didn't quite fit the economic or urban standard. As it happens, many of these students were interested in population, migration, inequality or race—the emerging social issues of America in the 1960s civil rights revolution. Because I was interested in these very issues, my emphasis quite fundamentally shifted in parallel with the students for whom I became responsible and with whom I became involved.

At the age of thirty (in 1964), I was no older than many of my students, so it was natural that our relationships were informal and went beyond the university to camping and hiking and long-term friendships. Because I was so ridiculously overcommitted, groups of advisees over the years would at times capture me to retreat for a day or so for advice and interaction—a practice I highly recommend. I'll admit that in the process I influenced students' political and social attitudes in a liberalizing to radicalizing way, and I'm glad it worked. Some students from these early days included Hal Brodsky, who studied urban transportation; Murray Chapman and Dick Wilkie, who studied migration in the Solomon Islands and rural Argentina respectively; and David Stallings, who studied the efforts to preserve Seattle's Pike Place Market. I can't help but observe that fieldwork, interviews, and appreciation of alternate value systems are not a discovery of the "new regional" geographers and social theorists, nor is political economy. In support of these research areas, I introduced the country's first course in social geography and later created courses in population geography and the geography of inequality. Meanwhile, I maintained my role in economic geography through developing a course in location and movement models.

At the same time that I was writing up my Swedish research, I also worked on projects on racial inequality and carried out my studies of the expansion of the black community in Seattle. Researched in 1962–63, published in 1965, and, I would note, fully aware of structural forces for inequality, this study was probably my most influential and most-cited contribution. I have done several follow-up studies over the years, the latest being my 1996 study of class and racial segregation in Seattle.

In 1962, my radical convictions were put to a personal test. Along with other university faculty and staff, I refused to sign the required loyalty oath, and we sued the state to overturn the requirement. Because I was untenured, I would certainly have been fired, but fortunately the dismissals were stayed pending the outcome of the suit. We eventually won in the Supreme Court, a proud moment in our lives. (I learned later that colleagues in Canada were prepared to offer me a position if we had lost.)

Outside the university, Margaret and I filled our lives with radical causes. We shifted to the Unitarian Church when we thought the Congregational Church was becoming too "Christian." I chaired the Unitarians for Social Justice for some years and with others was a founding member of the Congress of Racial Equality in Seattle, as well as working with the American Friends Service Committee on antiwar activities and on the reform of mental hospitals. I am not a religious person, being agnostic, but have belonged to the Unitarian Church for forty years not only because of its social action role in the Seattle area, but also because I enjoy singing in its outstanding choir.

In the summer of 1963, Torsten Hägerstrand brought a group of Swedish geographers to the United States. Margaret and I bought an old bus, and a good friend and I drove our visitors around some ten thousand miles of America—a fabulous experience for us as well as our visitors. Even I was astounded by the poverty and discrimination in the rural South. After the excursion, we sold the bus to John Lewis of the Student Nonviolent Coordinating Committee to take folks to the March on Washington.

Because Margaret was physically unable to bear children, we began an adoption process in 1963. The agency was not overly pleased with our radical associations and lifestyle, including our houseboat, but in the summer of 1964 we were approved and went to Spokane, Washington, to bring home Lee. That Christmas we went to Montana to be with my parents, and while we were at Flathead Lake, Margaret died of heart failure. Calling her parents was surely the most difficult task I ever had to do. The Morrill way of dealing with tragedy was to be even more immersed in university work and outside activities. Because of the risks of houseboat living with a one-year-old, we had just moved to a sort of commune with another couple and some boarders. The adoption agency wanted to take the child from me—men could *not* adopt children, and the adoption had not yet gone through—but agreed that the trauma to Lee had been so great that he needed some stability. A good friend, a new young lawyer, succeeded in helping us convince the court to finalize the adoption, so I was the first single male in state history to adopt a child. And in late winter, I wrote to a good friend of ours who was teaching in Illinois, Joanne Cooper, a University of Washington Ph.D. student in English, and encouraged her to come back to Seattle—to finish her degree, I said.

Joanne returned in the summer of 1965, and all three of us got better and better acquainted. In late fall, Joanne and I decided to get married during the Christmas break. Some friends thought this was a little hasty, but fortunately the two most critically involved persons—Margaret's parents—didn't agree with them. Rather, they were utterly supportive and proved to be all our children's most involved and loving grandparents.

## Research and Teaching

Although I've been energetic all my career, my most productive period was probably 1966–71 (when I was thirty-two to thirty-seven years old), when I wrote my two best books, *The Spatial Organization of Society* (1970) and *The Geography of Poverty in the United States* (1971), and many of my most effective articles; was promoted to professor and elected to the AAG council; made my most difficult climbs; and fathered two children.

Unpredictable forces were again at work. In spring of 1966, through the efforts of Brian Berry at Chicago, I was invited to become the director of the Chicago Regional Hospital Study, designed to assess the utilization of the hospital system and to evaluate issues of access by race and parts of the metropolis. Clearly I was no medical geographer, but, armed with excellent data, talented assistants, and competence in statistics and modeling, I became a leader in the health services aspect of medical geography. Many papers and several Ph.D.s (for example, Robert Earickson and Philip Rees) resulted from this project. I was certainly in a position to carve out a major niche in geography if I had continued to specialize in health services research. I did not abandon medical geography; I had excellent Ph.D. students in medical geography (such as Gary Hart, now a full professor of family medicine) and continued to write on related topics, such as studies of the diffusion of epidemics. However, I chose not to concentrate in this area but rather to move into broader areas of inequality.

I felt an extreme urgency to write a modern, theoretical text for human geography. Although by the mid-1960s, we "revolutionaries" had made great inroads in programs and in the journals, general texts seemed hopelessly descriptive. So I was (and am) very proud of *Spatial Organization of Society*, which appeared in 1970 and was slightly revised in 1972. A quarter of a century later it seems hopelessly incomplete but was at the time a statement that the revolution had taken place.

In 1969, I received a grant to study the geography of poverty. This research resulted in additional Ph.D.s—Ernest Wohlenberg and John Symons—and my third book, *Geography of Poverty in the United States*. I have maintained a strong interest in economic inequality and poverty, recently publishing an article on variation in income inequality across the states.

Studies of spatial diffusion were a major part of my research and writing in the late 1960s. Diffusion processes were at the core of my earlier work on urbanization and migration in Sweden, on the expansion of the black ghetto in Seattle, and on land conversion in the urban fringe of Seattle, but in the latter period I undertook some more formal and graphical analyses, as shown in "Shape of Diffusion in Space and Time." Later returns to the diffusion theme include studies of the spread of a smallpox epidemic and the 1988 monograph *Spatial Diffusion*.

When we returned to Seattle in the late summer of 1967, we bought our first house. Joanne had perceived a lack of opportunities for Ph.D.s in English and instead of finishing her degree opted to raise our family. Andrew was born in 1967 and Jean in 1969, so we kept within our politically correct quota of two of our own. In 1970, the whole family went to Glasgow, Scotland, for three months when I was invited—courtesy of Anne Buttimer—as a Sir John McTaggart Fellow to the Adam Smith School (ironic, because most of the faculty were good Marxists!). It was a challenge, for Joanne especially, to manage with three young children, but we enjoyed several outings and trips, including a visit to Margaret's remaining relative in the English Lake District. Raising a family and my increasing national involvements meant I had a little less time for local radical and social causes, but at the same time I became even more involved with service to the wider community and in finding intern and job opportunities for students. It was part of becoming middle class, too. In 1973, we moved to our current home, a large house, old for Seattle (1912), located in the inner city but within walking distance (twenty-two minutes) of my office at the university, wonderfully accessible to roads and transit as well as to many of Seattle's parks and libraries; we enjoy the parklikeness of suburbia and the convenience of the city. Old houses take a great deal of work, and I found it necessary and even fun to learn enough plumbing, electrical techniques, and carpentry.

Unlike the majority of professional families, we kept our three children in the Seattle public schools. Because we were in the inner city, most of the schools had a large minority population. Conditions were not always ideal for learning, and having the two boys attend these schools may not have been the wisest choice for them, but for our daughter, with wonderful peers, the Seattle school experience was great; she went on to Carleton College in Minnesota and is now finishing her doctorate in hydrology and atmospheric sciences at the University of Arizona.

In 1969, I was promoted to professor, the beginning of my heaviest formal involvement with the wider profession and leadership in the AAG. My early AAG activities were on the activist side: committees on geography and public policy, on the status of women, on the status of racial minorities. I was elected to the national council in 1971, serving for four years in a period when the main issues were institutional rather than conceptual. I was elected secretary in 1978 and served for a long three years.

In 1973, at the age of thirty-nine (a little unusual in those days), I was appointed chairman of the geography department, serving for ten years—it was much easier then than now, I suspect. I soon found that this position was less one of intellectual leadership than a job of protecting our interests against external threats and of dealing with personnel relations, from the all-important task of hiring excellent new faculty to the everyday problems of morale. I was only a fair chairman for internal affairs because I really preferred research and interaction with graduate students to administration, but I was pretty good at relations across the campus and with the outside community. Or perhaps the secret of my survival was my partnership with Susan Williams, the department's administrator, who could translate my horrible handwriting more accurately than I could and who saved me time and again from missing deadlines and appointments.

During both the 1970s and 1980s, my participation in national activities seemed constant—and probably excessive. For many years, AAG meetings were in late April, just at the time of my daughter's birthday, which made me and geography unpopular, and Jean vowed to "stomp" on the AAG. Finally, in 1984, when the meetings were in Washington, D.C., we took both Andrew and Jean out of school for a week. We all had a very rewarding time; the pandas were out and active at the National Zoo, and while I went to meetings, everyone else went to the Smithsonian museums.

My formal involvement with the AAG remained strong throughout the 1980s. I was elected vice president in 1981, but on Jim Anderson's untimely death became president immediately in 1982. This was a difficult period for geography, with attacks on the discipline as not up to university standards, but it also marked the beginning of more aggressive efforts to raise awareness of geography's contributions and to revitalize the AAG through the creation of specialty groups. I helped set up the population and political geography specialty groups, with which I've remained active, and participated as well in the medical geography, transportation, and other groups.

Teaching and preaching are not so very different, so my presidential speech in 1983 in Denver was on the responsibility of geography to apply our science to social and environmental problems, and in 1985 I delivered an exhortation on the importance of theory. Putting theory into practice, when Stan Brunn was chosen to become editor of the *Annals*, I worked with him from 1987 to 1992 as associate editor for human geography.

A parallel track of service was with the NSF. I served on the geography and regional science review panel from 1979 to 1982 and afterward served on a number of oversight panels for other programs, on National Academy of Sciences (NAS) and NSF panels for dissertation fellowships and on the NSF panel to award the National Center for Geographic Information and Analysis (NCGIA) initiative. Sometimes the trip from Seattle to Washington, D.C., for the NSF meetings had unexpected benefits: at 8:00 A.M. on the morning

Mount St. Helens erupted, the pilot of the Northwest flight to Dulles spent almost ten minutes (probably irresponsibly!) flying around the just-occurring eruption, affording us an incredible and unrepeatable view. By the mid-1980s at the age of fifty, I was already a sort of "elder statesman," so quickly had geography grown and changed, and my main role shifted to external department reviews and external promotion and tenure evaluations.

One advantage of reaching this stage of service is the invitation to participate in exchanges and in international conferences. Especially enjoyable and productive were conferences at Oxford University on segregation, race, and ethnicity; on political geography at Haifa in Israel; IGU commission meetings on government decentralization in Barcelona; on "postindustrial" geography at Osnabruck and at the International Institute for Applied Systems Analysis (IIASA) in Austria. And I also enjoyed side benefits, such as almost two weeks between Barcelona and Osnabruck, driving through southern France, Switzerland, back to Paris (and the legendary Louvre-in-a-day syndrome), on to Brussels, Liege, and Cologne.

Besides my work in the AAG, I have been active in the Population Association of America and the North American Regional Science Association, and have been especially committed to the Western Regional Science Association (WRSA), serving as president in 1983–84. The WRSA meetings are intellectually more satisfying than the AAG meetings because there may be as much as an hour per paper, and besides, as Joanne noticed, we meet in exotic resorts, often warm even in February.

I have to admit it. I have an addiction to data. I love and devour dictionaries and gazetteers and almanacs, lists and surveys, I love census reports and can hardly wait for the next year's local or national estimates so that I can manipulate and map them. One of my most enjoyable projects was creating, mapping, and analyzing a data set tracing the urbanization of the country from 1790 to 1990. I gladly spend days of volunteer work on revising the sets of census tracts, block groups, and census-designated places for the Seattle-Everett metropolitan area, and meet with other data addicts in the state and local governments bemoaning the lack of data we ought to have! Currently (1997–99) I am working on, and loving, a massive data-crunching effort to delimit the commuting areas of all urban places in the United States, using an intercensus tract journey-to-work file from the 1990 census.

In the old days before computer cartographic packages, I frequently made elaborate colored-pencil maps, as of all the counties in the United States. Fortunately, my daughter Jean, who inherited my love of data and detail, from an early age helped me with some of this scientific endeavor. I have a collection of hundreds of road maps, dating back to the 1930s. One goal for my later retirement is to undertake a study of the evolution of the U.S. road system. What could be better than data and maps!

## Public Involvement

A momentous and unexpected event intervened to change my life in 1972. The state of Washington was unable to complete the redistricting required by the state constitution, and perhaps because I had exposure in the community in speaking on population and settlement trends, the court appointed me as special master to redistrict both the congressional districts of the state and the Washington legislature. This experience was to bring me considerable notoriety in the state. My wife insists that it was with uncommon foresight that I arranged to be out of town the week the results were released! Out of the Washington redistricting experience came articles, my AAG Resource Paper monograph, and a series of consulting activities ever since. Although I have been involved in a great many local districting projects, as of school districts, both major parties in Washington have feared me, and since 1972 I have been purposefully excluded from county or state redistricting efforts, even as my activities in other states have increased! Most important among these activities was working with the courts and the Mississippi legislature in 1981–83 to redistrict the legislature, where race rather than population was critical. I worked in California as well in 1982, becoming part of the *Badham v. Eu* antigerrymandering case before the Supreme Court, and ten years later I worked in both California and in Illinois, emphasizing plans that would improve minority voting without trying to maximize minority representation. This project led to recent work on racial gerrymandering and alternative voting methods to overcome representation and redistricting. I also remain involved in the issues surrounding the census undercount and census adjustment.

In 1983, I received a Guggenheim fellowship and an NSF grant to study redistricting processes and consequences across the states, evaluating the effects of who did the districting and trying to measure resulting electoral bias.

The redistricting experience more fundamentally moved me into political geography generally, at mainly the local to state level. Since around 1972 I've become active in regional planning and urban development issues. I departed from the orthodox liberal establishment's espousal of centralized planning, maximum development of civic monuments, and downtown redevelopment because I believe that such planning and development is for the benefit of the liberal elite at the expense of ordinary, less-well-off citizens; I gradually became a skeptic about massive projects and new urbanist planning.

I was formally and strongly involved both in transportation planning and local-governance reform and in regional planning for the region. I served on the Seattle Metropolitan Study Commission, an early unsuccessful effort to create a Minneapolis-type metropolitan government, and later worked with METRO, the regional authority for water quality, sewers, and transit, on ways to modify its representation so that it could aspire to be such a government. My analysis led to the federal court's ruling that the METRO council, controlling transportation and water quality, was unconstitutionally constituted.

Although I am a dedicated geographer who professes to believe in objective science rather than in normative values, my interests in population, urbanization, transportation, and social inequality involved me from the beginning with planners and planning. My first job, in the summer of 1956, was as a planning helper in Billings, Montana. Over the years, I've taught joint courses with planning, served on dozens of urban planning master's and Ph.D. thesis committees, served on a score of planning committees and commissions for Seattle, King County, the state of Washington, and various regional bodies. Washington's governor appointed me to the Boundary Review Board for King County, a quasi-judicial body concerned with incorporation, annexation, special districts, regional planning, and environmental quality. I served for eight years, learning to survive long hearings in remote places. This experience led to my participation in IGU commissions in political geography and to several papers on local governance, regional planning, and growth management.

In 1986–88, I worked with the state's Local Governance Study Commission and in 1990 with the Growth Strategies Committee, leading to Washington's Growth Management Act. From 1992 to 1996, I was the initial chair and director of the Interdisciplinary Ph.D. Program in Urban Design and Planning. Even now in 1999, after my retirement, two of my four last Ph.D. students are in planning! As a radical concerned with inequality and as a professor (who certainly knows what's best for people and what the city ought to be like!), I am naturally drawn to planning as an alternative to the unbridled market. Yet I am profoundly uncomfortable with planning in practice. I shudder at many planners' ignorance of economics, at the elitist contempt for the preferences of ordinary people. So I often wonder if planning has done more harm than good and whether it aggravates rather than reduces inequality. This was the theme of my Golledge Lecture in Santa Barbara in 1998, "Reradicalization of a Liberal?"

From the 1980s to the present, *population* has been a prominent part of my work. Service activities often turned into research opportunities, and they also often became avenues to deal with issues of inequality. One of my most satisfying projects was my NSF grant in 1979 to study motivations for nonmetropolitan migration in the Pacific Northwest, especially the role of environment and community. Analysis of interviews and migration histories permitted modeling of the individual migration decision, and the project resulted in additional dissertations and articles (from Jeanne Downing, Bill Leon). As a result of this study, I came to appreciate the importance of age and gender differences in migration. For my presidential address to the WRSA in 1993, I examined patterns of age-specific migration in the Pacific Northwest, which proved also to be useful for the research I did for the University of Washington on enrollment projections and regional inequality in access to higher education. A large project reflecting my love of data, of mapping, and of generalization was my 1993 *Annals* paper on geographic variation in demographic characteristics of the American population—that is, variation in fertility, mortality, and their associations.

My consulting and applied projects have proven to be interesting and sometimes politically exciting as I measured the impacts of school closures, estimated the distribution and characteristics of special-needs students, evaluated the viability of tiny schools and school districts, and evaluated the effects of privatization of the system of state liquor stores.

Especially interesting and significant to my life and research has been my work on school busing. As a member of the Congress of Racial Equality and a social liberal, I helped design schemes for school integration as early as 1964. Mandatory busing was not implemented until 1977. In 1987, I was asked to study the effects of busing on the school district and the city, and to design alternative, less-unpopular busing plans. Discovering the severity of the demographic and social effects of busing—large-scale white flight, shifts to private schools, and greater income polarization—was a severe shock to my prior faith in liberal and especially geographic fixes to social problems.

In 1981, I was appointed to the Census Advisory Committee of the Population Association of America, serving several years in preparation for the 1990 census. From a geographic point of view, the major and dramatic accomplishment of this period was my creation of the TIGER files for the nation with national coverage of blocks and census tracts.

A most satisfying project, combining work on population and modeling, was to recommend a site for possible branch campuses for the University of Washington. Utilizing 2010 projections of population by age for the whole metropolis and a network-based optimum-location algorithm (ALLOC), I recommended two sites in the northern and southern parts of the region. The university and the state legislature accepted these results, and the new schools are in operation in precisely those sites. Such an experience is rare and makes up for many more studies that never saw the light of day!

In 1985, I was appointed by the dean of the University of Washington Graduate School to be one of two university representatives to the Technical Steering Panel established by the states of Washington and Oregon and the U.S. Department of Energy (DOE) to oversee the Hanford Dose Reconstruction Project. This multimillion-dollar effort reconstructed the volume, severity, and environmental pathways of radioactive releases to people from Hanford operations over the years. The task was so complex and difficult, especially exerting scientific control over DOE and its contractors, that it was not complete until 1995. We are still awaiting the results of the follow-up study, the Hanford Thyroid Disease Study, to assess the health of a large sample of persons exposed to Hanford fallout. My role as the "demographer," not geographer, was to help estimate not only the distribution of population by age by location by year, but also the details of lifestyle, farm practices, and movements of people and of farm produce, all of which could influence exposure. A side benefit for me has been a data set of migration histories for a sample of people born in the Hanford area from 1942 to 1945.

Since the completion of the Dose Reconstruction Project, I have been involved in a follow-up study called the Consortium for Risk Evaluation with Stakeholder Participation (CRESP), in which we are studying future land uses of the Hanford reservation. This situation turns out to be a classic case of political and economic conflicts across geographic scales and value systems, and it was the subject of my plenary address in political geography at the Boston AAG meetings in 1998.

In fall 1991, I returned to Dartmouth as a visiting professor, teaching my course on inequality. This experience was enjoyable because the undergraduates were so outstanding, and Joanne was able to visit briefly for a trip to the Maine coast. A valuable side benefit has been the attraction to Washington of wonderful geography undergraduates from Dartmouth ever since.

From this essay, one might get the impression that I am a pragmatist and an empiricist, mostly concerned with real-world problems. Still, part of me loves the abstract exercises in pure theory. I still feel some awe rereading Christaller and Lösch, and some of my most enjoyable projects have been escapes from the real world, as with the product of my 1970–73 NSF grant, "Experimental Derivation of Theoretical Surfaces," and as with other purely mathematical and graphic work on diffusion.

As a geography professor and educated northwesterner, I am of course an environmentalist. I was a founder of and professor at the Institute for Environmental Studies for twenty-five years. Along with Bill Beyers, I worked with the North Cascades Conservation Council and the Alpine Lakes Protection Society to create national parks and wilderness areas. Yet I place people before the environment, jobs before preservation. I'm a conservationist, not a preservationist, and this approach is consistent with my lifelong distrust of unequivocal commitments to causes.

## Hiking, Family, and Friends

When I was young, I loved to hike with my father, and I went on many great hiking trips at Dartmouth, but my passion for hiking increased after I came to Seattle and is unabated today. Many times I will get up at 4:00 or 5:00 A.M., drive two to three hours to a trail head, hike and climb sixteen to twenty miles, rarely stopping, and come home for a late dinner! Over the years, I've taken hundreds of hikes, probably half of them by myself, the other half with colleagues (dozens with Bill Beyers, including many scary climbs), with graduate students (every year since 1956), and with my family, ready or not. I've done almost every hike within two or three hours of Seattle, some as many as a dozen times.

I must really love hiking because I can remember many hikes better than I remember major academic events! One spring break Bill Beyers and I decided to hike a part of the Olympic National Park coastal strip. Just as we set up camp, a ferocious storm descended, totally demolishing our tarp shelter. This was not our best night, and we credit our survival to the foresight in bringing a fifth of whiskey. One of our climbing goals was a North Cascades peak, Eldorado, which shelters the largest glacier-snowfield in the lower forty-eight states. On our first attempt, we reached a high ridge at dusk in a deep fog and decided to rope ourselves to a small tree: wise, because in the morning sun we found ourselves on a knife escarpment and had to retreat from that class-five approach. The next year we tried a more tractable route and were snowed out (in July), but a couple of years later made it on our third attempt.

My children became avid hikers, too (as if they had any choice!). For several years, Andrew and I would try to get into some high viewpoint on June 21 (there's still much snow in the Cascades in "midsummer"). On many summer expeditions, Andrew would fish on some high mountain lake while I explored. While Lee was in Boy Scouts, for three summers I helped lead the Boy Scout fifty-mile hikes, once across the Olympic Mountains, once in the Glacier Peak wilderness, and once in the Goat Rocks east of Mount Rainier. Some of my trips were a little too adventurous, with swollen streams to cross, unforeseen cliffs to negotiate, or sudden summer storms. One long off-trail trek in the Montana Beartooths, with Andrew and Jean and Joanne's brother, became an extreme test of endurance but provided material for Jean to write a superb high school English essay on her experience.

Montana is our alter-state that we visit every year, sometimes more than once. Joanne was born and brought up in Montana, graduated from the University of Montana, and still has family in Helena, Butte, and Missoula, and roots in Billings. My mother grew up in the Flathead Valley, my sister married a wheat rancher in north central Montana, and I have relatives in Missoula, Libby, Conrad, and Loma. We spend two weeks every July at the Cooper family cabin near Yellowstone, at the edge of the Beartooth wilderness, where our agenda is reading when the weather is bad and hiking and exploring when it is good. In the early years, I took a great deal of work along. I still do, but the difference now is that I somehow don't get to it! Our son Andrew has recently become a Montana resident and has built his own cabin, although he circulates between Montana (fishing, snowboarding), Seattle (our home), and Maui (surfing)—a lifestyle 180 degrees from mine! Montana is our long-term antidote to the metropolitan rat race, but we could not live there permanently.

Montana is not the extent of our vacations and travels, though. In 1980, coming home from a meeting in Texas, I won a free trip in an American Airlines contest, so Joanne and I had a fabulous trip to Barbados, our first trip by ourselves without the children. The trip to Barbados was also the first time (Joanne's edict!) I went on a vacation without taking work! Now I can't remember why I was so reluctant to be without my briefcase. Since our daughter went to graduate school in Tucson in 1991 and we have become "empty nesters," we have managed to visit Tucson fairly often, usually in midwinter, to explore and to hike with Jean and her boyfriend. In recent years, we have also enjoyed wonderful September vacations (the University of Washington doesn't start its fall session until almost October), sometimes up and down the Pacific Coast, sometimes throughout the Southwest, to Mesa Verde, Santa Fe, and Taos, Arches and Canyonlands, Grand Canyon, Zion and Bryce, Canyon de Chelly and Rainbow Bridge. In February 1998, we enjoyed New Zealand so much that we hope to return, but in February 1999 we were in the Caribbean again.

In 1997, although I was only sixty-three, I calculated that if I retired, I would not only open the way for more faculty (two as it turned out) and get out of administrative task forces while being able to teach part-time, but be as well off as if I had kept working! So I officially retired as of September 1997 and have taught two courses per year, officially at 40 percent time. The tapering off has been gradual; I estimate my work week has only dropped from about seventy to about fifty hours!

In 1999, thirty years will have passed since I became a full professor at the University of Washington. And despite the ascendance of multiple perspectives, I believe as firmly as ever that the goal of geography is to explain human spatial behavior and the evolution of the landscape. I see virtually all the critiques as offering valuable, often necessary amendments to the process of science, but not serving as substitutes; and those perspectives based on a priori truths I find even potentially destructive.

A recapitulation of a career cannot avoid a confession of the beliefs that guide one. I define geography as the study of evolution of landscape—the physical and human forces that shape the earth's surface. The elements of study of geography as a social science are the *behavior* of persons, households, groups, institutions, and social systems that define and change the landscape. Geography as a social science tries to explain this behavior and analyze processes of change, including not only the effects of divergent values and perceptions of reality, but also "irrational" as well as rational behavior. But I do not believe social science can prescribe the "correct" values. Nevertheless, I defend social science simply as the only remotely fair way to attempt to discover a common understanding of the nature of the world, including its competing values. I prefer science to a priori belief because science can be self-correcting: the state of knowledge is always subject to revision.

Social theory—ideas about how social systems organize social relations—can be viewed in two ways: as theories, which social science should incorporate and evaluate; or as alternative structures for understanding the world, in which science may be viewed as product and tool of Western capitalism and where truths may be theorized (or theories defined as true). It will be no surprise that I reject the latter definition as no more than the assertion of faith over reason. On the other hand, social theory, including various forms of structuralism and realist modifications such as feminism, postmodernism, and others, all raise profound questions about social systems, human behavior, and balance of the individual and the collective—issues that an adequate social science must incorporate. My own exposure to social theory dates back to the 1940s and 1950s—I well remember hearing Levi-Strauss speak at Dartmouth—and I employed structural arguments in much of my earliest writing: for example, on the Negro ghetto, on the history of black America, on the geography of poverty, and in the series of health services articles.

Over the last thirty-five years, I have served on well more than two hundred graduate student committees, probably more than any other professor in the university, because I was asked and because I am able to work with students and faculty of all persuasions. I like to think that for those students most divergent from my worldview, I was able to influence their substance and methodology so that their work was social science, whether they liked the term or not. People are surprised, for example, that I was a significant advisor to such students as Anne Buttimer and Marwyn Samuels or that many of my own Ph.D. students would be considered more as humanists than as social scientists.

When asked to indicate what aspect of geography I specialize in, I have trouble because I have no narrow topical specialty but instead have chosen to pursue a wide variety of topics. I've taught at least twenty-two different courses. And despite the apparent diversity in these endeavors, there is underlying commonality. Almost all my work concerns spatial behavior—the location, movement, and settlement decisions of individuals, households, groups, and institutions (including firms and governments) within the constraints imposed by societal structures. Whether the topic is economic or social or political, I tried to explore the goals and strategies that result in the observed organization of the landscape. Because it is convenient and is where I live, I often used Seattle as a case study of divergent substantive topics. But I'm not a new regional geographer. Instead I have used Seattle to illustrate the evolution of landscape, which is inescapably physical, economic, social, and political, yet the product of actors pursuing goals of survival and self-realization, seeking or maintaining power, or improving well-being. Ironically, this defense comes perilously close to the very concept of geography as synthetic and integrating that we young turks decried in the 1950s as arrogant and hopeless.

My roles as a professor, a geographer, a scientist, a writer, a minor figure in the news, a hiker, and a husband and father have all been very fulfilling aspects of my life. The occupation of professor has been the ideal means for practicing both science and research (the compulsion to discover) as well as for writing and teaching (the equally strong imperative to show others what I've learned). Of course, I've been motivated as well by a drive for success and recognition, which sometimes has led me to act more aggressively and competitively than I am comfortable with.

The compulsion to work has been somewhat tempered by my love of hiking and of reading. I am incapable of sitting around doing nothing. Fortunately, my family has been understanding of these compulsions and yet been able to change me and free me for far more relaxation than I might otherwise have had, subverting me with murder mysteries, crossword puzzles, and domestic activities so that I could stop for breath and become a pretty good husband and father, too.

# Glimpses

## Gunnar Olsson _____

*Gunnar Olsson (b. 1935), shown here as a young boy in his father's arms, served in the Swedish army for eighteen months before receiving his first degree, the* Filosofie kandidat *from Uppsala University in 1960 and his doctorate in 1968. In between his first and doctoral degrees, he was a visiting fellow of St. John's College, Cambridge, a fellow of the American Council of Learned Societies at the University of Pennsylvania, and an assistant professor at the University of Michigan. In 1974, he became a collegiate professor at Michigan, before returning to Sweden in 1977 as professor and graduate director of the "consortium" university Nordplan, an institution supported by all the Nordic countries. In 1997, he became professor of economic geography and planning at Uppsala. He has also held visiting appointments at universities in twenty-four countries. Starting from strongly analytical perspectives, reflected in his first three books and monographs, he became concerned with the underlying methodological and ethical questions of planning, as developed in his most famous "double" book* Birds in Egg/Eggs in Bird, *published in 1980. After making a thorough study of various logics to see if they were suitable for the description of human spatial affairs, he moved to a genre of wholly original writing that emphasizes language, meaning, and power relations for their social implications. In addition to his seventeen books and more than two hundred research articles and essays, translated into fifteen languages, he writes feuilleton for newspapers to reach a wider audience regarding his concern for social planning and its consequences.*

It was Søren Kierkegaard who coined the phrase that we live forwards and understand backwards—a caveat that applies to every retrospective. The Romans knew it well, and as proof they invented the god Janus, the pivoting symbol of gatekeeping. In the fascinating company of that power-filled figure, I have learned a lot, less because he is equipped with eyes that make him see in opposite directions at the same time, more because he has a mind that allows him to merge seemingly contradictory categories into meaningful wholes. In the same evaluating glance of the present, he catches a glimpse both of those pasts that once were and of those futures that have yet to come.

How often have I not longed for a place inside Janus's head, a privileged position from which I might experience how he braids together ontology and epistemology, the world and our understanding of the world, the first sense of sight and the sixth sense of culture, body, and mind. What I would really like to grasp is how he managed to deal with double bind without going crazy, why he was celebrated as a wise god and not put away as a mad schizophrenic. Perhaps he knew that memory is nothing but a hypothesis not yet rejected and that remembering is a sanctioned technique for purifying individual and collective consciences. Since the Greek word *aletheia* means "true" and its opposite, *letheia,* means "forgotten," etymology suggests that by telling half-truths now, I pave the way for the whole lies that follow.

The temptations are too strong to resist, the echoes too loud to silence. Diana bathing in a stream of water, Actaeon spying from behind a tree. As William Shakespeare once put it (Sonnet no. 138):

> When my love swears that she is made of truth,
> I do believe her, though I know she lies,
>
> . . . . . . . . . . . . . . . . . . . .
> On both sides thus is simple truth supprest.
> But wherefore says she not she is unjust?
> And wherefore say not I that I am old?
> O, love's best habit is in seeming trust,
> And age in love loves not to have years told:
> > Therefore I lie with her, and she with me,
> > And in our faults by lies we flatter'd be.

And thus it is that in the evening of June 1, 1980, on my way home from an excursion to St. Petersburg, that white city temporarily called Leningrad, I caught a glimpse of what I had not seen before: we live horizontally and understand vertically. Not unlike the prisoners in Plato's cave, we are chained to the ground floor of the taken-for-granted, condemned to staring into the shadows captured by the perpendicular wall in front. The words of the prayer came automatically, admittedly helped along by the Edward Hopper-like mood of the airport bar:

> Oh, to sin is to trespass. To trespass is to cross a boundary. To cross a boundary is to break a definition. To break a definition is to create. To create is to be different. To be different is to sin. To sin is to live in self-reference. . . .
>
> So, Janus. Help me become a sinner. Let me understand how you break definitions. Show me how to erase what others see as irresolvable paradoxes. Teach me the equation of that third lense inside your head whereby you transform contradictory images into coherent wholes.
>
> Speak, memory, speak![1]

*       *       *

And when memory speaks, I hear the voice of Gerd Enequist, the lady who already in 1937—two years after I was born—defended her groundbreaking dissertation on the historical and cultural geography of Lule Valley in northern Sweden. In 1949, she was appointed to the new chair of human geography at the University of Uppsala, the first woman ever to hold a professorship at that institution. Eventually I came to know her as a dedicated scholar with much integrity, deeply religious yet blessed with an unusually open mind— one of those rare individuals who improve with age. I often think of her with gratitude and admiration, for even though she had little influence on my intellectual development, she always treated me with the same warmth and trust that I had gotten used to at home. It was in fact with Gerd as with my parents

---

1. Gunnar Olsson, *Lines of Power/Limits of Language* (Minneapolis: Univ. of Minnesota Press, 1991), 16 and 215.

that there was never a need for revolt or rejection, always a feeling that I was free as a bird encouraged to fly wherever the wings would carry. Since I never asked for permission, I was never punished; since I was never taught shame, I never learned to worry. Perhaps this explains why I have yet to meet a person of whom I feel afraid.

It is for the simultaneity of her presence and her reluctance to interfere—at the same time coming close and keeping a distance—that I now value Gerd Enequist most. As a telling illustration, I often recall that November morning in 1964 when I knocked on her door, opened it, entered, bowed, and officially submitted my licentiate thesis. The professor rose in honor of the candidate, received the tome, measured with her eyes, weighed with her hands. With mounting curiosity, she leafed through the pages. Thumb and index finger. Two minutes of eternity. Then, suddenly, the relief, among the finest words I have ever heard: "Oh, is this what you have written? That will be interesting to read!"

Whether she also read anything but the preface of *Birds in Egg* is more doubtful, but I can still feel the tears in her eyes when she caught the sentences that "what follows would be much different were it not for Gerd Enequist, who instilled in her pupils that nobody can tell you what to do, nobody can teach you how to do it, and the best way to learn is to learn how to learn. It is sad that in today's university a grant is worth more than an open mind." It is even sadder that since those words were written on the thirteenth of January 1975, the grants have gotten bigger and the minds have shrunk.

*

Of all this I had no idea when in 1955—at the age of twenty—I graduated from the gymnasium. Good marks, no plans for the future. Although innocent, I was not lost, for in many ways I had already learned to look after myself. There was in reality no alternative, as the Värmland village in which I happened to be born was too small to have even a secondary school, much less a gymnasium—a geographic fact that meant that from the age of thirteen I spent most of my time away. About others one learns a lot by riding a regular bus for three hours a day, six days a week, thirty-six weeks a year, for four years—a total of 2,700 hours and seventy thousand kilometers, countless two-öre poker games, occasional fantasies about the girl three seats away. About oneself, one learns even more during three years in a small room rented from a deaf landlady with nourishing lunches and a beautiful granddaughter. I liked it, and when my mother claimed that I was homesick, she was almost certainly revealing more about her own feelings than about mine. At school, the most important figures were not the grown-ups—although some were highly educated and excellent teachers—but the classmates, a creative mix of youngsters from a great variety of backgrounds.

People of diverse backgrounds were in fact such an integral part of my home environment that it took a long time before I understood how strange they actually were. The reason is again geographic, for in a small community it is inevitable that everyone knows everyone else. To this should be added that my mother was the local midwife, in storklike fashion delivering babies to virtually every household. Over time, their numbers increased to about five thousand. The boy's daily chore was to ski or bicycle to the bus station to fetch the urine specimens that the expecting women had sent for testing—not centiliters or liters, but hectoliters. Completely natural. No questions asked, no answers offered.

Even more taken for granted was the next-door neighbor on the north side of the creek, the poorhouse with its half-crazy inmates of dubious extraction. Many were drawn to my mother, who gave them coffee and cookies while patiently listening to their stories about jails and workhouses, trolls and fairies, witches and constables. They came with a special odor, and my friends often treated them with a cruelty born of fear. All wore funny hats even on weekdays, but for me they were as much part of daily life as the calico cat. It was nevertheless considered an act of great courage when one day I sneaked into the poorhouse attic to watch the naked madman locked into a cage of bamboo. I can still see his eyes and hear his screams. Sometimes he wakes me up. But how did they get the idea about the bamboo, a foreign material that normally was used only in ski poles and fishing gear?

It would be strange if my later work did not show traces of these childhood experiences, of growing up in a place where social problems were a matter of concrete human beings, not of abstract social systems. Perhaps it is in the crevice between gemeinschaft and gesellschaft that the seeds were sown for my subsequent attitudes to collective action in general and to politics in particular. Never shall I forget that winter day in first grade when the teacher—a card-carrying Social Democrat Party member—got so upset with a classmate who could not read that she grabbed her by the hair, lifted her up, slapped her face. A sob from the girl, buckets of tears from me. But the morning after, my mother asked that I bring Solveig with me home, where she was given the bath and the decent meal her parents could not afford, a weekly arrangement that lasted for years. I suspect the story is unfair to the teacher, but for the inconsolable boy it was formative. Thanks to the mother who through her actions showed that it is better to preach as one lives than to instil into others that they should live as they preach. The dinner parties were many and lively, driven by a generosity without intentions, a presence without pretense. There was a miniscule distance between the laughter and the tears.

My father was different. Born in 1895, he grew up on a small farm that now is mine. Like the other children from the village, he went to school for five years, three days a week, earned his own living at eleven, more intelligent and industrious than the majority of my professorial colleagues. At seventeen, he emigrated to the United States, where he spent three tough years in the forests of the Northwest, understood in time that he was going to be drafted and sent to death in Flanders, caught the last ship to Scandinavia vowing that he was never to be poor again. The war over and the times turning bad, he was not allowed to reenter the United States and got himself a job in a Vancouver sawmill instead. After another three years of saving, he figured he had enough, worked his way home as a stoker on a rusty steamship, set up his own business in the timber and building trade and—as far as I know—lived happily ever after. At ninety-two, he was still skiing; at ninety-five he died for no other reason than age.

When I had been in this life for only two months, he had the foresight of buying a tract of forest land that he gave to me and to me alone. Not enough to live on, yet sufficient to make a difference. Since my father's death, there has been much work with the axe and the chain saw, but also some rich moments in the solitude of the cottage. And in the evening, when the wind hums its lullaby in the tree crowns, it happens that I remember two other presents, both of them given to mark the end of the war: a pair of boxing gloves from my mother, a beautiful canoe from my father. The former got lost along the way; the latter is well kept and still around, albeit in the custody of my oldest daughter, who used to come along when I went canoeing. For many summers, the canoe was my home, a key to the land of freedom, a connecting link to the likes of Tom Sawyer and Huckleberry Finn, some of the best friends any boy could have. Of these two, Huck was by far the most important, for by being himself he showed that also outcasts are human beings. And through his relations with the slave Jim, he taught me first that the individual is worth more than the system, and then that I must never betray a runaway.

*

The importance of the Canadian canoe can hardly be exaggerated. The same holds for the Swedish army.

After three weeks of celebrations following graduation from the gymnasium, I resurfaced as a conscript in the student platoon, Second Company, First Battalion of the Royal Värmland Regiment. Since everyone had to obey, I had given little thought to the fifteen months ahead. But already on the first night in the barracks, I knew that this was not for me. As I woke up the next morning, I recalled that students of medicine, dentistry, and veterinary medicine could do their military service as part of their ordinary studies. The idea was appealing, the strategy self-evident. But which one of the three escape routes should I choose?

The answer was easy, for I could really not imagine myself as a lifetime explorer of gaping mouths. Neither did I feel the urge to cut into human flesh, and that is regardless of whether the knife is a surgeon's scalpel or a soldier's bayonet. To make a short story even shorter, it took less than twenty-four hours to assemble the necessary documents and send them off to the Swedish School of Veterinary Medicine. One day later Private 350911–621 Olsson stood at attention, saluted the flag, and bid the captain farewell.

The following fall, two weeks early, I arrived in Stockholm eager to commence my career as an animal doctor. An ambitious assistant took me around the laboratories and the stables, his sights set on a huge operating theater crowded with veterinarians dressed in greenish overcoats and black rubber boots. At home, I had heard much about the danger of Caesarean sections, but not until then had I imagined that the same operation could be performed on a horse. Blinded by the shock, I closed my eyes only to witness how the tied-down creature on the operating table changed into Jesus Christ on the cross, the surgical knives turned to lances, the sanitized clothes to sponges full of vinegar. Neither the moose hunts nor the village butchery had prepared me for this. Mulling it over on the next train home, I knew that I had learned a lesson I would never forget.

In retrospect, the mistake was elementary. For the first time in my life, I had intentionally tried to manipulate the world. But to no avail, for already in the operating room had I realized that in my attempts to change the world I had treated myself not as a living person but as a dead thing—a common fault more easily committed than excused. Yet I am fully aware both that I am not the only sinner and that there are many ways out. And, as if to prove the point, already Diderot had noticed how "peasants turned lackeys to escape the conscription, just as in our own days they turn priests."[2] As a variation on the same theme, my own father had chosen the poverty of Sweden over the trenches of France. In a similar fashion, Bill Clinton did not go to Vietnam as a soldier but to Oxford as a Rhodes scholar, eventually ending up in the White House as commander in chief. I myself was rescued by a crucified horse.

When I realized what had happened, the academic year had already started. Now that I was lost and had nothing better to do, grades and family connections made me a substitute teacher in a secondary school nearby. Seated in the front row of the graduating class was a shy little girl called Birgitta, too cute to approach, too forbidden to leave alone. It sounds more scandalous than it actually was, but seven years later we were married. Now she is looking forward to retirement and to the first walking steps of our newborn grandchild.

At the end of the school year, I returned to Karlstad for the remaining fourteen months of military training. I did not like it this time either, but I was not about to make the same mistake twice. It was not easy, but later, in August 1957, I walked through the regiment gates with a sense of newborn freedom, hoping never to be in uniform again.

2. See *conscription* in the *Oxford Educational Dictionary*.

Without much ado, and certainly without noticing, the boy was on the verge of becoming a man. The contours of a personality had been drawn at twenty-two years. An empty mold was waiting for the steel to be poured.

\*

And thus it came about that the sergeant climbed the train to Uppsala, two suitcases in his hands, new haircut, pinch in his stomach. Five hours later the freshman arrived in the city of learning, all set for the academic studies that I pray will last as long as I live. The first in the family ever to be so privileged.

\*    \*    \*

The future lay open; hence, I had only the vaguest ideas about how to approach it. Although I was certain that I wanted an academic degree, I knew neither why nor in which subjects. In lieu of a better alternative, I set out for a combination of literature, economic history, political science, and geography. With the thought of saving the best for the end, I began with geography.

It was during the second term of the first year, in March 1958, that I ran into a text as tempting and forbidding as anything I had ever imagined—the licentiate thesis of Esse Lövgren, a brilliant man obsessed with the idea of translating the vagaries of human behavior into the precise language of mathematics. I understood hardly a word, yet enough to understand how little I understood. Most crucial was the mysterious sense of beauty, a feeling similar to what I encountered in the sculptures of Jean Arp, which I met for the first time during the same spring. What was so attractive with both Arp and Lövgren was their minimalism, the Blakean attempt to capture the whole world in a grain of sand, the artist in a piece of polished marble, the social scientist in a set of well-formed equations. Forty years later, this aesthetic desire has yet to be satisfied.

For Jean Arp, the means of condensation lay in the touch of his hands; for Lövgren, they were in a series of correlation and regression models, linguistic expressions in which all information is distilled into a set of parameter values. My own stumblings in the same direction came in the fall of 1958, when I submitted an undergraduate thesis on the historical out-migration from my home parish. In hindsight, I realize that this was a juvenile attempt to capture an aspect of my father, but its lasting value is that it introduced me to the social gravity model. It is difficult to imagine a formulation more central to geography as a social science, for what is the glue of the social if it is not human interaction, and what is the fundamental variable of geography if it is not distance?

The undergraduate years passed quickly, a creative mixture of efficient studies during the day and intensive student life at night. In the spring of 1960, I graduated as a *Filosofie kandidat* in human geography, economic history and political science. I cannot remember that I ever consciously decided to go on to a higher degree, but I do recall the happiness when Gerd Enequist telephoned that the Faculty of Social Sciences had awarded me a two-year fellowship.

*

I knew from the beginning both that I wanted to pursue the theme of distance and human interaction and that the approach should be quantitative. Expressed in its most general form, it was thus assumed that

$$I_{ij} = f(d_{ij}),$$

which, through various specifications and mathematical manipulations, leads to the classical gravity model

$$\log I_{ij} / P_i P_j = a - b \log d_{ij},$$

where $I_{ij}$ = the interaction between places i and j;

$$P_i \text{ and } P_j = \text{the population sizes of places i and j;}$$

and

$$d_{ij} = \text{the distance between i and j.}$$

Translated from the story of the equation to the picture of the graph:

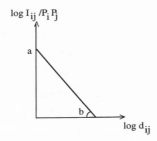

Embarrassingly simple. Extremely elementary. Seemingly not the nuts and bolts of an advanced degree but a trivial repetition of sixth-grade mathematics. Yet, and perhaps for that very reason, a well of insights so rich that no matter how much I have scooped up, it has never dried out. One reason is that in 1962 I followed Herman O. Wold's lecture series in econometrics.

This complicated little man—the only Swedish social scientist worthy of the Nobel Prize he was denied—had a lasting impact on my thinking. In particular, he showed how correlation and regression techniques are soaked in philosophical problems—on the surface about epistemology, causality, and logical inference, deeper down about human action, intentionality, guilt, and punishment. The time-bound tie between the two levels was in the practice of social engineering, the utopian vision that a better and more just society could be constructed on the basis of the exact knowledge of human behavior. As in the case of medicine, it was not enough to describe; the goal was to cure.

The basic point of Wold's highly technical teaching was that correlation measures covariation, whereas regression is a child of causation; to be specific, the correlation coefficient remains the same even when the $x$—and $y$-axes are interchanged, whereas the regression coefficients (the $b$-values or the slope of the line) are entirely dependent on which variable is put on which axis. In the vocabulary of perspectivism, the world looks like this when I watch it from here, like that when I view it from there. The geographic question is whether spatial variation is a matter of cause or of effect. More pointedly: Is distance a function of human interaction, or is interaction a function of distance? On analysis, however, these questions show themselves to be reformulations of more fundamental relations between form and process, picture and story, geometry and history.

Already during the academic year 1960–61, these difficult issues came into focus. As mentioned, I had just been admitted to the graduate program with a fellowship and an office of my own. One early morning Gerd Enequist stormed into the room with the message that also I needed company. A few days later the unknown stranger arrived from Cambridge, presenting himself as David Harvey. The professor's decision to put the two boys in the same room was nothing but remarkable, for it took us only five minutes to be-come friends for life. But before then—already after four minutes—both of us knew how little we knew. Along that same path, we have continued to move ahead.

David and I lived well together, socially as well as academically. The major scientific question we shared in common was exactly that of form and process, better known as the geographic inference problem. Is a phenomenon "thus" because it is "here," or is it "here" because it is "thus"? The question reaches into the depths of geography as a discipline, for since times immemorial the geographers have marched together in a kind of methodological two-step: *first* the investigated phenomenon is translated into the graven image of a map; *then* that idol is interpreted in terms of generating processes. Picture turns to story, matter to meaning, here to thus.

*

Packed with curious questions of this sort, Birgitta and I stepped off the plane at Idlewild Airport in New York on Labor Day 1963, greeted by a mass of well-armed policemen, a trickster taxi driver, and a saunalike heat. Behind the locked door of the hotel room, she broke into tears, insisting that this was not for her and that we must go home at once. By the next morning, the shock had subsided, and we continued by train to Philadelphia, where I was to spend a year in the regional science department of the University of Pennsylvania, fi-nanced by a prestigious fellowship from the American Council of Learned So-cieties (ACLS). The arrangement suited me perfectly, not only because of the trusting combination of freedom and responsibility, but also because of the practical help and social assistance provided by Julian and Eileen Wolpert, whom we knew quite well from their time in Uppsala.

The intellectual and social highlights of that time were many, but nothing compares to the year-long lecture course with Walter Isard, the best (perhaps only) teacher I ever had. His performances were always well rehearsed, regularly scheduled for Friday mornings, three hours without a break. What the small audience was offered was an excursion through the master's own writing, rhetorically intensified by the knowledge that the explorer was not inventing stories about others, but telling the truth about himself. As stops along the way he had selected a set of studies from his own bibliography, which he went through in mathematical detail, every step written out in symbols on the blackboards that covered the four walls of the room.

It usually took until Thursday evening before I had filled the empty spaces between the equations and got the comforting feeling that perhaps I might have understood. Then came the next Friday morning, the students and prominent visitors in their seats armed with sharpened pencils and stacks of note paper. Sound of footsteps and the PROFESSOR enters through the backdoor, already chewing on a piece of chalk, white foam around his mouth, ready to play yet another variation on the same theme. "You remember last week," he said. "You remember how we then started with this$_1$ particular set of assumptions and how we eventually arrived at this$_2$ conclusion. How nice it looked. And how good I felt when we parted! But the derivation has left me with no peace. Already that same evening I sensed that there was something strange hiding in the seeming clarity. And then, suddenly, I saw what it was. This particular expression. We must surely be able to do better. Let us try." And off he went into another lecture that took the poor Swede another six days to grasp. On and on, not once or twice, but every week. For two semesters! The most persuasive demonstration of the Hegelian principle of self-conscious reevaluation that anyone could ever wish.

From the ACLS year, two other events stand out. Both were National Science Foundation (NSF)-sponsored summer schools, the first in regional science held at the University of California, Berkeley, the other in spatial statistics at Northwestern University. We had a wonderful time, a constant party where I remain uncertain about what was most intoxicating, the lectures, the drinks, or the people themselves. Leslie King was there with his wife, Doreen, with whom we struck a lasting friendship; Akin Mabogunje was there, too, the funniest bowler ever to be seen. Roland Fuchs appeared every morning dressed in yet another of his Hawaiian shirts. With the two Swedes, Olof Wärneryd and Roger Malm, we shared many jokes that no one else could grasp. Among the youngsters were Mike McNulty, Bill Beyers, and Larry Brown. Add to this group the two lecturers Michael Dacey and Duane Marble, and there is a full bouquet of young promising men.

From Berkeley, Birgitta and I drove on to the other NSF-arrangement in Evanston, a hilarious trip that took us through a Beyers feast of clams in Seattle, on to Vancouver, where we picked up David Harvey, who was summer teaching in Edmonton, invited by Bill Wonders, who had met him in Uppsala and grasped his genius. Eventually at Northwestern, everything was in the computer hands of Michael Dacey, with whom I had shared an office during the year at Penn. It was this remarkable workaholic and political conservative who single-handedly adopted the cell-counting techniques of biometrics to geography, thereby opening the way to a deeper understanding of the relation between form and process, *the* problem that in a few years would blow his own dream of mathematical geography to pieces.

\*

Throughout these years—in Sweden and in the United States—I kept working on the licentiate thesis. It eventually came to consist of three parts: the first on the spacing of central places in Sweden, written jointly with Åke Persson; the second a review and bibliography, published by the Regional Science Research Institute; the third an extensive empirical study of Swedish migration. Of these three, the second got most attention, but it was the third that was the most important. With the aim of testing whether migration and central-place theory may be treated as special cases of a more general theory of human interaction, I formulated two sets of regression models, one in which distance was entered as the independent variable, the other where it was the dependent. The influence of Herman Wold was evident both in the care with which I handled the mathematical assumptions of the least squares technique and in the experimentation with the axes.

The crucial questions were addressed to the $b$-values, the data set condensed into its most minimalistic form. Are the slopes of the gravity lines stable over time and space, or do they exhibit systematic variations? The short answer is that they vary, especially in relation to the hierarchical level of the places of in- and out-migration. Out of the ashes of that finding rose a set of new problems, for the results revealed more about the spatial distribution of opportunities than about migration per se. Put differently, the *b*-values do not speak directly to the question of people's movements. Instead they furnish a sophisticated map of the spatial prison within which the human exchanges occur. And with that Isardian conclusion, the question changed; no longer was it about migration but about the location of places. Gerd Enequist's wonderful reaction has already been mentioned. In due time, I was made a *Filosofie licentiat.*

*

The steam was up, the process of self-conscious reevaluation started. But from the very beginning it had bothered me that the gravity model is based on a conception of man alien to my inherited sense of freedom and independence; for instance, it is well known that the model is a deterministic structure appropriate for the analysis of aggregate behavior but less suitable for the study of individual values. As a solution I pursued the idea of translating the various interaction models into the language of probability, an idiom in which variance plays a more prominent role than the mean.

The results were made public in a doctoral dissertation that I defended at the University of Uppsala on May 22, 1968. The work was well received, and I was duly promoted to *Filosofie doktor* with the highest possible marks. This was an extraordinary distinction, and the subsequent dinner party was well worth the apartment we sold to pay for the feast. Best of all was nevertheless the telephone call that brought the good news to my parents. Hearing my father's voice break into tears is the most valuable of all gifts he ever gave his son.

The dissertation was in essence a study of the geographic inference problem. To that end, I conducted a series of experiments with a range of deterministic and stochastic models in which the aggregation levels were allowed to vary such that distance information was entered sometimes as the dependent and sometimes as the independent variable. What was thereby illuminated was in reality nothing less than Kenneth Arrow's impossibility theorem, itself the social scientific variant of Kurt Gödel's impossibility proof in mathematics. A crucial move, for Arrow's important point was that without the intervention of a dictator, social preferences cannot be derived from individual preferences; Sweden in a nutshell, Plato revisited.

It says a great deal about the Swedish establishment that when Arrow was awarded his Nobel Prize, the impossibility theorem was not even mentioned. Yet the relations between the irreconcilability of aggregation levels, on the one hand, and Swedish reality, on the other, are so striking that only a Swede can fail to notice. Thus, it was clear already in the 1960s that the welfare state showed many signs of counterfinality; rather than solidifying the good intentions of liberty, equality, and fraternity in bricks and paragraphs, the results were often the opposite. A modern case of classical tragedy, for what is good for the state is often bad for its citizens. For the young geographer, it was an added memento that much of the actual planning was based on practical applications of the gravity model. Whether this alliance between the social sciences and politics is a sign of dishonesty or of deficient scholarship remains an open question. At any rate, the situation brings forth memories of my mother, of Solveig, and of the poorhouse inmates.

Gradually it became clear that the issues were not methodological but ethical. And then, suddenly, in the fall of 1968, it was proven that the traditional two-stage approach to geographical description and interpretation is fallacious. The various stochastic models—à la the principles of logical positivism doubly anchored in the spirit of reason and the matter of observation—showed that the same spatial form can be generated through drastically different processes. It follows that even the most perfect description of a spatial pattern cannot be used as the basis for inferences about how and why it came about. Although it is sometimes possible to reason from process to form, moving in the opposite direction is never appropriate; even though two plus two normally equals four, four can equal anything, including two plus two.

The methodological conclusion is so radical that most practitioners have yet to fathom it: the disciplinary ship *HMS Geography* was hit below the waterline. Repairing the damages has proved more difficult than anyone could wish, perhaps because information and understanding never are one and the same. In the meantime, much effort is being spent on the development of sophisticated geographical information systems (GIS). The only crux is that the majority of the GIS techniques draw on observed autocorrelations not on insights into causal relations. Much money and many Ph.D.s have been burned on the altar, institutionalized sacrifices as morally corrupt as the geographers' role in Swedish politics.

Double fault: (1) according to the principles of utilitarian ethics, purposeful action should be judged in terms of its consequences; (2) according to the principles of causal theory, the consequences of actions based on observed autocorrelations cannot be foreseen. Match lost, players disqualified, for some tricks are too tricky to be excused. Reasoning from aggregates to individuals is one of them; making inferences from form to process is likewise.

\*     \*     \*

Although the impetus for studying the inference problem came from my attempts to understand the contradictions of Sweden, I carried out most of the actual work at the University of Michigan. The universities were expanding everywhere, the jobs were many, and in September 1966 I arrived in Ann Arbor accompanied by Birgitta and Ulrika, the latter only ten months old. I had never set my foot in that town before, I had met none of my colleagues, and all I had read were some discussion papers from the Michigan Inter-University Community of Mathematical Geographers. The latter used to meet in a tavern in the nearby town of Brighton, the point of minimum aggregate travel from the three universities in Ann Arbor, East Lansing, and Detroit. John Nystuen set the agenda at beer time, Waldo Tobler's soft clarinet and Bill Bunge's blaring trumpet dominated the intellectual jam sessions.

It tells more about the times than about me, but barely three weeks into the job I was promoted to associate professor with tenure. The English-Swedish dictionary gave little guidance to what these foreign terms actually meant, and, judging from the body language of the well-wishers, I could not really ask. In quick succession, I was then made full professor and eventually given a collegiate professorship. The Värmland boy shudders at the thought of serving as an American assistant professor or a Swedish docent. Yet another stage in the normal socialization process avoided. Lucky me!

*

The inference problem was a part of the times. I spent many nights discussing it with others, especially with Reg Golledge, Leslie King, David Harvey, and John Hudson. Most important was nevertheless Leslie Curry, the Beckett-like figure admired by everybody, understood by no one. Most appealing was his conception of the random spatial economy as an undulating surface, a virtual landscape in which the mist-enveloped hilltops are rising and the impoverished valleys sinking. In voluptuous waves of three-dimensional Fourier functions, the world was depicted as sucking on itself, the rich getting richer, the poor getting poorer. But the initial problem remained: Is the geographic inference problem real or merely a phantasm produced by faulty reasoning?

As the panic spread, the members of the informal group set off in different directions. Golledge went into the laboratories of behavioral psychology; King disappeared into the corridors of university administration; Hudson boarded the historical train to the pig farms of the Midwest; Harvey began the intellectual journey that eventually made him the world's most creative Marxist. For my own part, the search led to the theory of science: first to the relations between theory, model, and reality; then into alternative logics and fuzzy sets, ethics and aesthetics. Quite soon I found myself deeply entrenched in investigations of the equal sign, that intellectual glue that keeps the two sides of the equation together, the symbol that couples cause to effect, the magic wand with which ideology is touched to stone, stone to ideology. Five of the best years of my life.

The background is unusual. All my friends kept talking about Karl Marx, and it would have been strange if I had not followed suit. But once I did, what had happened before happened again. I understood not a word that I read. I tried again. Again and again. And then, suddenly, I understood why I did not understand: Marx's conception of the equal sign was drastically different from the multiplication table I had been taught and hitherto taken for granted. Even such a central word as *capital* had one meaning on page 37, another on page 180, a third on page 306. Not because the text was badly written, but because the phenomenon under discussion kept changing as the story progressed. Revolutionary indeed, for it is in the methodological manual that one cannot have a science about a subject matter that hops capriciously about.

In the now familiar fashion, it became clear that although I had thought I was investigating one particular problem, I was in fact grappling with quite another. The key was given to me by Bertell Ollman, who in his turn had got it from Vilfredo Pareto, the Italian sociologist whom nobody could call a communist. Pareto's point was that Marx's writing is impossible to understand because it is written in a confused language, an idiom in which words and meanings are combined in a manner that violates the standard rules of logical and scientific conduct. Instead of being clear and well defined, the Marxian words are batlike; viewed from above they look like rats, from below like birds. On one level of abstraction, a question about appearance and reality, on another about the relation between word and object. On still a third, a problem of intellectual fix-points.

G. W. F. Hegel—the master dialectician whom Marx was determined to turn upside down—was already well aware that the issue of word and object is a special case of the relation between identity and difference. In minimalistic concentration: How do I recognize something when I meet it again? How do I know that what you see is what I see as well? What are the relations between the five senses of the body and the sixth sense of culture? How and why do you believe me when you hear me say what you never heard before? Which are the rhetorical principles that govern the unconscious?

Some of the answers can be found in the taboo-ridden interface between epistemology and ontology, in the conjunction of what I say and what I say something about. Indeed, it became the major conclusion of *Birds in Egg*, published in 1975, that what I happen to say in my allegedly objective analyses usually reveals more about the language I am talking *in* than about the phenomena I am talking *about;* the fisherman's catch furnishes more information about the meshes of his net than about the swarming reality that dwells underneath the surface. Kant's Copernican revolution in a new context, for it is not the words that mirror the world, but the other way around. Language is not mine alone. What better verification could anyone have wished than the message from the Library of Congress that the batlike geography book had been shelved under the label "ornithology"! A few days later a letter arrived from the Dutch Chicken and Poultry Institute. Twelve copies ordered.

In a similar situation, Barnett Newman—a painter from whom I have learned much—once remarked that aesthetics is to the artist as ornithology is to the bird, perhaps another instance of the Kierkegaardian dictum that we live forwards and understand backwards. It should be added that the title *Birds in Egg* was coined during a visit to the Art Institute of Chicago, where I saw a wonderful exhibition of Max Ernst, including several of his birds in eggs and eggs in birds. It was immediately clear to me that what he tried to capture in his pictures was exactly what I was grappling with in my own text. The manuscript was baptized on the spot, Ernst its godfather, Allan Pred the only witness. As an added benefit, I saw an opportunity to include Ulrika's drawings as well.

*

From 1970 to 1977, I gave a collegewide graduate course entitled "Thought and Action." Much of its content eventually found its way into *Birds in Egg*, circumstantial evidence that I have never felt any conflict between research and teaching, writing and lecturing. In fact, just the opposite, for there has always been a symbiotic relationship between work and leisure, family, friends, colleagues, and students. Countless are therefore the seminars and guest lectures held in our living room at 2128 Geddes Avenue, telephone (313) 761–3195, events that invariably progressed into some kind of symposium. The prominent guests were many; most loved were nevertheless the students. With Birgitta's assistance, a functioning home was built, not only for the small family in exile but for a number of gifted graduates as well. The first was Stephen Gale, now himself teaching at the University of Pennsylvania, the last was Michael Watts, currently at the University of California, Berkeley. In between were many, some eventually wandering off into the wider world, others remaining within the confines of academia: Jack Eichenbaum, David Russell, Bob Douglas, Roger Ulrich, Chris Smith, Adrian Pollock, Peter Hoag, Widdicomb Smith. We still enjoy each other's company whenever the opportunity arises, the former pupils now both older and wiser than their teacher at the time.

As an integral part of these interactions, I managed to arrange a set of seminars together with like-minded people from two other universities: the University of Toronto, where Bernard Marchand (whom I had met for the first time in John Hudson's Evanston kitchen) held a visiting appointment, and Ohio State University, where Reg Golledge was the prime mover. When times eventually got rough, Reg and I grew very close, a personal and professional friendship that has withstood both the constraints of distance and the assaults of time. But there were others, too, especially Leslie and Doreen King, with whom we shared many relaxing holidays away from home. Ineffaceable are the four-clover memories from Reg's thirty-seventh birthday: a windswept beach on the west coast of New Zealand's Southern Island, bottles of brut Veuve Clicquot, bags of newly cooked crayfish, the huge Australian, the thin Swede, the New Zealand king, and the Greek archbishop, Dr. Papageorgiou himself. Open minds and open charge accounts. Thirty-five years later the white little stone from the shore of the Tasman Sea is still with me, touchable proof of untouchable bonds.

To complete the Michigan story, mention must be made of Jim Clarkson and Peter Gosling, in my estimation the brightest of all my departmental colleagues. Singly and together they taught me much about the university, some quantitative geographers included. In addition, Jim introduced me to cultural geography, bluegrass music, German wine, and the lonely whiff of marijuana that ever came my way. Peter, on his part, first helped the immigrants buy a house and then took the entire family on an unforgettable summer of fieldwork in Thailand. The latter experience had a lasting impact, for, in addition to solidifying my Swedish scepticism about social engineering, it made me appreciate that it takes a lifetime to get into another culture; on Thailand I wrote nothing, on North Armorica either. For the same reason—and contrary to prevailing rumors—I had nothing to do with the eventual elimination of the Michigan geography department.

*

Completing *Birds in Egg* brought a number of intellectual and existential issues close to the breaking point. Perhaps the situation would have been easier if the various parts of my life had been less tightly interwoven. But in the bat-like world of the Tally-Ho Tavern, it became increasingly difficult to tell the

difference between the silver-tongued devil and me; as my friend Kris Kristofferson—another Rhodes scholar—once put it, "All he is good for is getting in
trouble / And shifting his share of the blame / And some people say he's my
double / And some even say we're the same." As retold in his "Border Lord":

> Breakin' any ties before they bind you
> Runnin' like you're runnin' out of time
> Take it all—take it easy—till it's
> over—understanding
> When you're headin' for the border lord
> You're bound to cross the line.

And thus it became painfully evident that just as knowledge is an issue of
naming, identity, and existence, so life itself is in the art of being prisoner and
warden at the same time, of shouldering the burden of freedom, of having the
courage to be what I am. Very Precious. But what really remains to be studied
after five years in the constant presence of the equal sign, that ultimate symbol
of the word *is*? Creativity running amuck, fettering fetters, anxious anxiety
that what is left to do has already been done. How do I grasp the ungraspable,
touch the untouchable, say the unsayable? Ylva was born as the independent
joy she has remained; I was appointed to the chair of economic geography and
planning at the Nordic Institute for Studies in Urban and Regional Planning
(Nordplan) in Stockholm, arriving there in May 1977. As in Ann Arbor eleven
years earlier, I knew nobody, and the first time I went there, I had to take a taxi
to find the way.

The decision to return to Sweden had nevertheless been easy, for the alternative was to die in exile, to continue doing for yet another quarter of a century what I had already done for more than a decade. In addition, I had never
thought of the Michigan appointment as a regular job, but merely as an exceptionally valuable postdoctoral fellowship. As initially planned, we left the
land of plenty in style, sailing first class on board the *Queen Elizabeth II,* almost missing the departure for the surprise party that a group of pupils had
arranged in New York.

In retrospect, I am happy that I had resigned rather than taken a leave of
absence, for returning to Sweden proved to be the most traumatic experience
of my life; if the opportunity had been there, we would almost certainly have

gone back to the United States. The sad truth is that Sweden is not the paradise that the political propaganda makes it out to be; the mixture of fiddling, wire pulling, and hypocrisy makes a foul brew. When recent immigrants assure me that I am the most un-Swedish Swede they ever met, I am therefore moved and flattered, accepting their remark as the compliment I know it is meant to be.

But even though Swedish life presented (and still presents) many difficulties for return migrants, it offered (and still offers) a wealth of opportunities for the independent social scientist. Just as Karl Marx might not have written *Das Kapital* had he not been an alien in England at a time when capitalism had not yet disguised its contradictions, so Sweden, especially in the 1980s, was the world's foremost laboratory for anyone interested in the rise of postindustrialism and the decline of the welfare state. Nordplan was in fact an ideal observation point, for even though the five Nordic countries are sufficiently similar to experience the same problems, they are different enough to have chosen slightly different solutions.

*   *   *

As *Birds in Egg* was rapidly going out of print, Allen Scott convinced John Ashby that Pion Ltd. should publish a new edition. which is also what eventually happened, but not until 1980, after a new section, more explicitly devoted to the issue of linguistic self-reference, had been completed. The old and the new were then bound together in one single volume to be read first in one direction then turned upside down and approached from the other end. In order not only to say but also to show, the self-destructing text was given the chiastic title *Birds in Egg/Eggs in Bird*. For the same reasons of reflection, the old and the new—the chicken and the egg, the egg and the chicken—were made to meet in a reproduction of Marcel Duchamp's Large Glass, *La Mariée mis à nu par ses célibataires, même*. No whimsical coincidence, for Duchamp is the only person apart from my wife with whom I have had daily discourse for almost four decades.

With the completion of *Birds in Egg/Eggs in Bird*, the formal education and the informal postdoctoral were brought to fruition. At the age of forty-two, I was finally ready to proceed to what I consider my major contribution: twenty years of advanced transdisciplinary and transcultural teaching. The major lesson from the Uppsala and Ann Arbor experiments was that the words of ordinary language are not rich enough to express the emotions I wish to convey. As it was for the French poets of the nineteenth century, the challenge was therefore to learn to think-and-write in such a way that the text is not *about* something but that it *is* that something itself. The inspiration came from the poetry of Stéphane Mallarmé, the prose of James Joyce, the drama of Samuel Beckett, the (post)philosophies of Roland Barthes and Jacques Derrida, all creative explorers at the limits of language. Like Piet Mondrian—the foremost artist of silent lines—I came to dream of a work in which the expression of things gives way to the pure expression of relation. The study objects nevertheless remained the same: the power-filled relation between individual and society, the crises of representation, the concepts of intentionality and purposeful political action. The specific questions also remained: How do I know the difference between you and me, and how do we share our beliefs in the same?

\*

I reported the explorations of the 1980s in a series of articles, some of which eventually were collected in two books, the Swedish *Antipasti* from 1990 and the English *Lines of Power/Limits of Language* from 1991. Whereas the former consists of a portfolio of watercolors from the Swedish landscape, the latter is a retrospective of abstract oils and genre-crossing installations. In constant focus are the classical issues of POWER, which for the first time became acute with the geographic inference problem and then reemerged again in the braiding of *Birds in Egg/Eggs in Bird*.

The overarching idea of these studies is that POWER is a game of ontological transformations, a magical performance in which visible things are turned to invisible relations, untouchable mind to touchable matter. The paradigm is as old as the first chapter of Genesis, in which it is told how reality pops out of the LORD's mouth: "Let there be!—And there was." The Almighty did not make the world, he uttered it. As a latter-day illustration I have been drawn to the story of Czar Peter and his St. Petersburg, that enthralling city of white light, the concrete outcome of a series of ontological transformations, where ideology is transformed into huts and palaces, serfs and princes, Gogol's nose,

Andrej Belyj's senator, the bronze of Pushkin's horseman, the granite of St. Peter's rock. But my most violent critiques have been directed at the closest target: the case of Swedish social engineering, the institutionalized hatred of everything different.

The difficult question is this: How do I find my way in a world I do not know? Not in the well-ordered universe of rivers and mountains, hotels and airports, but in the cultural contradictions of hopes and fears, promises and obligations. The easy answer is a dressed-up variant of what I learned from the madman in the poorhouse attic. As a consequence, some of the subsequent analyses are firmly lodged in the unconscious, impossible to explicate. Others are more directly related to the points and lines of the regression models of the 1960s and thereby to Wassily Kandinsky's book *Point and Line to Plane,* first published in German in 1926. That text used to be suppressed as well, for years kept in the dark, too dangerous and too important to be read.

And yet. When the moment eventually came for putting together *Lines of Power/Limits of Language,* I could no longer ignore Kandinsky's book. Shocking discovery, for when I finally dared to open the text, I found that this man, nine years before I was born, had "quoted" a number of passages from me, who so stubbornly had refused to read him. Not merely general ideas but exact phrasings, not just once but repeatedly. A leery experience, if one ever was. As a way of righting the wrongs, I composed the introductory chapter as a collage of citations, some lifted from Kandinsky's book, some from my own articles. More important still: as the Russian artist once had reacted to Claude Monet's *Haystack,* so the Swedish geographer began to wonder if it would not be possible to get farther in this direction, to push the relation between expression and impression to its limits.

*

Once in that dematerialized void, it becomes clear that every sign (regardless of whether a word, a gesture, a dollar bill, a gene, a regression line, or anything else) consists of two inseparable ingredients, Signifier *(S)* and signified *(s).* The full sign can consequently be written as

$$\frac{S}{s}$$

where the most crucial aspects are in the fraction line, the untouchable symbol that at the same time separates and unites. If the equal sign was in focus for everything in the 1970s, the same holds for the fraction line in the 1980s.

About the point of that line and about the plane onto which its shadow is projected, much can be said. Not, however, here and now. Suffice it merely to note that whereas the Signifier *(S)* by necessity assumes some kind of material form, the signified *(s)* is always invisible. It follows that the full sign within itself melts together two drastically different forms of being, such that the *S* dwells in the five senses of the body, the *s* is in the sixth sense of culture. To be possessed by POWER is in that context to know the magical art of ontological transformation so well that I am believed when I perform it. And so it is that creativity always involves a struggle over categorial limits, a guerilla war fought against the silent majority. Marcel Duchamp provided the perfect example when in 1917 he went to the New York hardware store, bought the infamous porcelain urinal, carried it to the exhibition hall of the Society of Independent Artists, turned it ninety degrees, signed it "R. Mutt" and baptized it *Fountain*. The scandal was immediate.

*

The sign formed the glue also of my Nordplan teaching, an adventure that lasted from 1977 to 1997. Throughout this period, I alone carried the responsibility for an unusually ambitious doctoral program. Approximately seventy-five high-level politicians, administrators, university teachers, and researchers with degrees in a range of disciplines went through the system. Because the institute was not Swedish but a functioning part of the Inter-Nordic government cooperation, the participants came in equal numbers from Denmark, Finland, Norway, and Sweden. The obligatory program ran for three years and ended with a day-long oral examination, after which the candidate could move on to the writing of the dissertation. When time eventually came for the defense, the majority were in their late forties, several much older. The traveling was extensive, with three-fourths of the teaching abroad; for two decades I spent a little more than 180 nights a year away from home.

The teaching and the dissertations ranged over a wide spectrum of topics. Yet everything was steeped in closely related questions of power, ethics, aesthetics, and rhetorics, where planning was treated as a special case of human action located in the taboo-ridden interface between individual and society—Kenneth Arrow was never forgotten. The overriding challenge was to approach the outer limits of understanding, to explore the no-man's land where different taken-for-granted elements are pitted against each other, stripped bare by their critics, showing off their hidden secrets.

Personally I took it as an encouraging sign that the program was under constant attack, especially from Swedish and Norwegian social democracy. It took the evasive enemy longer than expected, but in December 1995 the last dissertation was defended. One year later the institute was thoroughly reorganized, all teaching abolished, ongoing research turned into extensions of state bureaucracy. For me, the Nordplan mission was completed, an occasional poem to be read many times, written only once. In the process, my chair was transferred to the University of Uppsala, in all respects an ideal solution.

But prior to the official closing, there was a big feast, a celebration of what had been done and an indication of what was to come. As part of the transition rite, I asked everyone with a completed dissertation to write a personal paper on what he or she had learned. The resulting testimonials are collected in the volume *Chimärerna: Porträtt från en forskarutbildning*, a title that shall be interpreted as the teacher's homage to those courageous individuals who had the stamina to complete the program.

The Swedish word *chimär* comes from the Greek χιμαιρα, an iconic sign that at the same time says and shows; as can be seen, its first letter is χ. More specifically, a chimæra is a fire-breathing monster often depicted as a lion with a goat's head on its back and a serpent as tail—a frightening creature of imagination, an incongruous union of parts that must not be united. Since every monster by definition is a child of the limit—and in that sense a symbol of Nordplan, perhaps of myself—the chimæra must be eliminated. That task was given to Bellerophon, who set off to complete it riding on Pegasus, the winged horse that once played a prominent role also in Bertrand Russell's theory of types. But so wise is Greek mythology that it made Bellerophon the grandson of Sisyphus, the existentialist figure who in life was rewarded for his cunning shrewdness and in death was punished for exactly the same characteristics. Forever doomed, sentenced to pushing the rock that refuses to be pushed. But *chimæra* is also the term for a biological organism that contains tissues of two or more distinct parental types, yet preserves the genetic characteristics of the parts unaltered; a branch of an Åkerö apple tree grafted onto a stem of Astrakan continues to yield Åkerö apples—the same Åkerö, except of a higher quality than on the original tree.

Such is my own dream of teaching. A chimæra, a phantasm, a combination of that which must not be combined. Yet another reality impossible to leave alone, at the same time forbidden and extremely beautiful. Neither a scientific laboratory nor a government-sponsored Center for Brainwashing, rather an open environment in which the grafting proves to be so successful that everyone can develop one's own personality. The point is not to produce pre-ordered copies but to allow some genuine originals the freedom they need to be what they are.

By fortunate circumstances, the dream became reality. No wonder the apparatchiks got terrified, for in everything their attitude is different. Sharpened finger in the cyclopes' eye. Nohbdy tricked them.

*

Aristotle observed that he who uses a friend thereby shows that for him the friendship is over. Keeping that warning constantly in mind, I am most grateful for the generous help provided by a number of individuals from the geographers' jet set: Bernard Marchand, Guiseppe Dematteis, Franco Farinelli, Derek Gregory, Michael Watts, Allan Pred, Allen Scott, Ed Soja—all of crucial importance for memorable excursions to Paris, St. Petersburg, Cambridge, Milan and Venice, Berkeley and Los Angeles. In the Nordic countries, somewhat similar roles were played by Olavi Granö in Turku, Georg Henrik von Wright and Erik Allardt in Helsinki, Arne Næss and Per Andersson in Oslo, Johannes Møllgaard in Copenhagen. Within Nordplan itself, I learned most from José Luis Ramírez y González, the Spanish exile who devoted his dissertation to the praxis and rhetorics of Aristotle. In addition, he introduced me to a menagerie of artists and philosophers from his home country, especially to Antonio Serrano de Haro, for a period Spain's ambassador to Stockholm. The only Swede to matter was Johan Asplund, the reclusive social psychologist, who in reality is not a Swede at all, but an immigrant Finn. Much the same applies to Hans Olsson, my moose-hunting brother in the flesh. More recently,

Ali Bensalah-Najib has offered a sip of the pride, honor, and generosity so typical of his Berber family.

Most important is nevertheless the informal, nameless, and constantly evolving group of international geographers that for two decades has met to share ideas, food, and wine. One early meeting was held in the stunning Villa Serbelloni in Bellagio, where they served the best scrambled eggs in the world, eventually to be followed by a row of gatherings elsewhere in Italy, Switzerland, California, and Denmark. With few exceptions, the discussions have been outstanding, sometimes resulting in a book, sometimes not. Key participants include Peter Gould, Dagmar Reichert, Ole Michael Jensen, Allan Pred, Verena Meier, Ulf Strohmayer, Reg Golledge, Michael Watts, Derek Gregory, Beppe Dematteis, Alessandra Bonazzi, Jette Hansen-Møller, Guilia di Spuche, Enzo Guarrasi. But nobody compares to Franco Farinelli, the only geographer with a significant impact on my own thinking. The first time we met was on a bridge over the railroad tracks in Geneva, an event quite similar to what happened twenty-four years earlier, when Gerd Enequist burst into my office, David Harvey in tow.

What I find so fascinating about the Italians—as about the Spaniards—is the living symbiosis of the Greek and Roman classics, on the one hand, and the (post)modern avant-garde, on the other. It has indeed been a tumultuous experience to discover how several of the paradoxes that the established professor has forged for himself are well known to every Latin fourth grader. And yet, despite the differences, the cross-cultural meetings have always turned into mutually rewarding exchanges; while the autodidact from the periphery is charmed by the perfectly polished formulations, the exegetes from the center are stunned by the skewed associations and the never-heard-of combinations. Constantin Brancusi and Joseph Beuys running into each other, discovering that they are working on the same sculpture.

For anyone interested in the history of the taken-for-granted, there is no better guide than Franco, the obsessed pursuer of Odysseus and Moby-Dick. And in that exclusive company, the old question resurfaces again: Where do they come from and to where are they heading, all the self-evident premises and all the confirming conclusions? Restating what is worth restating: How do we become so obedient and so predictable?

The answer to these forbidding questions is embarrassingly simple, hence extremely complex: we find our way in the invisible by treating it *as if it were* visible; we handle social relations *as if they were* material things. The differences are nevertheless decisive, for in the world of things we are guided by maps and compasses that themselves are material, while in culture we are pointed the way by tropes of rhetoric, especially by the likeness of metaphor and the closeness of metonymy. Modern reason is in deed a form of cartographical reason, a mode of understanding inherited from Immanuel Kant, *the* explorer of the boundary zone between the sensibility of the body and the intelligibility of understanding. Two centuries after his death we are finally approaching the moment of a fourth Critique, the explication of those principles that he himself lived and therefore never fully understood, a volume devoted to the rhetorics of how we make sense of the world and of how we share that sense with others. Its provisional title: *A Critique of Cartographical Reason.*

*           *           *

In the 1990s, a series of studies have appeared exploring these lines of thought. Of particular concern is the laying bare of the double roots of European culture, one deeply sunk into the polytheism of the Greeks, the other grown out of the monotheism of the Jews; while the former depicted their gods in sculpture, the latter are warned that the graven image is an idol not to be worshiped. In that vein, the key stories of Odysseus and Polyphemos, Abraham and Isaac, Esau and Jacob, Job's refusal have all received their heretical interpretations. Who knows what fate has in store? Perhaps there will be a book from the 1990s as well.

The principles of cartographical reason is the theme also of the pyramid that I am busily constructing with Ole Michael Jensen, a sculpture in glass and granite, a reformulation of Moses' first stone tablet, a novel rendering of the relations between the Almighty and his subjects. Perhaps a minimal answer to the question of how and why we become so obedient and so predictable.

*

All as exciting as ever. GO ON, GO ON. Cannot be stopped. Not with the fire in me now. And for that batlike reason, these autobiographical glimpses will be neither summarized nor concluded. No one should judge himself fortunate until after his death. Too late by then.

And yet. For being a confessed solipsist I must be blessed with an unusual number of close friends. And for being a devout nonpolitician with a conservative bent, I am spending a great deal of joyful time with leftists, albeit coming from a variety of congregations. Although firmly believing that human action is structured as a tragedy—everything perfectly right in the beginning, everything horribly wrong in the end, no mistakes in between—I have rarely wanted to be dead.

The idea of thought-and-action is too tempting to resist. On the wings of geographical imagination, I am therefore returning to the dim-lit Casablanca bar, where one single bottle of Black Label costs three cubic meters of fine timber. Say it again, Sam! But once you do, you must remember this: a kiss is just a kiss, a sigh is still a sigh. The fundamental things apply, as time goes by.

## Selected References

Farinello, Franco, Gunnar Olsson, and Dagmar Reichert, eds. *Limits of Representation*. München: Accedo, 1993.

Gale, Stephen, and Gunnar Olsson, eds. *Philosophy in Geography*. Dordrecht: Reidel, 1977.

Golledge, Reginald, Helen Couclelis, and Peter Gould, eds. *A Ground for Common Search*. Santa Barbara: Santa Barbara Geographical Press, 1988.

Gould, Peter, and Gunnar Olsson, eds. *A Search for Common Ground*. London: Pion, 1982.

Olsson, Gunnar. *Antipasti*. Göteborg: Korpen, 1990.

————, ed. *Att famna en ton*. Uppsala: Acta Universitatis Upsaliensis, 1998.

————. *Birds in Egg*. Ann Arbor: Michigan Geographical Publications, No. 15, 1975.

————. *Birds in Egg/Eggs in Bird*. London: Pion, 1980.

————. "Chiasm of Thought-and-Action." *Society and Space* 2 (1993): 279–94.

————, ed. *Chimärerna: Porträtt från en forskarutbildning*. Stockholm: Nordplan, 1996.

————. "Complementary Models: A Study of Colonization Maps." *Geografiska Annaler*, Ser. B, 50 (1968): 115–32.

————. "Distance and Human Interaction: A Migration Study." *Geografiska Annaler*, Ser. B, 47 (1965): 3–43.

————. *Distance and Human Interaction: A Review and Bibliography.* Philadelphia: Regional Science Research Institute, 1965.

————. *Distance, Human Interaction, and Stochastic Processes: Essays on Geographic Model Building.* Avhandling för filosofie doktorsgraden, Uppsala Universitet, 1968.

————. "Heretic Cartography." *Ecumene* 1 (1994): 215–34.

————. "Invisible Maps: A Prospectus." *Geografiska Annaler,* Ser. B, 73 (1991): 85–91.

————. *Krisens tegn/Tegnets krise.* Nimtofte: Indtryk, 1989.

————. *Lines of Power/Limits of Language.* Minneapolis: Univ. of Minnesota Press, 1991.

————. "Misión imposíble." *Anales de Geografía de la Universidad Complutense* 17 (1997): 39–51.

————. *La senza ombre: La tragedia della pianificazione.* Roma: Theoria, 1991.

————. "Towards a Critique of Cartographical Reason." *Ethics, Place, and Environment* 1 (1998): 145–55.

Reichert, Dagmar (Hrsg). *Räumliches Denken.* Zurich: Vdf Hochschulverlag an der ETH, 1996.

# Sliding Sideways into Geography

## Forrest R. Pitts

*Courtesy of William E. Feltz*

*Forrest ("Woody") Pitts (b. 1924) served in the United States Navy during World War II, studying Japanese at the language school at Boulder, Colorado. He received his B.A. in Oriental languages and literature, an M.A. in Far Eastern studies, and a Ph.D. in geography (1955), all from the University of Michigan. Before his retirement in 1989, he taught at the Universities of Oregon, Pittsburgh, Hawaii, and Seoul National, and was a visiting professor at the Universities of British Columbia and Washington. Emerging from a traditional curriculum in cultural geography, he was one of the few geographers so educated to see the potential in more formal quantitative approaches, particularly in the area of spatial diffusion, where (with Duane Marble) he advanced the programming of Monte Carlo simulation models. He served for eight years as executive director of the International Geographical Union Commission on Quantitative Methods. Despite such formal approaches, his research is marked by meticulous fieldwork in Japan, Okinawa, and Korea, and he is one of the few geographers forming an intellectual bridge between East and West. For many years, he served as editor of* Korean Studies, *and in Korea he is known as "the father of the hand tractor," being personally responsible for the early promotion of the large-scale introduction of this technological innovation into Korean farming. His research also includes several studies of the structure of Korean cities and further studies on the political unrest among young adults in these rapidly growing centers. He is professor of geography emeritus at the University of Hawaii.*

A t the end of June 1989, I retired from the University of Hawaii, ending a rewarding career of thirty-five years of university teaching and field research. Becoming a geographer was something that I did imperceptibly, well prepared for it, but not having the field as a goal until events showed the way.

## At the Foot of the Grand Mesa

My father's ancestors came from England and settled near Winston-Salem, North Carolina, sometime after 1680. My great-grandfather, Cadwallader Jones Pitts, was born there into a poor Mennonite family, and was "bound out" at the age of eight to another family. That family moved to Kentucky, and when the contract term was up, he went to the Minnesota Territory to trap animals with a friend. On his return, the two were walking down a country lane near Vinton, Iowa, where they tipped their hats to two young ladies. One of these said to the other, "Did you see that tall fellow? I am going to marry him." Not long afterward she became my great-grandmother. At least three of their sons moved to Cripple Creek in Colorado to make their fortunes in digging gold. When that did not work out, they settled in 1886 in the Collbran-Molina-Mesa area, at the foot of the high lava plateau, the Grand Mesa—now a national forest. My grandfather, Thomas A. Pitts, homesteaded a farm at Mesa, where he and Grandmother Sylvia planted alfalfa and raised eight children.

My mother's ancestors in this country were Dutch, from Haarlem, Manhattan, where in 1732 Christine van Zandt was born on a dairy farm. My maternal grandfather, Moses Thompson, was a glassblower in Newcastle, Pennsylvania, when he got the westering urge. They wintered in Broken Bow, Nebraska, where my mother was born in 1893.

Grand Junction was my birthplace, in a hospital. My mother had been a practical nurse in the Great War and had traveled widely. Her marriage to a sailor came to an end with the Armistice, when he wrote to her on a postcard, "My ticket don't read home." E. Frank Pitts saw his chance, and after their marriage they operated a grocery store with a cousin of my dad, who ran the meat counter. The business was not a success, and the couple moved to Mesa to live at his father's farm, in the original log cabin built by the former gold hunter. Later on, Granddad had built a larger, two-storey house a short distance across the farmyard.

The farm as it evolved consisted of a twenty-acre orchard, mainly apples of many varieties, apricots, pears, peaches, and cherries. Tom Pitts followed closely the experiments of Luther Burbank of Santa Rosa, the town I chose for our retirement, owing to the fact that my first cousin, Glen Price, a retired chemist from Shell Oil, lived there. He was ten years older than I and told me much about the early 1920s, when I was on the way to being born. I might not have been born but for the act of a faithful dog. As a child, my mother, Mabel, and several of her brothers and sisters were playing in a sandpile one day, when brother Elmer saw the hair rise on the back of the family dog. Before he could wonder why, the dog leaped over the children and behind Elmer treed a mountain lion. The older boys sent the girls for Moses, who came and shot the lion.

On the farm was a large five-bedroom farmhouse, with a root cellar for winter storage of potatoes, onions, and the like. A part of the root cellar was later used as space for an acetylene-generating apparatus that provided gas for the lighting system. There was also a smokehouse, the apple-packing house, a blacksmith shed (where my father lost an eye at the age of eight), an icehouse where winter ice cut from a pond was stored for summer use, and a milking shed. There were also 150 colonies of bees, bought primarily to pollinate the fruit trees and alfalfa, but supplying honey as a by-product in sufficient quantities to pay for a new car each year. Aside from coffee, sugar, flour, and tea, very little in the way of groceries was bought with cash.

A great deal of the work on the farm when I was young fell upon my dad and his brother Walter. At harvest time, more hands were hired, and Grandma Pitts, along with Mabel and Aunt Pearl, had to cook large dinners for the temporary help. Then, as always, dinner at noon was the substantial meal. Supper at about six in the evening was cornmeal mush poured into a plate, served with a bowl of milk on the side. Both Brian Berry and I recall the noon and evening meals as dinner and supper. None of this effete urban business of lunch and dinner!

My younger brother and I—he grew up to be a fighter pilot—were only sixteen months apart and were a worry to my mother, who often entertained the Ladies Aid Circle of the Methodist Episcopal Church. At one meeting, my brother grabbed an ax and I a pitchfork, and we chased two Foster brothers onto the county road, locked the gate, and wouldn't let them back in. The missing boys were the object of a great search, and when they were found, my brother and I were spanked on the spot. We boys lost our mother to pneumonia in 1932.

I had a well-remembered run-in with our grandfather about the age of seven. He had asked me to fill a tin pail with ripe apricots lying on the ground in the upper orchard and take it down to feed the pigs. I was afraid of the many bees and hornets humming around and filled the bucket with alfalfa before I covered it with two layers of apricots. This ruse was quite successful over several trips, until he visited the pigpen and saw the alfalfa, which was starting to make the hogs sick. Another spanking ensued.

After outliving two wives, Granddad married a part Ute lady. The chemistry between her and my now-single father was not the best, and she asked—or ordered—him to leave the farm. After our mother's death, Aunt Pearl and Walter Pitts took us in for a year, while my father sought to reestablish his family. Our new mother was but twenty-one when the families were united, consisting of her daughter (her husband had died in a car wreck), we two boys, and later another sister. Our new family lasted until our parents' deaths in 1973. I had been on the farm when my grandfather dropped dead of a heart attack and was with my own father when he died of an aneurism in a Santa Cruz hospital.

In the 1930s, we were caught by the Great Depression, in which the reconstituted Pitts family was often needy, sometimes on relief, but never hungry. The experience left me a lifelong Democrat. In 1934, we rented a ten-acre farm, and our phone number was 24-J. To this day, that local phone system is separate from the larger network, a result of both isolation and local stubbornness. In a two-year period, we lived in eight or nine houses or apartments, one of which was a deep rural two-bedroom house that had been used to store wheat. We caught dozens of mice before we could move in.

My father, with his relatives and hired hands, often played for dances on weekends, and I have many memories of going to sleep in a corner as he called square dances. He was a natural musician who never found an instrument he could not coax a tune out of, but he never learned to read music. With an eighth-grade education and one eye, he went further than most men in those mountains. Other memories of rural life include milking a cow (which I did not like), berry picking in the early dawn on the Severson farm, and climbing the numerous mesas along Plateau Creek. These climbs made me aware that the mesas had at one time been joined, and I appreciated at age thirteen the erosive processes that formed them. The lava rockfalls on the slopes of the Grand Mesa provided similar insights.

Often I rummaged around in the attic of an aunt's parents. There I spent hours looking at stereographs of the Civil War and famous places. Later, when I studied aerial photography, I found that I could place photos side by side and look through them to produce a three-dimensional image. I was the envy of my class for what I thought everyone could do.

The last farming my father did was at Groundhog Gulch in the summer of 1938, where he planted nine acres to wheat at almost the upper limit of wheat growth. I spent an idyllic summer hiking in the sagebrush hills on one side of our cabin and exploring the aspen forests on the other. Later I recognized this as an ecotone. So many of my natural experiences, for which I found names in graduate school, made geography a welcome study.

## High School Days

On our return to Grand Junction in 1939, my brother and I got jobs in a bowling alley, where we spotted pins by hand. It was a great improvement when the boss built a new and larger alley with racks into which we could toss the pins and push the placing mechanism down. This was my first side job as I attended high school.

I studied two years of Latin and pined for the other two that were offered. But my stepmother was a "practical" sort, and she talked me out of taking more Latin. By finishing Caesar's *Gallic Wars* ahead of the second-year class, I managed to squeeze in three weeks of Greek, owing to the kindness of another teacher. Geography as such was not something I yearned for. When someone urged me to take a commercial geography course, I responded, "I took the last geography course I am ever going to take in the fifth grade." The highlight of my high school career was learning to write, both essays and verse, taught by Miss Emma Groom, a dedicated lady of Sherman tank proportions. I graduated seventh in a class of 168 and found that the first 6 had received college scholarships.

At ages sixteen and seventeen, I flirted with the far right. I jumped ship before I did too much damage, though I lost my best friend, a Jewish lad. This experience inoculated me, and I came independently to hold the Stephen Jay Gould position on the natural distribution of abilities, in whatever lands and climes. In the summer, after my junior year in high school, our geology teacher, Edward Holt, took me and three others on a three-week fossil dig not far from Green River, Utah. We found the scapula and coracoid of a large dinosaur (now called an *Apatosaurus,* I believe) and carefully wrapped them for shipment to the Field Museum in Chicago. In return, Professor Holt was sent plastic models to use in his teaching at Mesa Junior College.

After leaving the bowling alley, I worked as a carhop in a drive-in at twenty-five cents an hour. Together with many tips, my wages went to help support the family. My stepmother viewed my contributions as a loan, but I knew better and did not resent giving up the money. When our parents' estate was settled in 1978, each of us four children inherited $2,000.

## College and the Navy, Intertwined

Upon my graduation from high school in 1942, our family moved to Sandpoint, Idaho. After a summer's job as a plumber's assistant helping to build the Farragut Naval Training Base on Lake Pend d'Oreille, I took my $950 in wages and entered the University of Idaho at Moscow. Only $50 remained at the end of the spring semester. I enrolled in geology courses, mostly paleontology and geomorphology, but the scorn poured on my head when I let it be known that I believed in Wegener's "floating continents" was enough to put me off geology on my postwar return to college.

During my freshman year, where fifteen credits were the norm, I took twenty-one and got good grades, even in remedial algebra and Algebra 1, which I took concurrently. During this time, I enlisted in the navy, where I later went through a course in celestial navigation and differential calculus. After a year at Pocatello as an apprentice seaman, I entered the navy V-5 flight program at St. Olaf College in Minnesota, only to be transferred once again. The navy was losing fewer pilots than anticipated, so all who wanted to go to a midshipmen's school were released. I reached Throgs Neck in the Bronx (where a bridge now exists) for my training. Toward the end of my four months, a Commander Hindmarsh visited, and I was selected for Japanese-language school in Boulder, Colorado. He gave me the spiel, "With you, Pitts, we are reaching the bottom of the barrel." I learned only a few years ago that that was what he told everyone he recruited.

While I was in the Naval Language School in Boulder, in June 1945, I met a young lady at a Friday evening picnic. I asked Miss Jones to go with me to Elitch's Gardens in Denver the next day, and at the end of that day I proposed to her. She wanted to think it over, and on Sunday night she said yes. We were married three months later, in September, just after the war ended. We have had three daughters, who live in Santa Rosa near us, and a son whom we lost to leukemia at the age of six months. We celebrated our fiftieth anniversary with a cruise through the Panama Canal, a long-time dream for Valerie.

Almost four years of navy service gave me another year of college, fifteen months of Japanese-language school, considerable experience in teaching gunnery to people more warlike than myself, and several months in a Washington, D.C., document facility, which was later incorporated into the CIA. I left the navy in November 1946 with a commendation in hand from the admiral in charge of Naval Intelligence. The GI Bill had just become available, and I enrolled in Barnes School of Commerce in Denver for a few months to learn typing and Gregg shorthand. Without the GI Bill, which paid for my education, I would have spent a life snaking logs out of the forests of the Grand Mesa.

Valerie and I then moved to Berkeley for the 1947 summer school. The visiting professor that summer was Robert Burnett Hall, who taught a course on Japan and was a close friend of Carl Sauer. When Professor Hall learned of my language training, he persuaded me to transfer to the University of Michigan, where they were just starting their Center for Japanese Studies. My third year in college proved to be my senior year (I had been given a year's credit for my language stint), and in 1948 I entered the center at Michigan for an M.A. in Far Eastern studies. In November 1997, I was a guest speaker at the fiftieth anniversary colloquium for the center, reading a paper called "Japan: Twelve Doors to a Life."

Life at the Ann Arbor campus was interesting, especially for the annual

Easter-season arrival of disillusioned graduate students of Finch and Trewartha in Madison, asking us if we had room for them. Hearing we were full
up, they departed sadly to careers in government.

My own career plans were not fixed. After a semester and a half of studying
Mandarin Chinese, I went to the former Washington Document Center location and asked to be given language tests to join the CIA. On the Japanese
test, I had no trouble, but on the Chinese I scored seventy-two points. For
lack of three points, I was saved from a life at the agency. I never made it to
China, nor did I become a China watcher in Hong Kong. Returning to Ann
Arbor to attend field camp at Waugashaunce Point, I then wrote my M.A. thesis on a regional study of Mt. Fuji, translating materials from geological journals in Japanese.

### Graduate School and Japan

With Robert B. Hall as my mentor, I more or less slid into geography sideways. My background in geology and languages made it easy to pass the geography doctoral comprehensives. In the early 1950s, these exams were a
curious remnant of nonbureaucratic times, for they consisted in persuading
examiners in seven fields that one was competent. When the document containing seven valid signatures was presented, one was on the doctoral path
automatically.

Armed with a $7,000 grant from the Social Science Research Council
(SSRC) for field study in Japan, I arrived in Grand Junction to visit my former
English teacher, Emma Groom, to thank her for teaching me how to write.
Then to Okayama, where the Michigan Field Station was located, in January
1951. All the graduate students there applied their particular social science or
humanities expertise at the small hamlet of Niiike ("new pond"), and one portion of my dissertation includes material from that settlement, to which I have
gone back many times for nostalgic reasons.

Basically, the aim of my doctoral dissertation was to find the roots of rural
prosperity in the Inland Sea area of Japan. My application to the SSRC read:
"On the northern shore of the Japanese island of Shikoku lies one of the
world's most productive farming areas—the Kagawa plain. Centered in the
fertile Inland Sea region, it represents the pinnacle of Asiatic indigenous agri-

cultural development. Yet this great productivity, and the relatively high level of rural prosperity which accompanies it, have been achieved only in the face of adverse conditions. The Kagawa farmer lives in *prosperity amidst adversity.*" That and a further paragraph eventually became the first paragraphs of my dissertation. In trying to explain rural prosperity in the Inland Sea area, I sought evidence from the Heian period to 1950, a stretch of a thousand years. After the fifteen months of fieldwork, it took a while before I sat down and wrote the text.

One thing that Robert B. Hall's students absorbed was an emphasis on problem orientation in writing a dissertation. We did not take his frequent query "What is your problem?" in a psychological sense. The search for roots of rural prosperity brought me to a realization, documented in several important instances, that Japan in the Tokugawa era had been economically fairly advanced. Thus, one by one, academia's older myths began to tumble. I was only one of several graduate students at the Okayama field station who found such evidence. This trickling line of inquiry gradually swelled to a stream under the aegis of the Michigan (later Yale) historian John Whitney Hall. His writings and those of his students have forever changed the direction of historical and economic inquiry into pre-1868 Japan.

George Kish was a cochairman of my committee owing to the frequent absence of Robert B. Hall in Japan or Washington, D.C. George's wife, Vin Anger, was a medical doctor and arranged for me to have emergency gall bladder surgery. She then invited me to recover at their home, the "House of the Seven Pear Trees," along the Huron River west of Ann Arbor, and to do minor chores while I wrote my text. Living in her sewing room, I took a footlocker of notes, divided them into four categories, each category into relevant piles and each pile into subpiles of usable size. The actual writing took me eight weeks. (Other things took longer: I did not learn to drive a car until I was thirty-three.) After my committee saw the result, all I had to do was delete a quotation from a Chinese document alleging that the traditional first Japanese ruler was originally a Chinese bandit.

While I was writing my dissertation in the Michigan countryside, my reading included Wilbur Cash's *The Mind of the South* and Edmund Wilson's *To the Finland Station*. I was seeking answers to the cataclysm of our Civil War and the revolution in czarist Russia. Good writing was evident on every page, and their ways of arguing must have impressed me, as a later incident showed. One day, in the spring of 1955, Professor Hall asked me to come to lunch at his home in Barton Hills, where he told me of an English professor on campus who had long since ceased writing professionally and who read each dissertation in the social sciences at Michigan as soon as it reached the library shelves. This pedant had the habit of showing up in the dissertation committee chair's office and telling him in great detail of the errors and infelicities of the document. One day he showed up in Professor Hall's office, demanding, "Who is this person Pitts?" As Hall was about to explain my background, the visitor barked, "His is the first literate dissertation I have ever read," and walked out. Osmosis from the Cash and Wilson styles had influenced me.

While teaching for a visiting year at the University of Hawaii, I was invited by George Jobberns to spend two years at Christchurch, New Zealand. But first I had to spend nine months in Okinawa on a Pentagon-sponsored project to "find the causes of the tension between Okinawans and Americans," in order to earn enough money to take myself and family to Christchurch. Back in Ann Arbor, a sudden recurrence of gall bladder pains led to the operation mentioned above. The surgeon's work was free for university students, but in paying for the nurses and anesthesia, I spent all my travel money, and to this day I have never reached New Zealand.

My experience with rural life as a child and teenager made it easy for me to talk to Japanese farmers about their very intensive farming practices. My father had done custom threshing in the fall and plowing in the springtime. During threshing time, we boys had helped bag the grain as it came from the thresher. In Kagawa, I once asked to be permitted to cut a row of ripened rice plants. The result established great rapport between us because it was most sloppily done.

I wanted to record the annual rhythm of crops and farming methods, and to that end took endless rolls of color slides. As I roamed the countryside, I was able to tell people what I was interested in and said that I had grown up on a farm. The rumor mill made sure that news traveled ahead of me, and I met with a good reception everywhere.

Not all my work at Okayama was for myself. The anthropologist John Cornell had been working in Matsunagi, "place for tethering horses," and had no good map of this settlement on the rolling limestone plateau. With the help of

John and the young son of a Michigan art professor, I plane-tabled the entire village at a scale of one to two thousand. The houses were individually measured at a scale of one to two hundred and were entered on the map via a pantograph.

In graduate school, reading Hartshorne's *The Nature of Geography* was the most difficult thing I ever undertook. Throughout it seemed opaque, and though I later met and liked the author as a person, I never changed my mind about his book. Another outlet for my iconoclasm has been my creation, since about 1957, of *A Little Devil's Dictionary*, in admiration and imitation of Ambrose Bierce. One of the longer definitions is geographic: "Eratosthenes— Geography's first quantifier, who measured the distance between Syene and Alexandria, multiplied it by 50 [for good geometric reasons] and got a figure of 25,000 miles for the circumference of the earth; his results were reduced to 18,000 miles because *it didn't sound right*, by Strabo, geography's first qualifier."

## Oregon and Korea

Choice jobs in geography in the mid-1950s were few, but positions did exist here and there. Merle Prunty of Georgia had contacted me, and I agreed to go there after my doctoral defense. Then Joseph Russell of Illinois visited Ann Arbor and learned that I was looking for a job, with a preference for one in the West. He told me to get in touch with Samuel N. Dicken at Oregon and said, "I have recommended another person already, but I shall tell him that you are more qualified." After Dicken had offered me a job as instructor, I wrote Prunty that I was no longer available. Thus, I became an honorary member of the Ex-Georgia Society, composed of young men who published like mad until they got an offer elsewhere. In 1990, I had the happy task of writing a short sketch of Sam Dicken for *Geographers: Biobibliographical Studies*.

A few months after I arrived at Eugene, I went to the Seattle meeting of the Association of Pacific Coast Geographers. I gave a paper on Vietnam, whose main point was to criticize in detail the remark in Joseph Spencer's book *Asia: East by South* on the Geneva armistice conference's decision to place a boundary near the seventeenth parallel. He had written that "the present lines of demarcation correspond to nothing in either the physical or cultural landscapes." This run-in with Spencer was the beginning of a long friendship, mixed with methodological disagreements. He had learned his Chinese working for the Salt Commission in the 1930s. Years later, in Kansas, I talked with A. W. Kuchler, who reported that he, too, had studied Chinese as a young man and now regretted that he had never used it in his career.

At that same Pacific Coast Geographers meeting, I met William Garrison and his band of young model-oriented graduate students. The chairman at Washington, Donald Hudson, learned of my interest in East Asia and asked me to teach in the summer session of 1957. It was there that a crucial conversation in my life took place. I was telling Bill Garrison that I had used experimental and control areas in my study of Japanese farming when he interjected, "There is a better way of doing that—it is called analysis of variance." That comment started me on the road to quantitative analysis, which I taught at beginning and advanced levels for many years.

In my first four or five years at the University of Oregon, I taught every geography course in the catalog, not because it was good for me (as Waldo Tobler was told when interviewed for a Madison position), but owing to many courses and few bodies. Once geology and geography at Oregon became two departments, and a doctoral program was instituted, those days were over. Eventually I was asked to teach a course in geomorphology and chose Thornbury's text.

Because my audience was mainly geology students, I was able to start on a level somewhat above the usual introductory course. A preference for the Albrecht Penck theory of parallel slope retreat (over the progressive gentling of slope advocated by William Morris Davis) was reinforced by a chance seeing of an article in *Science* that used analysis of variance to test the two models. I was slowly learning a new language, that of modeling.

One day Sam Dicken asked me to become coauthor of a book he was writing on human geography. He had already written two-thirds of it, but wanted a different point of view on cultural geography. To be frank, I agreed because I thought my next promotion might be in jeopardy if I refused. As often happens, cowardice had its rewards. The chapters I wrote taught me much more than I would have learned by less-focused reading. In one chapter, I introduced a form-function ladder, where material and technique advance in alternate steps. In 1997, Anne Buttimer told me of being assigned the

Dicken-Pitts book when she was a student. Included was a two-page typology of Old World agriculture, relating types of agriculture to common tools, crop emphasis, typical crops, land pattern, dominant animals, ownership pattern, settlement type, economic stage, and population densities. Later I was pleased to learn that this typology prefigured many of Ester Boserup's ideas. A later edition of the book, with a different title, contained my "Languages and Dialects" chapter, which was great fun to write because it drew upon my long-standing interest in languages and linguistics.

Where do ideas for articles come from? Returning from Okinawa in spring 1954, I was struck on the flight home by the great contrast between the way the Japanese organize space and the way Americans treat it. This thought led to "Japanese and American World-Views and Their Landscapes" in *Proceedings of the IGU Regional Conference in Japan, 1957*. A closer look at Japanese land-use practices appeared in my 1960 piece, "A Mirror to Japan," in *Landscape*. The published version was a great disappointment to me because the editor, J. B. Jackson, had cut quite a bit of the text without consulting me. Although he had paid $50 for a professional drawing of a summer versus winter land-use transect, sea to mountains, an illustration that was to have been placed across the bottom of four pages, he saved space by excluding it. I made the full text available to anyone who wrote me about the article, as several did, but the beautifully detailed transects were not returned to me.

About the same time, I made a series of colored maps of land use in the Okinawan hamlet of Nanahan, where I mapped and interviewed in the eight months after September 1953. My colleague Clyde Patton taught me the use of California Inks, and the resulting maps were a joy to behold, totally different from the monochrome and Zip-a-Toned maps of land productivity in my dissertation. One Okinawa map was a hamlet-size version of von Thünen's productivity circles. At the time, I was exploring the von Thünen model in my cultural geography seminar and wrote a summary of what his model would look like when certain assumptions were progressively relaxed.

British Columbia held its provincial centennial in 1958, and I was invited to teach at the university. Charles Fisher of Malaya was their first choice, and when he could not come, I was contacted. The university then had a department of increasing fame, and I was delighted to become acquainted with Ross Mackay and many others on the staff. Ross, to my mind, deserves a Nobel

Prize for what he has done for geography—if such an award existed for
our field.

During the late 1950s, I wrote to Samuel E. Martin, author of *Korean in a
Hurry*, about what seemed to me to be the amazing similarity of grammatical
structures in Japanese and Korean. An army language branch had loaned
books and records (78 rpm speed) to me so that I could study Korean, so my
first tries at speaking it were close imitations of the records. I have a good ear
for reproducing phrases, whether or not I understand the grammar, whose
study can always come later. The records were a boon because the greater
complexity of sounds in Korean, as compared to Japanese, posed a greater
challenge. Meanwhile, on the Oregon campus other things were happening.
The Business School, under Dean Richard Lindholm, was bidding for federal
support of a University of Oregon advisory team to help the South Koreans
with their first three-year economic development plan. Several economists
were sent out with a retired dean as pro forma chief, but the search for an agricultural economist dragged on for several months. During this time, I frequently approached Lindholm about being sent as part of the team, citing my
language interests and my field experience with Japanese farming. Rather than
lose the funding, he finally caved in. My family and I arrived in Seoul three
weeks after Eisenhower had visited the new government. Syngman Rhee had
by then been exiled to Hawaii.

My interest in Korea was not only linguistic. In Japan, many people spoke
of Koreans in a disparaging manner that was a chapter-and-verse echo of what
many white Americans were then saying about our black citizens. I wanted to
see what it was about the Koreans that so irked the Japanese, and this was my
opportunity.

In Japan, I had studied the use of rototillers in Inland Sea area farming.
Over the years, they had been adapted from dryland tillers imported from
Geneva in 1927. By 1951, the tillers were much-changed and more sturdy
versions of the original Swiss machines. When I arrived in Korea, I found that
several small Sears-Roebuck tillers were being used for horticulture, and one
fine Japanese tiller was *hidden* at an agricultural experiment station—because
of Syngman Rhee's hatred of things Japanese (he even caused Korean clocks
to be thirty minutes off Tokyo time). The hand tractor was there, but was not
being used. After two or three week-long excursions to the countryside—dur-

ing which a young farmer urged me not to disrupt the farmers' ecological relationships with their work cattle—I started on a plan to introduce hand tractors into Korean farming. In this project, I had the public-relations help of
Mr. Won, the editor of the local *Food and Agriculture Journal*. In both Korean and English, he published my 1960 justification for a pilot program. In
April 1961, the national legislature appropriated the hwan equivalent of
$39,000 to start the hand-tractor program. In the summer of 1966, Duane
Marble and I visited the office of the minister of agriculture, and there was the
same Mr. Won who had published my mechanization plan. He told Duane,
"Mr. Pitts is the father of the hand tractor in Korea."

At the request of Dean Lindholm, I arranged a sister-city agreement with
Chinju, near the south coast of Korea. It was Korea's first sister-city arrangement. It was the dean's idea to start such an arrangement, but he left the city
selection to me. The opening ceremonies were held in January of 1961. I
went into a barber shop not far from where the famous *kisaeng* Nongae, a
courtesan, had gotten a Hideyoshi general drunk and then pulled him into
the river, where both drowned. As my hair was being cut, two young men
came in and started talking to the barber, who told them I was a university
professor. They started needling me about having so much education and that
they would never get a degree, but had to go into the army. I told them that
they were already graduates of one university. They asked, "Which one?" and
I told them, "Shikishima Taehakkyo" (University of Shikishima). They
laughed and became very friendly thereafter. *Shikishima* is an old word for
Japan and is the brand name for a popular condom in Korea.

The Oregon advising contract had been for two years, and I had little time
to study Korean formally, though I practiced common phrases in the street.
My coming aboard a year late meant that we all returned in May 1961. As my
family and I breakfasted in Tokyo's International House, we heard about the
military coup of General Park. We had left Korea nine hours ahead of the
takeover. My sabbatical was coming up, and I took a semester off early in 1962
to study marriage distances in Kagawa—an intellectual heritage from Torsten
Hägerstrand and his idea of a mean information field. I visited Torsten in
Lund on my way to Japan, spent three days in Beirut, and visited an archaeol-
ogist friend of his at a dig at a Portuguese Fort in Bahrain. In Japan, I also
toured key universities with Fukushichi Uyemura, trying to generate interest

in starting a Japan section of the Regional Science Association. This lure was successful, and when I attended a meeting in Tokyo some years later, I was publicly thanked by Professor Konno, its president.

Richard Morrill and I later put my Kagawa marriage-distance results in an article that we submitted to the *Annals of the American Geographical Association*. Joseph Spencer as editor was rabidly antimodel so far as human geography was concerned. Our paper on marriage, migration, and the mean information field was put on hold for a long time while we received letters from him asking, "What are you shysters trying to put over on us?" He then published the article precisely as it was submitted, two years later.

This incident requires me to return to the hiring of Duane Marble by Dean Lindholm, who wanted several social science components in the Business School, to the ire of the rest of campus and the disdain of Sam Dicken. Duane had told me of Torsten Hägerstrand's visit to Seattle and gave me a copy of his long-unpublished "Monte Carlo Simulation of Diffusion." I was immediately interested and asked Clyde Patton to teach me how to program a computer in FORTRAN, the only language then designed for scientific computing.

My adventures in the numbers trade led to a National Science Foundation (NSF) summer seminar in regional science at Berkeley, followed by two more NSF-supported summers at Northwestern University. At the first of the Evanston sessions, I was a student, but a year later, under Michael Dacey's wing, I was teaching FORTRAN and learning to sample point patterns. In the map room of the clock tower in the earth sciences building, I invented torus sampling so as not to throw away information found within half a random-sampling circle. Dacey's program TOINTS *("torus points")* later was included as a technical report in an Office of Naval Research (ONR) contract that Duane Marble and I were given for the programming and exploration of the Hägerstrand spatial diffusion model. My programs for Hager III and IV, released in 1965, were vastly improved by a short insert that Margaret Cliff programmed. Programs for calculation of the mean information field and a noncellular arrangement of receiving areas (small political units) were released in 1967. Many fine reports from graduate students supplemented our work, which the ONR geography program chairperson, Evelyn Pruitt, regarded as the best return for navy dollars during her overseership.

In 1964, I chaired an Association of American Geographers (AAG) committee on the use of stored data systems. Waldo Tobler chaired the renamed committee on geographic information systems in 1965. These two groups were two of several advocating the geocoding of the 1970 census. About this time, I moved from Eugene to the University of Pittsburgh to become a full professor at the age of forty. While there, I attended an Asian studies conference in San Francisco and was surprised to be offered a position as associate director of the reconstituted Social Science Research Institute at the University of Hawaii. A quick visit affirmed instant tenure in the geography department, and we put our Mount Lebanon house up for sale. For a few weeks in the autumn of 1965, we owned houses in Oregon, Pennsylvania, and Hawaii.

### Hawaii and More Korea

At Hawaii, several professors had deep interests in Korea, and over many lunches we talked about forming a Center for Korean Studies. At length, the administration took notice, and I was asked to formulate a plan, with attached documents. As in the original three bears story, the first document was too long, the second too short, and the third one "just right." The center was approved and funded early in 1972, while my family and I were in Seoul on an NSF-funded grant for study of the internal functions of Korean cities. Thus, I became a founding member of the center and edited its annual journal, *Korean Studies,* for three years and much later for another six years. While in Seoul in 1972, I engaged a personal tutor from the Myongdo Academy to coach me in Korean. Because we both spoke Japanese, and the grammars are similar, as are the words adopted from Chinese, I was able to finish a year's work in afternoon sessions over four months.

Korean and American scholars in the mid-1960s formed a research consortium called the International Liaison Committee for Research on Korea (ILCORK), funded by the U.S. Aid Mission. I was its secretary until its dissolution in 1973. We sponsored a trinational multidisciplinary research summer in 1969 in Taegu, where each foreign scholar was paired with a Korean scholar in the same field. The resulting book, *A City in Transition: Urbanization in Taegu, Korea,* contains my chapter "A Factorial Ecology of Taegu City." Later, several of my students worked on periodic markets in Korea, spurred by the ground-breaking article by James Stine in a 1961 monograph on economic development and urban change, which I edited at Oregon. This work culminated in a sociology dissertation at Hawaii and a geography dissertation at Seattle.

In my first year in Korea, I wandered around taking many pictures of each type of business or amenity, which led to my NSF-funded study in 1971 of the internal structure of thirty-two Korean cities, for approximately 1,350 precincts. Still later, the same source funded my study, with historian William Henthorn, of the economic conditions in Kyongsang Province in the fifteenth century. For the study, we were able to microfilm several rare historical royal annals from the Seoul National University Library.

On several recent visits to Korea, the last in 1996, I noticed that the most visible improvements to the landscape have been the massive reforestation of the hills and the mechanization of farming. I am proud to have been instrumental in the second of these improvements.

In 1968, I went to the International Geographical Congress (IGC) meetings in New Delhi and was asked to become the executive director of the International Geographical Union (IGU) Commission on Quantitative Methods. The commission assembled a large number of corresponding members from many countries and launched a program of one international conference each year. When the eight-year stint of conferences was finished, we could boast of many influential publications. Despite many criticisms from the touchy-feely antinumbers crowd, we built a basis for many extensions of quantitative methods into areas of traditional concern to geographers. Years later I deposited all the letters and documents from these eight years with the Royal Geographical Society in London.

One of the more interesting aspects of working in anthropology and geography is the opportunity to restudy many years later those field areas you researched when you were young. In the summer of 1983, I got a travel grant from the National Geographical Society for a restudy of two of my six doctoral townships. "Thirty Years of Change in Two Rural Areas of Western Japan" appeared in the last volume of *National Geographic Society Research Reports* two years later.

In 1982, I worked on a proposal for a multisummer space-time budgets study in Manila and Honolulu. A prominent time-geography reviewer of grants claimed (incorrectly) that I had never seen most of the five hundred references in my very detailed proposal, and so funding was denied. In the re-search, I intended to use canonical correlation analysis, a technique ready-made for geographers, but one that most of us sadly neglect, despite a few very perceptive studies. Individuals were to be interviewed daily on their time of travel and destination for a two-week period, an approach that had been field tested in Korea in 1967. Data would be for 300 Filipino individuals, 150 in Honolulu in the summer of 1984, and 150 in Manila in 1985. In each place, respondents would be divided into those born and raised in the city, those who migrated to the city earlier, and those who were recent migrants to the city. Some information would concern characteristics of areas where the respondents lived. Canonical correlation would have been performed pair-wise on three data sets: the movements in space (behavioral) data set would be related to the background data set to produce *person-specific ecologies.* The be-havioral data set would be related to the psychological data set to produce *person-specific syndromes.* Finally, the environmental data set and the psycho-logical data set would produce *person-specific mindscapes,* or topophilias and topophobias.

Instead of going to Manila for the proposed second summer of study, I un-dertook a study of hand-tractor use and financing on farms near Taegu, Korea. A few minutes after I delivered the copied results, I had to go to the hospital with a mild heart attack. The fact that I was booked into the teaching hospital of Kyongsang National University gave me the superb care of a top Korean cardiologist. Had I been in Manila at such a juncture, I might not have survived. Thus, I belatedly appreciated the review board's turning down my proposal. The two-week stay in hospital cost my medical insurer, Kaiser-Permanente, $1,400.

Since 1961, I always have left a copy of my study notes or filled-out sched-ules with my foreign counterparts, for whatever use they wish to make of them. Nowadays this procedure has justifiably become the legal requirement in Thailand and many other areas. To me, it has always been a matter of main-taining the dignity of colleagues.

## Thoughts on Geography

It is time to speak of my interests and basic outlook in geography. I expressed these ideas in the July 1972 issue of the *Geographical Review*. A specific piece of research tends to be normative in tone or phenomenological in emphasis. The former starts with sparse models and adds new variables only as needed. The latter looks at as many aspects of a problem as possible in a multivariate framework to classify families of variables or to reduce many variables to a smaller set of dimensions. My own work has oscillated between the two approaches. In all of this work, I have been guided by the three questions posed by Stanley Dodge, "Where? Why there? Why *just* there?"

I view attempts to answer these questions as involving *(a)* locating a phenomenon, *(b)* studying the macroprocesses of general localizing causes, and *(c)* following up the general concern with microprocesses that lead to specific locations. Answering "Where?" involves patterns in two dimensions (as a dot map), three dimensions (if an isopleth map), or four dimensions (if a time series of isopleth maps). Answering "Why there?" involves the great underlying processes of weather and climate, resource availability, levels of organization, technical skills of a population, and major secular or historical trends. Answers to "Why *just* there?" involve a fine-tuned concern with such variables as local factors (sometimes called uniquenesses) and historical accidents. The latter is almost a misnomer because there seems to be an inverse relation between the hours spent in archives and the number of such "accidents" left unexplained.

The very best regional monographs combine good distributional information, delineation of how macroprocess brings about macropattern, and how microprocess (intermittent and local "vetoes" of macroprocess) account for micropatterns. These works are both a joy to read and a frustration to imitate. Being a good regional geographer is an exceptionally difficult task and not the soft option that some deem it. To sum up: geography involves a curiosity about two-dimensional patterns, a commitment to find the processes bringing those patterns about, and the ability to explain convincingly how the patterns change through time. We are devotees of a discipline in two-space presented in time slices.

Let me explain an instance of my preference for a normative approach to a problem. In an informal multidisciplinary staff seminar on frontiers at Eugene, held in our homes once a week, I became annoyed at the arcane approaches to the problem of why Moscow had become of preeminent importance in medieval Russia. Sparked by an article William Garrison published on the interstate highway system in a regional science journal, I decided to explore graph theory as a possible approach. The first result of this exploration appeared in *The Professional Geographer* in 1965 and rapidly became well known in other fields, such as history and anthropology. Nicolas Rashevsky, in his *Looking at History Through Mathematics,* remarked: "Forrest R. Pitts (1965), using methods of the theory of graphs, studied how the problem of how many cities can be accessible by rivers from a given city and how many different *river pathways* there are between a given city and another one. He introduces the notion of *connectivity* which characterizes a city with respect to these two properties, and finds that in Russia, with the exception of Kolomna, Moscow ranks highest in connectivity. . . . It may perhaps be further generalized by considering not only the topological distance between two cities but also the actual metric distance. Such a combination of relational and metric approaches seems to be particularly promising."[1] A second and much more germane study of the same data, using real distances rather than incidence data, appeared in *Social Networks* in 1979.

When my former graduate school companion, David Kornhauser, retired from Hawaii, I was asked to teach his Japan course and seminars. My own Japan notes were two decades out of date, so I began a crash course in reading the latest materials, in particular those in archaeology. Once again I came up against a wall of accepted but outdated opinion about the early Japanese state (Yamato) and its unpolitical predecessors, the Jomon and Yayoi (periods named after pottery styles). It was a joy to learn of new discoveries and to pass along to my students accurate information and interpretations rather than tattered and worn outlooks.

An iconoclastic bent had shown in me early on when I submitted to the AAG meeting in 1959 an abstract on the myths of isolation and stagnation in Tokugawa Japan. The abstract was accepted, but I was not permitted to present the paper. Forty years later this little piece of stupidity still rankles.

1. Cambridge, Mass: MIT Press, 1968, 133–34.

Many times in my career I have been an editor, both of journals and mono-graphs, and of student theses and dissertations. Allied to editing is bibliogra-phy. I once compiled a 150-page list of articles and books on computer cartography, which was used by local and distant colleagues, and a bibliogra-phy on simulation methods occupied my spare time for several years. Being in a research institute, I had free rein to be more of a social scientist if I so chose. During the early days of the AIDS scare, I was a founding member of the Hawaii AIDS Task Force, which coordinated all the monthly meetings and other AIDS-related efforts in Hawaii. Interviewing people with the disease or those perhaps prone to be exposed is difficult. I developed a double-blind protocol for personal interviews, in which the interviewer did not know the precise question being asked or the specific answer. The National Institutes of Health were apprised of the technique, and I received a letter of thanks.

Though I have done only a moderate number of book reviews, I see re-viewing as a minor art and have always taken great care with them. The longest I ever wrote is a review of Hermann Lautensach's magisterial *Korea: A Geography Based on the Author's Travels and Literature*. I have kept a "Books Read" file since 1951. I averaged a book a week during my teaching and re-search years, and twice that in the past decade. But it is now time to speak of a more relaxed time of life.

### Santa Rose and Retirement

Retirement to the mainland meant that we could take car trips longer than forty miles from home and that we could ride trains. One of our first acts was to take Amtrak from Oakland to Grand Junction for a family reunion at Mesa (population in the 1930 census was ninety and appeared even less in 1989), then to Denver for another reunion.

What does one do in retirement? Reading one's old correspondence is an anodyne for boredom. By habit, I have kept a bound file of my typed letters since 1949, and whenever I made up handouts for my classes, these handouts were included as a sort of professional diary. What our three daughters will do with the great mass of these letters after my demise, I cannot predict.

Writing to retired colleagues is a pleasure. Teaching summer sessions at Oregon on occasion and regular semesters at Sonoma State University when asked has helped to augment our income. Just living a long life brings rewards, as were the invitations to read a paper ("The Spatial and Metaphoric Extent of Ryukyu") at a Harvard symposium on Okinawa in 1994, at the fiftieth anniversary in 1997 of the founding of the Center for Japanese Studies in Ann Arbor, and at a conference of "veteran Asianists." I was also asked to write on U.S. relations with Korea in the twenty-first century. This latter project was particularly interesting.

One of the founders of U.S. Steel endowed the eponymous Donner Foundation with funds for primary attention to Canada. Its staff once a year are allowed to plan a conference on any topic of their own choosing. In 1993, the foundation invited many aging Asian-area scholars to prepare papers on various areas of Asia concerning the Pacific century and the possible range of U.S. interests in Asia. I wrote one of the two papers on Korea. We met, in autumn 1994, at a posh hotel in Laguna Niguel, California. No one working at the hotel knew the translation of the name, and it was not until I got to a large Spanish dictionary that I learned it was "Chigger Cove." The proceedings were published in 1997.

While attending the AAG meeting in San Diego, I was interviewed for *Geographers on Film*. A few minutes later I interviewed a somewhat nervous Waldo Tobler, who had to be coaxed to appear before the camcorder of Maynard W. Dow and his wife.

Every person has a few regrets along the way. Mine include that I never took the time to study Hawaiian and that data from some of my projects were not published or were printed in shortened form. Over many years, I amassed some four thousand articles and chapters on diffusion, with the aim of writing a book tentatively titled *Cultural Diffusion: Models and Reality,* for which I had a publisher's contract. These materials line my living room. Bill Garrison encouraged me to develop my manuscript "A Dignity Revolution in America: The Regaining of Self-Respect, 1960–1985." Both remain undone, although I work on them intermittently.

The area in Santa Rosa where we live is Valle Vista III, a family-management association in a buried-utilities city subdivision, where each family owns its own large mobile coach (ours has 1,780 square feet) and the land beneath it. We are a self-governing group, with no resident manager or employees. The association owns in common a recreational vehicle yard, an office building with laundry facilities, a clubhouse, a swimming pool, and a large gardening area. As I write, I am serving my seventh year as secretary of Valle Vista III. I am happy to report that academic life prepared me well for the backbiting and snidery of a small minority of our association members. I feel right at home.

# Ma Vie

## Waldo Tobler

Courtesy of Waldo Tobler

*Waldo Tobler (b. 1930) was photographed during a field trip to China conducted by 优美的球状和谐 (Beautiful Spherical Harmonics). Professor Tobler enjoys dual American-Swiss citizenship but at eighteen chose to enlist in the U.S. army in Europe for a tour extended to nearly four years by the Korean War. During this time, he served in intelligence in Austria and learned Russian to add to his interpreter skills in German and French. Returning to the United States in 1952, he enrolled initially in the University of British Columbia, but received all three degrees (Ph.D., 1961) in geography from the University of Washington. After teaching at the University of Michigan for sixteen years, he joined the newly formed Department of Geography at the University of California, Santa Barbara (UCSB), where he is now professor emeritus. For many years, he has been a member of the National Academy of Sciences and has served on several committees of the National Research Council. He has received many awards for his research, and in 1988 the University of Zurich awarded him its doctorate* honoris causa. *His imaginative and pioneering research has made him the major analytic cartographer of the twentieth century, not simply for the ingenuity he has displayed in generating map transformations and for summarizing large quantities of spatial information, but for his concern to provide graphic expression for these analytic procedures. His work on transformations, spectral analysis, animated simulation, spatial prediction and interaction, bidimensional regression, and pycnophylactic reallocation have opened up a series of methodological ap-*

293

*proaches to his fellow geographers. He is presently engaged at the global level in building a latitude- and longitude-oriented demographic information base two orders of magnitude better than currently available.*

## Growing up in America and Europe

I was born in Portland, Oregon, thus making me an American citizen, of a Swiss consular employee, thus making me a Swiss citizen. These aleatoric facts have, to some extent, shaped my life. My paternal grandfather was a farmer from Appenzell in northeastern Switzerland and quite short, as most Appenzellers of that generation were. I first recall meeting him years later when he had left farming to manage an apartment house in St. Gall. I do not recall meeting my maternal grandfather but may have done so on a trip to Switzerland at age one. He was a Swiss chemical engineer who worked in pre-revolutionary Russia, escaping back to Switzerland with a suitcase full of worthless rubles. My mother recounts how he used the family bathtub for chemical experiments that eventually led to a small factory in Switzerland. I never knew either of my grandmothers.

Through my grade school years, my immediate family lived in Seattle, where my father was the Swiss consul for Alaska, Idaho, Oregon, and Washington. We lived at the top of Queen Anne Hill, and the school, to which I walked daily, was at the bottom. I was back there last year, and it is still a big hill! Pleasant memories persist of summer camps in the San Juan Islands—among other things sliding down the sandy cliffs. In one camp contest, I won a prize using an inexpensive box camera with no controls except point and click, while at another I received a fifteen-inch wooden sculpted arrowhead with various kinds of affixed medallions obtained from some forgotten boyhood camp achievement. It hung on the wall in my room for years. Returning home at age seven from one camp, I found I now had another sister. We have an old home movie from about that time of me dressed in native Appenzeller attire—white shirt, red vest, leather suspenders with brass cows, yellow leggings, high white stockings, black shoes, fancy leather skull cap, and a pipe with a convex metal cover so that it could be used upside down or in windy weather. I also remember picnicking at Mount Rainier and outings that included Swiss American dairymen from farms in the vicinity of Seattle.

The attack on Pearl Harbor in 1941 found me in a softball game on that Sunday afternoon, and I recall all of the adults becoming very serious. With the outbreak of World War II, father was transferred to Washington, D.C., where we stayed until the spring of 1945. We took the trip by train, stopping for a sunny month-long vacation in southern California, then on to Chicago via the Santa Fe train arriving in a blizzard. From there, we went on to the capital, where we stayed briefly in a hotel across from the White House and then moved to a house on the outskirts of Washington, D.C. Junior high school in Chevy Chase included participation on a baseball team, a course in Latin *(Brittania insula est,* the first sentence in the text, is about all I remember), and a paper route. Father's duties included visiting prisoner-of-war camps in the southwestern part of the country. Switzerland reported to the Red Cross on the condition of German POWs in these camps, and my father usually came home annoyed that their main complaint was that they did not like having to eat white bread. He contrasted their treatment with the German atrocities in countries that they occupied.

On the first civilian transport ship after the cessation of hostilities in Europe, we steamed, in May 1945, from Boston to France, embarking via landing craft to the ruins of Le Havre. The ability of little children to speak French amazed me—I thought they must really be smart. We went by train from Le Havre to Paris and then on to Switzerland. After a short vacation in the Swiss Alps (it snowed in July) and some hiking in the mountains around St. Moritz, I was sent off to a boarding school south of Zurich to learn German. The school was not very far from Zurich so I could visit the family on occasional weekends. I enjoyed the train ride and felt quite grown up to be traveling alone. Dad was stationed in Budapest, but my sisters stayed in Zurich with Mom. One sister eventually became a stewardess for Pan American, the other a doctor, and both later married doctors. On one visit to the big city, I saw *Das Land des Lächelns,* a Lehar operetta, my first, in the Zurich Opera house and was enthralled. I still love the music, although at that time I did not realize that it is a sad story.

At school, the headmaster's daughter and I had exactly the same birthday, and that made us good friends. Naturally everyone at the school, except the other foreign students and myself, knew how to ski, but we all learned on the normal Swiss school outings. My most memorable incident was becoming a minor hero by knocking out one of the local bullies at the school in a teenage fight. Of course, I did it quite by accident with a lucky blow. I also played soccer (not very well) and took a summer bicycle trip through Switzerland, which included sneaking into large estates after dark in order to sleep on the grounds at the edge of Lake Geneva. A school friend from Chexbres, near Lausanne, and I spent a warm summer afternoon in the local wine cellars sampling the new grape juice, not yet quite fermented. I also somehow got copies of Upton Sinclair's Lanny Budd books and devoured them.

After a year at the boarding school, I moved to a public high school, but boarded with a Swiss family near St. Gall. The rest of the family was then stationed in Casablanca. My English class (required of all students) allowed me to quibble with the teacher because I was a "native expert." I recall taking classes in chemistry (the Bohr atom is the extent of my current knowledge of the subject), having fun in algebra, and being bored in French, although I seem to have learned a bit. The class put on a German-language production of the *Mikado,* so English, German, and French were the languages I grew up with at this time. The whole school participated in community work, which included pushing wheelbarrows and other menial tasks to help to build a village for orphans. The Pestalozzi Dorf is still there, now much expanded. I made many friends, including the eight of us in the boarding house.

In the winter, we went sledding down the main highway (rather risky, but everyone did it). This was the time of the year when the Gasthaus next door cooled some of its liquor supply between the outer and inner windows (which could be opened from both sides), which is how I developed an early aversion to gin. Across the street was a butcher shop that made sausages from pigs, which one sometimes heard squealing. Amusements included bicycling to a small, extremely isolated neighboring village to see some of the consequences of inbreeding, racing by bicycle down the very long, steep hill to the Rhine River, or biking over to gawk at a girls' finishing school on a nearby estate. The little town of Trogen is connected to St. Gall by a rickety rail line that takes about ninety minutes, with tracks alongside the main road and through the lovely hills of Appenzell. Goiters were still common in older people, something I had never seen in the United States. I spent one summer break at the École de Commerce at Neuchâtel, having been sent there to learn more French. My afternoons were free, and from former GIs living on the GI Bill (at a time when you got five Swiss francs for a dollar) I learned to appreciate "une blonde" at the local outdoor café. Another possibility was swimming at the lovely Neuchâtel plage.

After contracting dysentery in Morocco, my father was transferred back to the home office of the Swiss Foreign Service, and I went to live with my parents in their apartment in Bern. There I attended the local gymnasium just across the river from the capitol buildings. My only recollection of this time is going through the state-required track-and-field exercises, all recorded in a little government-supplied booklet. From Bern, a short hike up a hill on overcast days enabled one to climb above the clouds for a gorgeous panorama of the Bernese Alps. The American embassy gave Fourth of July parties for resident compatriots in rooms filled with nut dishes and free whiskey, a potent combination for a youngster like me. A few summer weeks as a worker in a machine shop and on an outdoor construction site in Bern brought my stay in Switzerland to an end.

## Army Service

When I turned eighteen, the Swiss army was eager to draft me into their compulsory service. To avoid the draft, I went to France to enlist in the American army in Europe. The process took a month, allowing me to explore Paris, where I borrowed an apartment near Pigalle from an American embassy employee on temporary leave in Africa. One lunchtime a place to eat caught my eye, and so I wandered into Maxim's. I was pretty naïve and did not recognize it, but it soon became obvious that it was a rather fancy place, so I ordered pheasant under glass from the menu. At the next table sat an American banker and his lady, and we had a pleasant chat. I recall this experience every time music from Lehar's *Merry Widow* hits my ears. Paris is still one of my favorite places to visit.

The U.S. army had somehow assembled an entire company of assorted Euro-Americans, and for six months it trained us to march, to shoot, and to hide in small gullies near Marburg, Germany. The Browning automatic rifle was the fun weapon. Those were the days when a carton of cigarettes or a chocolate bar would yield quite a bit of local currency. After basic training, because I was fairly fluent in German, I was shipped off to an intelligence unit in

Austria, first to Vienna, then to Salzburg. Our main task was to interview former Austrian prisoners of war returning from the Soviet Union about their
Russian employment experience. We used German maps and Wehrmacht aerial photographs of the factories in which they had worked to elicit details
about their workplaces. It was little more than industrial espionage. Because
we had no actual authority to order them to attend our interrogations, our in-
ducement to come for an interview was an invitational letter with many offi-
cial-looking imprints and imperious wording. It worked!

Our unit wore civilian clothes, and we lived away from the barracks, at the
Pitter Hotel near the Salzburg railroad station. The office was located on
Mozart Platz, where bells played *Eine Kleine Nachtmusik* every evening at
five. For a while, our unit's activity was also carried on in St. Johann-im-
Pongau, near Zell-am-See, some kilometers south of Salzburg. The army had
requisitioned a villa and used it as a residence and office. I learned to drive in a
jeep on the narrow mountain road between Salzburg and St. Johann, my
teacher being an American civilian named Alfred Schwarz employed by our intelligence unit. There were a number of these civilians, all recently discharged
from the military, but of European ancestry and competent in the German
language. They now did the same work as before but at higher pay. Rest and
recreation were provided for U.S. troops in nearby Berchtesgaden and
Königssee. German army jeeps, subsequently marketed in the United States as
"The Thing," had a speed advantage over U.S. jeeps of about one mile per
hour—as we learned by racing on the autobahn between Salzburg and
Berchtesgaden.

At that time, Vienna, like Berlin, was occupied by American, English,
French, and Russian troops, and the command rotated monthly between
these four powers. The army had sent me to Oberammergau for six months to
learn Russian, and thus I got the job of interpreter for the American general's
train (although not for the general). Once a month he went from Salzburg to
Vienna and back for the change in command. I went along to do the Russian
paperwork for his train, reporting who was on board, the purpose of the trip,
and so on to the Russians at their checkpoint on the Danube at Linz. For this
job, I wore my army uniform. The checkpoint was surrounded by armed
Russian soldiers and manned by a sergeant. Photographing Soviet troops was
not allowed, but this did not deter some of the MPs on the general's train. I

did not mind anything except that I was out there surrounded by the Soviet soldiers. On one occasion, when we stopped at the border, the Soviets were inspecting an Austrian train full of passengers, and I had to hurry between the two trains to the guard post. Two Soviet soldiers got off of the passenger train behind me with their Kalishnikovs and saw me running away from them. That did scare me, but nothing happened.

The general traveled on a train with dining and sleeping cars, a radio car, and a car to transport his automobile. There was an additional railroad car for some aides, a few MPs, and myself. During the three-day changeover cere-mony in Vienna, I was free to do as I wished but slept on the train. Usually I went to the opera, which shows my long-standing enjoyment of this form of theater. On another occasion, I was assigned, with an American officer and several Soviets, to return a Russian deserter-criminal by car from Salzburg to Vienna. As a matter of courtesy, we exchanged cigarettes, but I do not recom-mend driving after smoking a Russian cigarette—they are worse than Gauloise. The Korean War extended my army tour, but I was eventually dis-charged in Seattle, never really having seen any combat in three years, seven months, and twenty-two days of military service. The return trip passed through Pisa, where the Leaning Tower was visible from the train, then by ship through the Pillars of Hercules, with Europe on one side and Africa on the other. After seven days on the Atlantic Ocean, with dolphins accompany-ing the troopship, we arrived in New Jersey, and I went on to New York and then Seattle.

## University Life

Now out of the army, I spent an enjoyable summer in Seattle carrying mail for the post office in the mornings, watching hydrofoil racing on Lake Washing-ton in the afternoons, and taverning with longtime friends in the evenings. The Red Onion tavern is still there, but our usual haunt has been turned into a small shoe store. At the time, my father was Swiss consul in Vancouver, so I moved there, lived with my parents, and entered the University of British Co-lumbia. An army high school equivalency test got me admitted without hav-ing ever graduated from a high school. The Varsity Outdoor Club at that time had a cabin on Mount Seymour, so skiing was only an hour away. Without quite knowing why, except perhaps that we had traveled a bit and I enjoyed reading maps (I still have an Imhof school atlas from Switzerland), I decided to major in geography. I pestered Ross Mackay into letting me attend his advanced cartography class, which he did, albeit reluctantly. Later he offered me the position of summer camp cook for one of his Arctic expeditions. To my complaint that I did not know how to cook, he responded, "Anyone can cook better than an Eskimo!"

I did not take the chef's position but instead went to work as a mucker half a mile underground in a gold mine in northern British Columbia. The only way in was by seaplane, but the pay and food were good. It was a mine site with nothing but barracks, a few houses for married couples, a dining hall, and a reducing mill, all in an isolated wilderness. Animal tracks, reputedly of grizzly bears, could be seen in the mud a few hundred yards from the barracks. The cage (shaft elevator) held six people on its three-thousand-foot descent. The working face was farther into the mine, reached by wagons pulled by a small trainlike conveyance. The mine operated three daily shifts, extracting quartz containing about one-half ounce of gold per ton. One could occasionally see tiny flakes of the gold in the quartz veins. The underground miners drilled into the vein and set off explosive charges that created a rubble of debris. Muckers shoveled this debris into small inclined shafts that emptied into large carts at a lower level. The carts were then transported out of the mine on the narrow gauge tracks and up to the reducing mill. I believe cyanide was used to extract the gold.

It was quite hot at the bottom of the mine, and I was eventually assigned to help a carpenter seal off old tunnels because they stole air from the ventilation system, which consisted of huge fans blowing air down a thirty-foot-diameter shaft to cool the lower-level tunnels where the miners were currently working. Because the old tunnels, dug some twenty-five to fifty years earlier, also intersected the air shaft, they diverted the cool air from its intended target. The solution was to block them off, which the carpenter did by building wall-to-wall wooden forms to close them. The frames were doubled so that concrete could be poured inside to block the tunnel from the air shaft. These sealing walls were built about six feet from the edge of the three-thousand-foot deep air shaft, a really awesome hole. One difficulty was that parts of some of the old tunnels had caved in, and the only access was via the steep narrow shafts formerly used to dump the debris into the ore carts. So a part of my job was to move fifty-pound sacks of cement or sand to the narrow shafts, crawl into these two-foot-high shafts with a sack on my shoulder, and then go down the eighty-degree incline on a narrow wooden ladder to get to the next lower level. Once there, I had to move the cement bags, sand, and wood to the air shaft so that the wall could be built. The only illumination came from a small helmet-mounted lamp powered by a battery worn around my waist. In the old tunnels, the timbering holding up the ceiling was rotted and covered with mold or fungus. Not the best kind of summer job for a college student.

The carpenter, however, was a nice older man, a refugee from eastern Germany who had emigrated to Canada with his entire family. He was a master carpenter and taught me many useful woodworking tricks. Three of his sons, he told me, currently also worked underground in the mine. In Germany, one had been a surgeon, another a concert pianist, while the youngest had no profession. An American company operated the mine, and the foreman once complained to me that he had never had such troubles with indigenous labor in Arabia or Indonesia as he had here in Canada—referring, of course, to the Canadian miners. Had I a talent like that of Van Gogh I would have sketched some of the miners. One really looked like a gnome from a Disney movie.

My reading for the summer was intended to be Dante's *Inferno* in the original Italian because I assumed that it would be much lovelier than the translation I had read in class. I never got around to it, although I did get to some of Byron's works that I had also brought along. Frustratingly, cockroaches ate all the glue off the sheet of stamps I had brought for correspondence, and the sprockets in my camera kept tearing the film, so I have no written record or pictures of the mine site, the miners, or the lovely northern British Columbia terrain. I read later that the mine was closed after the price of gold dropped.

Had I taken Mackay's Arctic trip I probably would have become a physical geographer. When at Christmas I took a bus trip east as far as Omaha to visit a former army friend, I had just finished Mackay's physical geography course. As a result, I took great delight in observing the countryside with greater understanding, especially at The Dalles. Then it was back to the university, where the second-year English teacher, an elderly lady, let me write any kind of term paper I liked instead of those assigned to the others when she learned that I was a veteran. I had had enough of organizations so refused to join a fraternity.

The GI Bill funded my undergraduate work, and for the third and last year I moved to the University of Washington in Seattle, where I was able to afford a tiny basement room near the university. Apparently the procedure used to clean the apartment before a new occupant moved in was simply to apply a new coat of paint. Clearly there were many layers. I also purchased my first car, a two-door 1941 Chevy. By then, skiing was available at Snoqualmie Pass and I got away on a few occasions. I had a part-time job parking cars in a garage in downtown Seattle, and the next summer I was a fire lookout in Seattle's Cedar River watershed. Fire watchers spent ten lonely days in the high tower, ten days at the base camp clearing brush or constructing roads, and three days off—usually spent in Seattle. The view from the top of the lookout tower was spectacular, but scary during thunderstorms because we had to spot and report by radio the estimated coordinates of each lightning strike. One of the interesting features in Seattle's watershed is a fairly large cement dam that never held water. Apparently it was constructed on a bed of porous sandstone.

## Graduate School

After graduation from the University of Washington with a bachelor's degree in geography, I was asked to stay on as the resident cartographic assistant, and it was this offer that allowed me to enter graduate school. Three friends and I rented a delightful old house on a pier over Lake Washington, at the foot of Madison Street. We could dive right from the porch into the water. This quaint pier has now been turned into a huge high-rise apartment building. Each week one of us was in charge of cooking, and the nuclear physics student was easily the worst. When it was his turn, he made up a huge pot of pinto beans, and that was our food for the week!

At the university, John Sherman became my adviser, which worked well for me because he was very patient and conscientious. My nominal field of study was cartography, and many of my publications include specially devised illustrations. Scientists, as a look at the journal *Science* will confirm, freely use graphs and pictures. This practice can be contrasted with the verbal and logical presentation found in many books on mathematics or philosophy, books that tend to denigrate graphics and diagrams as inappropriate and misleading in the search for clear thinking. A recent mathematical exception is Tristan

Needham's *Visual Complex Analysis.* I passed the foreign-language examinations in German, French, and Russian, the latter somewhat to my surprise because I did not think that I knew it that well. Occasionally the Russian has been useful, for example to look things up in the *Referatny Zhurnal* or to skim Soviet books on geography, cartography, or map projections.

Professor William Garrison was an interesting teacher. The way he taught was to get his students angry, at least that is how I reacted. I would not let the so-and-so confuse me with what seemed to me to be exotic terminology, so I went out and learned the material. This teaching method is quite contrary to most theories of education, but challenges and puzzles have always drawn me. He coerced several of us into taking Arnold Zellner's course on econometrics, using the newly published textbook by now Nobel laureate Larry Klein. I don't know what Zellner thought of having a bunch of geography graduate students in his class. Possibly he was happy because it was a very small class, and we made up more than half of the group. Anyway, he was very kind and gave us all passing grades.

Duane Marble was the best at using the computer, then located in the attic of the chemistry building and available for geography students to operate (literally, we ran it) at two in the morning. The programming language was Symbolic Assembly Programming (SAP), later SOAP (adding *Optimized*), all before FORTRAN, and with your best coding you could get two pieces of data on one revolution of the IBM 650's "huge" two-thousand-byte rotating storage drum. I found programming and mathematics pretty simple and took naturally to them. Differential geometry, offered by Carl Allendoerfer, then chairman of mathematics, has been invaluable for my work on map projections. This course work also enabled me to read the literature of relativity theory and to learn about various mathematical notions of space, crucial knowledge for a geographer. Allendoerfer later served on my doctoral dissertation committee.

After two years, I left the University of Washington with a master's degree in geography, having completed a thesis on hypsometric colors on maps. Locally, as I tried to find a job, I was surprised to learn that I was "overqualified and would be bored" with any work available. In retrospect, these assessments were probably correct. Fortunately, both Rand McNally in Chicago and the Systems Development Corporation (SDC, then splitting off from RAND) in

Santa Monica were interested, the latter possibly because Professor Garrison had encouraged us all to take mathematics courses and to learn how to use a digital computer in addition to the mechanical desk calculator. On a desk calculator, a regression involving only four variables took about a week's work to invert the matrix, so using a computer was obviously advantageous.

The SDC became my choice, and the group I was associated with assisted in the preparation of computer-printed maps used in air defense system simulation exercises. I stayed there two years and at that time also got married to my first wife, Dorothy, whom I had met in Seattle. One spring the Association of American Geographers (AAG) held its convention in Santa Monica, and I was able to arrange a tour for the geographers of the unclassified North American Air Defense System's immense computer facilities. This must have been my first attendance at an AAG meeting. At SDC, Dick Kao gave some internal lectures on map projections using wedge operators, but unfortunately these lectures have never been published.

Some travel to SDC system sites was also involved in this job. On a company trip to Tacoma, I applied for and subsequently obtained a position with the Pierce County planning commission. Dorothy and I moved to the pleasant village of Steilacom, and from there I commuted to Tacoma. I was well qualified for the work, which consisted mostly of using colored pencils to shade areas on large land-use maps. After another two years, a multiyear National Science Foundation (NSF) fellowship enabled me to return to the university to attempt a doctorate. I had interviewed at the University of California, Los Angeles, which offered the possibility of obtaining a foreign area fellowship, then spending a year in South America, then another year writing up the results. *Et voilá*, a Ph.D.! According to the UCLA geography chairman at that time, this was the only acceptable path.

In order to save a year of study, I returned to the University of Washington. I knew what my thesis topic was going to be before returning to school, having been intrigued by a master's thesis by Carlos Hagen—a fellow student at Washington and later the curator of maps at UCLA. Thus, I finished my doctoral work in a fairly short time, having been influenced particularly by John Sherman and William Garrison. Torsten Hägerstrand was a visitor at this time, but I did not attend his lectures, much to my later regret. Another visitor was Ross Mackay from the University of British Columbia, who compared the methodology of statistics with that of cartography. Chairman Donald Hudson was good at bringing such speakers in as visitors. He had previously been with the Tennessee Valley Authority as an administrator, and because of this experience he undoubtedly made a better chairman than do most academics. He would frighten the graduate students by loudly yelling at them down the hall: "Tobler come here!" But he also greatly encouraged all of us. He himself wrote all of the brochures advertising the department, and he made sure to emphasize the amount of financial assistance available from the department.

Financial aid is always of great concern to students, and Hudson would list all of the graduate students by name, together with the type and amount of financial aid they received. He once told me that he got his first academic position by immediately applying for the post vacated by the person who actually got the position that he, Hudson, had also applied for. I enjoyed Ed Ullman's urban field course, although he was not a very conscientious teacher. He was often late for class and came in smoking his cigar. He would put a map of U.S. railroads up on the wall and talk for hours using the map only as an aide-mémoir; most of what he talked about was not actually on the map. He was interesting to listen to, throwing out many ideas, and occasionally he would even talk about urban geography—the nominal subject of the course. It was a summer field course, normally held in the San Juan Islands, but on this occasion it was held in Seattle. George Kauders, a student from New York, now a consultant in Frankfort, objected to our being charged field course rates when we were in town, and he got us all a sizable refund from the administration. My term project was to visit all of the parking lots in downtown Seattle and record the hourly parking rates, presumably to investigate distance decay of some sort around the central business district. Ullman also had an Office of Naval Research project, and several of us worked on making the maps for this study, which eventually came out as a book.

Donald Hudson had the sense to put all of the graduate students at desks together in a big room, and as result of our interactions I learned just as much or more from fellow graduate students as from the faculty. Students at that time, the late 1950s, included Howard Albano, Bob Alexander, Brian Berry, Bill Bunge, John Campbell, Michael Dacey, Art Getis, John Kolars, Julian Minghi, John Nystuen, Duane Marble, Richard Morrill, Richard Preston, Bill Siddell, and others. The night janitor was a former member of the International Workers of the World (the IWW, also known as "the Wobblies") with amazing tales from the 1920s and 1930s. Some labor picketing was currently going on in Seattle, and doctoral candidate Bill Bunge bought himself an expensive suit and tie so that he could dress up to go on the picket lines.

In the late 1950s at this Washington geography department, the academic concern was for rigor and models rather than for numbers as such. This approach included interest in nonnumerical mathematics, scientific method, and the development of theories—theory being the most compact form of data compression yet invented. Verified theory is really the most practical thing there is—not useless, as some seem to think. This period is often called the "quantitative revolution," somewhat mistakenly, but it was a most exciting time. Walter Isard's book *Methods of Regional Science* was published, as was Bill Warntz's work on macrogeography. From both, we learned that an important aspect of the geography of Seattle was its situation with respect to New York City, something that was not readily apparent in the field at any particular site in the Seattle area. This approach was in contrast to the then-current notion that in situ field studies were the only way to do geography. Peter Haggett's book *Locational Analysis in Human Geography* had marvelous maps on which cities were identified by names such as $\alpha$, $\beta$, $\gamma$, $\delta$, and so on, so that one could easily conceive of an abstract geographical study. Ullman and Chauncy Harris had just brought Walter Christaller's *Central Place Theory* to the attention of American geographers, and I also read August Lösch's wonderful *Economics of Location*. I still refer back to this book from time to time and recommend it to students. Tord Palander's book *Beiträge zur Standortstheorie* was also available. The Regional Science Association was established at about this time, and I later became one of its counselors, only to learn that it was a paper organization with no legal status. To go further back into the history of the geography department at the University of Washington, before my time, one can now turn to a web page: http://www.weber.u.washington.edu/~geogdept/history.

My Ph.D. thesis, "Map Transformations of Geographic Space," concerned some problems introduced by modern technology so that maps scaled in time or dollars, rather than in kilometers, could be constructed. In the thesis, I also spelled out for the first time the partial differential equations that govern the class of map projections known as anamorphoses (also known as cartograms) and described how they could be used for analytical purposes. Similar solutions of technical cartographic problems have been typical of much of my later work. Dorothy's and my first child, Eric, was born the day after I turned in the dissertation, which was incredibly good timing on my wife's part.

## Michigan Days

Thesis done, John Nystuen recommended me to the chairman of geography at Michigan—those were the days of immense hirings (cohort counts!) and before affirmative action—and Charles Davis did hire me. I felt good about my decision to go to Michigan, especially after being told by the chair at Wisconsin that new faculty would be expected to teach every course in the department at least once so that they would become well-rounded geographers and being informed at another university that the cartography professor would be expected to do the maps for the other faculty's publications. After joining the faculty at Michigan, I spent a summer session working with Bill Garrison in the Transportation Center at Northwestern University in Evanston on the modeling of geographic transportation relations. Emilio Casetti was there finishing up his dissertation, and I also worked with Dave Boyce, recently graduated from the University of Pennsylvania in regional science. Charles Davis had obtained funding for cartography students from the National Defense Education Act, one of the many impacts of Sputnik, and Phil Muehrcke, now a professor at Wisconsin, attended Michigan as a student under this program.

Professional meetings tend to be a mixed blessing but do provide opportunities for travel and to meet colleagues. At a meeting in San Francisco, Ed Horwood from the University of Washington established the Urban and Regional Information Systems Association (URISA). Bob Alexander, Jack Dangermond, Andrei Rogers, Art Getis, and I, as well as others, became founding members and presented papers at that first meeting. In spring 1968, the AAG hurriedly transferred their annual meeting from Chicago to Ann Arbor to protest against the riots held during the Chicago Democratic Party Convention. We all had to work on this relocation, and my task was to rent space for commercial and academic displays. Despite the last-minute arrangements, the meeting seemed to go well.

The National Research Council (NRC) ran several committees on which I was invited to serve, with meetings in Washington, D.C., or at the Advanced Institute for Behavioral Studies at Stanford. The Mathematical Social Sciences Board, which I chaired for a short time, put me in contact with many outstanding scholars. The High School Geography Project, directed by Gilbert White, convened a meeting in Boulder, and I took the opportunity after the meeting to explore a bit of western Colorado. A conference in Seattle also enabled me to explore the northern portion of the Olympic Peninsula. One early meeting on the application of satellites to cartography was held at the National Academy of Sciences (NAS) Woods Hole estate, and our family rented a Cape Cod cottage at Falmouth for a pleasant month. Some summers Professor Rhoads Murphey invited me to help crew his sailboat off the east coast from Cape May to Monhegan and back. The east coast of the United States, a submerging coastline, is much more interesting than the west coast, an emerging shore with more cliffs than harbors. Not only is the physiography more interesting, but the country was settled from the Atlantic, and from the sea one gets a completely different impression from that which one gets by driving on land. During another summer, I attended a Gordon Conference on Theoretical Biology in the hills (they call them mountains) of New Hampshire, as the lone geographer in with the mathematical biologists. Later, back at the University of Michigan, student sit-ins and strikes occurred but evolved more gracefully than elsewhere, with teach-ins and nothing more threatening than seemingly endless discussions and meetings.

A notable event was the organization of the Michigan Inter-University Community of Mathematical Geographers (MICMOG), which involved three Michigan universities—the University of Michigan at Ann Arbor, Wayne State University in Detroit, and Michigan State University in East Lansing. From 1962, the seminars met once a month or so in the back room of a tavern in Brighton, just off a freeway intersection approximately equidistant from the three schools. Faculty and students from each of the three institutions attended these sessions, which often also included faculty visiting one of the universities. Among them were Stig Nordbeck from Sweden and Peter Gould from Penn State. Informal discussion papers had been a tradition at the University of Washington—the chapters in Bunge's Ph.D. thesis "Theoretical Geography" had come out as such papers—and I did several on map projections. Thus, a series of MICMOG discussion papers was begun, funded in part by each of the university geography departments—usually off budget. John Nystuen agreed to serve as editor. The magnifying-glass map projection on the cover, used as a distinctive signature of the series, was mine. These papers were distributed widely, and the general feeling was that the series as a whole would have a cumulative effect and that the wide distribution would enhance the reputation of the group and the sponsoring universities, as well as invite scholarly interchanges. In fact, Gunnar Olsson once told me that his decision to come to Michigan was prompted by this discussion paper series. Indeed, so well known did the MICMOG papers become that the negotiations with Leslie King regarding the founding of the journal *Geographical Analysis* at Ohio State University Press involved halting further publication of the MIC-MOG discussion papers. At the time, we had difficulty getting articles published in the traditional journals, such as *The Professional Geographer* or the *Annals of the Association of American Geographers,* because, we were told, their typesetters did not have any mathematical symbols! Professor Joseph Spencer from UCLA, the editor of the *Annals* at that time, was definitely unsympathetic to any submissions from "quantitative" geographers. This situation was another reason for starting a new journal. It was about this time that John Kolars and John Nystuen also started the monograph series *Michigan Publications in Geography.* Such series are common in many European universities, but had been started at only a few in the United States. I had two monographs published in the series, one a translation of Lambert's 1772 classic on map projections and another filled with listings of computer programs. The latter proved to be particularly popular overseas.

While at Michigan I developed my analytical (as opposed to visualization) cartography course to include methods geographers used in research involving information contained on maps. I also made available a number of computer programs, two of which included the first computer algorithms for the finite difference calculation of cartograms. I presented material on this subject at a New York Academy of Sciences conference, to Howard Fisher's computer graphics group at Harvard, and at a National Institutes of Health conference. Other programs I developed were concerned with "winds of migration." In Nottingham, England, my presentation at the NATO Advanced Studies Institute was on the use of computers for the spatial filtering-smoothing of census data. The concept of geographic information systems (GIS) had not yet been invented, but many of the functions detailed in my programs were later included and sometimes copied in these systems. At Michigan, the university computer facilities were very advanced for their time, and interactive computing became possible at an early period. The Michigan Algorithmic Decoder (MAD) language—an Algol derivative—perhaps did not have quite as dramatic an impact as BASIC developed at Dartmouth, but it was almost as easy to use.

Those were the days when faculty had to write their own computer programs, and we did a great deal of it. Ken Knowlton at Bell Labs had written some routines for computer movie making, and with several others I tried my hand at this in the early 1970s. The interaction with faculty, including computer graphics engineers Richard Phillips and Bertram Herzog, was of great help when I produced the computer movie simulating the population growth of the city of Detroit, calibrated using data that Donald Deskins had collected. The paper describing this movie contains the oft-cited "first law of geography"—that near things are more similar than things far apart. One undergraduate student collected the date at which streets in the city of Ann Arbor were established and was able to produce time series maps of the city's growth on a computer screen. Another program simulated continental drift, and Hal Moellering produced his movie of automobile-caused deaths in Washtenaw County using the screen's phosphor decay to mimic the magnitude of the event. A colleague, Henry Pollak from geophysics, later presented his computer simulation of the down-cutting of the Grand Canyon at a computer conference in Lawrence, Kansas. When he started the movie, I was able to surprise him and the audience by playing a recording of Ferde Grofe's *Grand Canyon Suite* in the background as Henry's computer movie played on the screen. On the trip to Kansas for that meeting, I had stopped at a small grocery at an isolated road intersection on an extremely hot day. Upon entering, I noticed a woman with a small child, in addition to the clerk. When I asked to purchase a beer, the woman did not actually shriek, but she looked like she wanted to, and, grabbing the child by the hand, she ran out of the store. I had hit a dry county and had to settle for a soft drink, most of which I detest.

Ann Arbor was also one of the sites where the engineering department established an early form of the Internet, initially for interaction between the three Michigan universities. Being at a good university really makes a big difference, with outstanding colleagues not only within your own department but throughout the university, and with good library, computing, and support facilities and infrastructure. Especially important is an understanding of the scholarly endeavors, attitudes, and needs on the part of the university administrators. I was able to interact with faculty in the geology, archaeology, and statistics departments, and to a lesser extent with psychology and botany. On one occasion, observing a similarity in the pattern of major veins in a maple leaf and the arterial roads leading out from the center of Detroit, I went to the botany department to ask how one might use a leaf to simulate automobile travel-time isochrones. My idea was that one could feed a plant an alternating variety of minerals to get patterned leaf coloration. I ended up giving a lecture to the botany group on leaves—how the vein intersections on the main stem of a leaf resemble what one might expect from Snell's Law. For example, the angle at which the secondary veins intersect the main veins decreases from the stem to the tip as the main veins decrease in size. The overall shape of leaves is related to this phenomenon. The smallest veins visible to the eye also resemble the street pattern in modern subdivisions. The botany professors recommended water-borne radioactive isotopes to do the simulation, but I have not tried it yet.

Also in the early 1970s, the opportunity arose to teach a spring term at the University of Minnesota in Minneapolis. It turned out to be a pleasant experience, even though it required commuting by plane for a while because the sessions at Ann Arbor had not yet finished. Jim Johnson, a geographer from England, and I rented the house of a computer science professor who was on leave in Israel. The geography faculty at Minnesota were a congenial group, then in new quarters, and springtime is the best time to be in Minnesota. Another year I presented German-language lectures during the summer semester at the University of Zurich, then under the direction of Hans Bösch. Needless to say, my German improved very quickly! I painfully had to write out each lecture in full in advance. I do not remember what I taught but probably still have the handwritten notes. Kurt Brassel, recently chairman at Zurich, was one of the students in my class, working on his development of cartographic hill shading, following the method of Eduard Imhof but using a computer and a line printer with a special print chain. I was not much help but certainly encouraged him. I also visited Imhof at his house outside Zurich because I have always made it a practice to try to meet people whose work I admire when I have the chance. I also got to know Osborne Maitland Miller of the American Geographical Society in this manner during a trip to New York City. Before our return to Michigan from Switzerland, Dorothy and I spent a few days in Malta and London, and of course took the hydrofoil down the Thames to Greenwich. There we all straddled the two hemispheres divided by the Prime Meridian.

An interesting process occurred during that summer in Europe and probably repeats every summer throughout the world. At one point, strawberries from Italy come on the market in Zurich (I think the price then was about five Swiss francs for two kilos), but after about a week the price slowly starts dropping. Suddenly it jumps up again. The reason is that the Swiss strawberries come to market as a subsidized crop, and the Italian berries are no longer allowed into the country. These latest prices again slowly decline. Then suddenly the German strawberries appear, and there is another price change. Now suppose that instead of just looking at the Zurich market, we could collect the berry prices in every one of the European cities every day. We could now draw daily isotims (lines of equal price) across the map of Europe and then link them up to make a computer movie. Thus, we could watch the economic "weather" as it moves across Europe. I was not able to complete this project despite collecting some of data for it, but I am convinced that August Lösch would have appreciated it. In a similar project, John Nystuen and I at Michigan funded an undergraduate student (with a rented movie camera and a bit of film), who made a movie of the geography of tomato production in the western United States. He used data from Nystuen's master's thesis and represented tomato quantities by proportionally sized stacks of red Lifesavers at the county locations week by week. For the frost line, he used a piece of white string positioned according to the appropriate weather reports. He placed all of this on a colorful Wenschow relief map of the United States, which he photographed from above on a stepladder. In the movie, those stacks of tomatoes chase the frost line north in the spring and flee south from it in the fall! This is the sort of student enterprise that makes being a geography professor fun. At last account, the movie was being shown as background in a disco in Berkeley, California, complete with strobe lights.

During the 1974–75 academic year, I was able to take a sabbatical leave at the International Institute for Applied Systems Analysis (IIASA) in Laxenburg outside Vienna. Howard Raiffa had just taken over as director of this new institute, and Harry Swain from Canada headed the urban group with which I was affiliated. There were occasions for discussions with some outstanding Soviet, Italian, German, Austrian, and American scientists. John Casti gave some beautiful lectures on the calculus of variations, and I learned about penalty-function methods of optimization from a Soviet mathematician. At one seminar, I was able to meet Tjalling Koopmans, the Dutch economist and Nobel laureate, and we had a delightful hike together. Gilbert White also visited for such a seminar. I presented a lecture on *"analytische Kartographie"* at the University of Vienna and one on *"schachtbrett Geographie"* (cellular geography) at the Institut für Raumforschung. These lectures were later published in English. The time at Laxenburg also gave me a chance to work with Ross MacKinnon and Martyn Cordy-Hayes. It was a very productive time for me and really advanced the foundation for my continuing studies of human migration.

On one occasion, a group of us from IIASA took the hydrofoil down the Danube from Vienna to Budapest for a regional science meeting. Only the geographers were out on the open deck of the boat in the light rain during the whole trip, with short visits to the bar in the cabin for bottles of Egri Bikaver (a.k.a. bull's blood) to keep us warm. The return was by train, and I recall border guards looking under the seats for who knows what—but probably defectors. Ross and I also made a foray into Prague in his new car, paying a "fine" along the way to some local county police carrying submachine guns. In Bratislava, we met with a geographer, now a dean, Joseph Krcho and some of his colleagues. My family and I were also able to ski at Präbichl, travel to the Dolomites, and visit my parents in Zurich. Swain and I hiked the Schneeberg several times, and once I enjoyed a solitary but tiring walk all the way down from the top to the valley. On a quick trip back to Michigan, I was able to stop briefly in Amsterdam and visit the Rijks and Van Gogh museums. There was also a computer conference in Rättvik, Sweden, almost at the Arctic Circle, so close in fact that when I awoke at 2:00 A.M., I was certain that I had overslept. Andrei Rogers, having learned about IIASA from me, joined the group just as I was leaving and spent many years there. He is now at the University of Colorado.

### Santa Barbara

Five years after my sabbatical leave, UCSB, in the form of David Simonett, called. Geography at Michigan was in good shape, but they had made some poor decisions. Several of the faculty had affiliations with area centers or other units and were thus de facto only part-time members of the department. Three senior geographers—Gunnar Olsson, Melvin Marcus, and I—left at the same time, and the department was able to replace these positions only with nontenured assistant professors, a mistake that weakened the department. Few of the faculty served on campuswide committees, another weak point. From my service on the Dean's Promotion and Tenure Committee, I was aware that a historian on that committee was antagonistic toward geography, arguing that Harvard did not have a geography department, so Michigan did not need one. All these factors, combined with a financial crunch in the state of Michigan, led to the closing of the department. In my case, having grown up on the West Coast, I was in the mood to return there, and a pending divorce cinched the decision. I joined the newly formed geography department at UCSB. Simultaneously Reginald Golledge came from Ohio State, and he and his students and I subsequently collaborated on measuring the rubber sheeting transformations implicit in mental maps.

When I arrived at UCSB in 1977, the computer resources were terrible because they had been assigned to a financial administrator. In contrast, the library was and is, in spite of declining resources, quite good in contemporary materials, although, being in a new university, the holdings did not go back very far. Fortunately I had read at Michigan nearly all foreign materials and those of interest from the nineteenth century. My Santa Barbara colleagues in 1977 included Jeff Dozier (a Michigan graduate on whose Ph.D. committee I had served and today the dean of a new college), Jack Estes, and Terry Smith. They were and are bright and hard working. Simonett was in many ways an outstanding chairman, supportive and encouraging but with high standards. His support for assistant professors—light teaching loads, no committee assignments, and inclusion in research grants—had the intended effect. You hire bright young people and get them tenure quickly. Thus, the department has built its reputation to near the top of the profession in less than fifteen years.

At one point, the Swiss National Science Foundation sponsored Dr. Guido

Dorigo for a year's stay in Santa Barbara, accompanied by his wife Marianne. He and I collaborated on a long paper eventually published in the *Annals* with revisions sent back and forth at a furious rate. This was before e-mail and on-line collaborative editing, and also before word-processing programs were popular, so we kept the best departmental typist, Teresa Everett, busy going through at least twelve revisions. The research involved the depiction of geographical migration by a continuous vector field driven by a partial differential equation representing the attractivity potential. The interesting thing about spatial interaction, as studied by geographers, is that it is asymmetrical, though only slightly so. The notion that the classical spatial interaction model can be used to generate vector fields portraying the asymmetry or (assuming symmetry) can be inverted to predict distances between places is most useful. We later extended these ideas to include interaction relations on a spherical earth, and I find it curious that most spatial analysis done by geographers uses a flat earth.

About this time, Peter Gould and I engaged in our experiment in geographic coding. He convinced some thirty-four geographic colleagues throughout the world to mail letters to me addressed only by my name and the latitude and longitude (degrees, minutes, and seconds) of my house. Four of these letters actually arrived, via rather circuitous routes, but were delivered to my university office because the "professor" title was included. The purpose of the experiment, of course, was to demonstrate that there are many names given to geographic places and that they can be considered aliases of each other, for use in different contexts.

Before the official release of the Topologically Integrated Geographic Encoding and Referencing (TIGER) system by the Census Bureau in 1980, I organized three well-attended annual conferences on the subject organized by the California State Department of Finance and its Population Bureau. The geographer from the Census Bureau, Dr. Robert Marx, was one of the principal speakers. Somewhat later the NSF-sponsored National Center for Geographic Information and Analysis (NCGIA) was established at Santa Barbara under David Simonett and Michael Goodchild. I played only a minor role in this organization, although nominally I had the title of chief scientist. Hikes with the Sierra Club in the local hills kept me physically active and led to meeting my future wife, Rachel. I also walked to the top of Mt. Whitney.

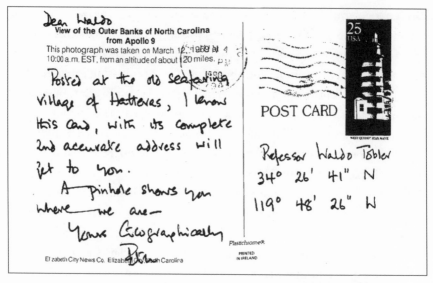

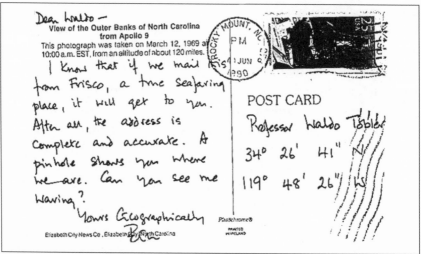

2. Peter Gould and Waldo Tobler's experiment in geographic coding.

As is the custom in the University of California, I was able to take sabbatical leaves, and in this way I could build up my competence in the history of cartography, a field in which I was doing research and teaching. One trip involved going to several libraries within the United States: the Library of Congress; the libraries of Brown University, Yale University, and the University of

Wisconsin; and the Newberry in Chicago. The next set of trips were to the Vatican Library (where I also walked around the perimeter of this country, something every geographer should do—that is, walk completely around the outside of a country), the British Library, Zurich's Zentral Bibliotek, Florence (several libraries), and the Bibliothèque Nationale in Paris. All the time I was looking at original pre-1500 A.D. Portolan charts.

More recently I have had the opportunity to see Rio de Janeiro with Reg Golledge, his wife, and my son Stephen, at an International Geographical Union (IGU) regional conference. My first IGU meeting had been in London in 1964, where Imhof amazed me, and I think everyone else, at the business meeting by rapidly switching between languages. He then rebuffed the parodic successive complaints of the Soviet bloc Europeans because scholars from the German Democratic Republic had not been granted visas to attend the conference. I always advise students and colleagues to attend the IGU business meetings, which a sensible person would expect to be quite dull. They are miniature United Nations squabblings and fascinating to watch. The London IGU meeting was followed immediately by the first International Cartographic Association meeting in Edinburgh. Ray Boyle was demonstrating his automatic scanner/plotter at this meeting, and I gave a paper called "Automation in Thematic Mapping," after which I headed for the hills. Actually just one hill. Ross Mackay and I seemed to have the same idea, and we played hooky by walking every morning to the top of Arthur's Seat (a basalt intrusion and a very prominent feature of Edinburgh). Mike Goodchild and I did the same thing years later at a joint U.S.-British GIS conference.

In contrast, the 1982 Rio IGU meetings were a bit embarrassing. At the first meeting, at the National University in a cement block classroom with terrible acoustics, I listened to talks by Reg Golledge from Santa Barbara, Ross MacKinnon from Buffalo, and Allen Scott from UCLA, and gave my own talk, all of us having traveled who knows how many miles to hear each other. The one non-U.S. speaker for that meeting presented a political harangue. From the program, it appeared that virtually all of the South American geographers had similar political rather than scientific themes. I found this trend rather disappointing and depressing. The next day my son and I took the inexpensive Rio bus and subway system to the university, where I was refused admittance by a soldier with an automatic weapon—the military was still in charge in Brazil at the time—because I was wearing shorts. He explained that this was a university and not the beach. So we went to the beach. We had taken hotel rooms at Copacabana Beach, rather than at the convention hotel in downtown Rio, which turned out to be a smart choice. There were fabulous fruit breakfast buffets and, amazingly, the hotel bill went down every day! This was the result of inflation relative to our payment in dollars. According to the hotel employees, their pay was adjusted upward once a month. What a way to have to live, but somehow people seem to survive.

The surf in the ocean at Copacabana is rather abrupt, and on one occasion Reg Golledge had to haul me out after I had been tumbled around quite a bit and was completely out of breath. We did most of the tourist things—up the cable car and by bus through the jungle to the lookout with the big cross. Most enjoyable was jogging along Copacabana in the early morning or walking to Ipenima and back along the beach, stopping at little cafés along the way for light refreshments. The only Portuguese we needed was *obrigado*. Fruits, fish, and meat dishes were common, but salads were difficult to find—as both Allison Golledge and I discovered. There were also some less-attractive features: not least the *flavelas*, the persistent and aggressive hustlers near the hotels, and the fake gems for sale along the beachfront at ridiculously low prices. Purse snatchings on the beach were said to be common. Golledge saw one and grabbed the man, who then pleaded that he not be turned over to the police because he expected that he would be summarily shot.

The GIS community seems to love conferences, so I have gotten to Hawaii and other places several times with colleagues such as Roger Tomlinson and David Rhind. A few years ago Andrew Frank invited me to give some lectures at the Technical University in Vienna, and while I was there also met with Manfred Fischer and some of his students from the Economic University, so I got to revisit Vienna all for the price of a few lectures. On another trip, Bob Stimpson escorted me around several Australian cities in exchange for a keynote address to the Australian Regional Science Association. From there, I traveled to Bangkok and Hong Kong on my way to parts of the People's Republic of China.

The trip to China was sponsored by the Academica Sinica and was arranged by Professor Shupeng Chen. It included talks at Kunming, Chengdu, Xian, and Beijing. In preparation for the trip, I had attended an evening Santa Barbara Adult Education class in spoken Chinese, which enabled me to begin the lecture with the Chinese phrase "I'm sorry that I am not able to speak Chinese," and that always got applause. Using an interpreter turns a one-hour lecture into a whole morning's session, particularly when the compulsory tea break is included. Student interest in my talks peaked at a discussion of how geography is taught in the United States. From my knowledge of Soviet cartography books and methods, I was able to deduce that many of the Chinese cartographers whom I met had been trained in the Soviet Union. My interpreter had indeed studied cartography in Kharkov; when his English faltered, we were able to converse using my limited Russian. After my presentation at each site, I was escorted to the outstanding places of interest. In Kunming, we went to the Stone Forest, an amazing limestone landscape. Outside Chengdu, my hosts, interpreter, and I visited an amazing two-thousand-year-old water control and diversion project. I was also given a demonstration of map drawing using the delicate Chinese *maobi*, used for calligraphy. The County Museum in Xian contains the "Forest of Tablets," displaying centuries of treasured calligraphy, and a pair of world-famous maps from the 1100s engraved in stone.

In 1994, the AAG meeting was held in San Francisco, and Ned Taaffe from Ohio State University organized a session on the impact of the so-called quantitative revolution in geography, which he had done at previous AAG meetings and was to do at subsequent meetings. Emilio Casetti, Fred Lukermann, and Eric Sheppard were among those who gave presentations at these sessions. I did not attempt to fathom the meaning of the quantitative revolution but instead presented a substantive paper, my philosophy being that doing it is better than talking about it. My presentation was on human migration and referred to the work of Ravenstein, Thornthwaite, and more recent scholars; it was subsequently published in the journal *Urban Geography*. I had gotten interested in migration and its cartographic depiction, and so programmed a computer to draw a variety of flow maps. I was motivated by that fact that although movement of all sorts is the basis of most change in the world, too many maps depict static situations; that is, they show state rather than flow information.

## In Retirement

Part of the motivation for my first big effort after retirement was the same as that enunciated by Torsten Hägerstrand many years ago—namely, to render geographic observations susceptible to the same kind of objective analyses as are performed on time series data. Since my graduate school days, possibly because of my reading in geodesy and related fields, I have observed that earth scientists generally deal with data in *(x, y, z, t)* space. And so do geographers. Much of modern chemistry also requires spatial analysis, so it makes sense to use methods if not identical, then at least similar to those used in the physical sciences, especially geophysics, but applied to the subject matters that interest human geographers. My working definition of geography is "the study of processes that affect the arrangement of people on the surface of the earth." This is my guide to topics to be studied, somewhat akin to the dictum "the proper study of mankind is man."

The Global Demography Project involved the creation of a roughly thirty-megabyte database of world population in five-minute latitude/longitude quadrilaterals, rather than by the conventional country units, which are in any event ephemeral. This was twelve-squared times better resolution than existed previously. A corollary to the use of quadrilateral data is that large amounts of physical data are now recorded and transmitted by earth satellites in a similar pixel format. If geophysicists can describe the arrangement of magnetism (or of gravity or whatever) at the surface of the earth using spherical harmonics (a form of Fourier series adapted to a sphere), why can't geographers do this for the arrangement of people? So I did it. This attitude is consistent with most of the work I did during my career and published in about 150 papers.

But there are problems. One is that the fields studied in geophysics are continuous, whereas people are not. Yet population density is considered a continuous field, albeit one that is a little nebulous because it depends on the denominator used. Second, the geophysicists use these forms of mathematics to interpolate where they cannot measure, and no one can measure everywhere. In the case of population, our censuses break the world into little chunks and then count people and record their attributes within these chunks without observing what is going on at a finer level—except that somebody else always wants a different or finer chunkation. Thus, there is no apparent need for interpolation between chunks; rather, one must learn how to convert data between different chunkations. Third, geophysicists are often concerned not with the scalar phenomena themselves, but rather with their spatial or temporal derivatives, which they use to study related phenomena of interest. These derivatives are easy to compute from spherical harmonics. Spatial population analysts are rarely, if ever, explicitly concerned with spatial derivatives, although these derivatives are often implicit in the types of analyses performed, and there is no agreement as to which properties of interest are related to these possible derivatives. Finally, spherical harmonics were developed, back at the time of Gauss some two hundred years ago, for continuous phenomena that are "harmonic" in the technical sense that the value at any location is the average of nearby values. Thus, the fields are smooth in that the spatial derivatives in all directions change slowly and continuously. In what sense can this be said to be true of the arrangement of population on the earth?

It is clearly the case that nearby people resemble each other more than they resemble people located at great distances—in part because people are gregarious and mutually influence each other, and also in part because they are sufficiently mobile to do such influencing. We also know that there are genetic and cultural ties between peoples. Thus, some of the spatial derivatives can be expected to change slowly, and the data display geographic auto- and cross-correlation. Some people, respectful of human diversity, abhor this convergence of culture to a global mean. But it is stretching things to say that population density satisfies the harmonic assumption, so probably other methods might be better. Many physical fields that seem continuous are not really so when one looks more closely—for example, at discrete molecules in gases and liquids, moving particles in electric fields, or temperature as the motion of molecules. But these elementary items are so numerous and so tiny that few question the fictional hypothesis of continuity in the work of physical scientists. And, at one level, the equations "work" as approximations, whether or not the phenomena are really continuous.

To continue in the same vein, Guido Dorigo and I have modeled the aggregate asymmetric migratory movement of humans using Poisson's equation and have used the related Laplace equation to reallocate population in a density-conserving manner (pycnophylactically) within or between the chunks I referred to above. These mathematical methods are similar to those used by physical science practitioners. Is this approach reprehensible? I think that these methods yield only crude approximations, linear and usually isotropic, but they seem to give definite insight into the processes being studied. The main basis for their rejection seems to be that they are considered "positivistic." Such objections do not interest me, my philosophy being that the only relevant and best criticism is better work—that is, improvement in the specific task at hand. To go off in some other direction is not to my mind a useful criticism. And to a limited extent, the equations also "work" with human populations. Science, after all, consists of approximations that attempt to achieve increased understanding with a limited numbers of variables.

## Looking Back

From all of the foregoing, it can be seen that my childhood experiences were perhaps somewhat unusual but not extraordinary. During my career, I consider that I have been quite lucky. I could have lived in some unfortunate place with civil wars or continuing riots or without a good educational system and infrastructure. I have had excellent colleagues and students at excellent universities, have received several honors, and have traveled a good part of the world. Numerous students have taught me much. In retirement, I have a good wife and a good computer at home and do not miss committee meetings. I continue to use the university library, publish papers, work with an occasional student, and give a few presentations at conferences. There are still plenty of interesting problems that need my attention. All in all, my career choice has served me well.

# A Life of Learning

## Yi-Fu Tuan

Courtesy of Yi-Fu Tuan

*Yi-Fu Tuan (b. 1930), shown here as he was about to go for his first job in 1956, attended University College, London, but received his B.A. and M.A. from Oxford in 1951 and 1955. Specializing initially in geomorphology, he went to the University of California, Berkeley, and was awarded the Ph.D. in 1957. The University of Waterloo awarded him its doctorate honoris causa in 1985. He taught at the Universities of Indiana, New Mexico, Toronto, and Minnesota before becoming the John Kirkland Wright and Vilas Research Professor of Geography at the University of Wisconsin, where he stayed until his retirement in 1998. He has received many professional awards, including those from the Association of American Geographers, the National Council for Geographic Education, and the American Geographical Society, as well as literary awards from library associations. In 1986, he was elected a fellow of the American Association for the Advancement of Science and a year later received the Collum Medal from the American Geographical Society. He is noted as a remarkable teacher and in 1992 was not only elected "Best Professor" by the Wisconsin Student Association, but also received a certificate of appreciation for "Special Contributions to Student Lives." He is noted for creating a unique genre of geographic writing, accessible not only to his colleagues and students in geography, but also to a wide audience far beyond the academic walls. Many of his books, such as* Topophilia, Space and Place, Landscapes of Fear, The Good Life, *and* Passing Strange and Wonderful, *have become classics.*

The invitation from the American Council of Learned Societies to give the Haskins Lecture came to me as a great and pleasant surprise.[1] All honor should seem undeserved—a surprise—but this one is way off my normal chart of hopes and aspirations. Among geographers, I believe only Don Meinig of Syracuse University has given the Haskins Lecture. The honor to him is as well earned as it is predictable, for Don is not only a distinguished scholar and humanist, he has also labored in an established field—the history and historical geography of the United States—that is itself distinguished and honored in the sense that many of the country's best minds and scholars have chosen to cultivate it. My own area of study is quite otherwise. It is off center even in geography. Students who take it up tend to be regarded as impractical loners with uncertain job prospects after graduation. True, I now enjoy a modicum of recognition in geography, but it is more a consequence of my longevity—my longtime card-carrying membership in the profession—than of any marked intellectual influence. And so I return to my surprise and pleasure at the invitation from the council, a venerable institution with a superb record of support for the highroads of learning, which this year chooses, boldly if not wisely, to acknowledge a maverick.

What is it then that I do? My answer is human geography—more precisely, a subfield within it that might be called—though rather inelegantly—systematic humanistic geography. And what is that? Let me try to provide an answer, drawing on my own experience and work. A good way to start is to envisage a faculty social gathering. At such a gathering, a historian is unlikely to be asked, "Why are you a historian?" Yet I have repeatedly been asked, "Why are you a geographer, or why do you call yourself one?" My unimposing physical appearance may have prompted the question, for people even now tend to see the geographer as a robust explorer in the mold of Robert Falcon Scott or Indiana Jones. As a matter of fact, when I was an undergraduate, the professors of geography at both Oxford and Cambridge were explorers.[2] The question

---

1. The Charles Homer Haskins Lecture was given in Benjamin Franklin Hall, Philadelphia, on May 1, 1998, and is here reprinted with kind permission of John H. D'Arms, president of the American Council of Learned Societies.

2. Kenneth Mason at Oxford and Frank Debenham at Cambridge.

"Why do you call yourself a geographer?" may also have been prompted by the titles I have given to some of my books. People don't immediately see how *Morality and Imagination, Passing Strange and Wonderful, Cosmos and Hearth,* and *Escapism* can be the works of a geographer.

To those who have wondered about my vocation, I respond in three ways, each geared to a different level of seriousness. At a social gathering, when people are not at their most attentive, I will probably say, "As a child, I moved around a great deal with my family, and there is nothing like travel to stimulate one's appetite for geography." Sad to say, this lazy answer nearly always satisfies my inquirer. It is what he or she expects. My second and more thoughtful response is: I have always had an inordinate fear of losing my way. Of course, no one likes to be lost, but my dread of it is excessive. I suspect that more than physical discomfort is at stake. To be lost is to be paralyzed, there being no reason to move one way rather than another. Even back and front cease to be meaningful. Life, with no sense of direction, is drained of purpose. So, even as a child, I concluded that I had to be a geographer so as to ensure that I should never be disoriented. Geographers always know where they are, don't they? They always have a map somewhere—either in their backpack or in their head.[3]

A childhood hero of mine was the great detective Sherlock Holmes, and one reason why I admired him and considered him a superb geographer was that he always could find his way, whether in the backstreets of London and Chicago or in the wilds of Utah and Tibet. Moreover, Holmes was always socially at ease: he knew how to behave whether the place was a duchess's drawing room, a Mormon meeting hall, or an opium den.[4] That too was most admirable, for, like many young people, I feared social disorientation quite as much as I did geographical disorientation. This fear of losing my way has strongly affected my environmental preferences. Unlike many people, I prefer American towns with their geometric street pattern to Old World towns with their maze of narrow alleys. Old World towns are unfriendly to strangers, who must stay a long time to feel spatially comfortable. By contrast, the open-grid

3. Yi-Fu Tuan, "Orientation: An Approach to Human Geography," *Journal of Geography* 82, no. 1 (1983): 11–14.

4. Yi-Fu Tuan, "The Landscapes of Sherlock Holmes," *Baker Street Miscellanea* 45 (spring 1986): 1–10, 57.

characteristic of many American towns says right away, "Welcome, stranger." I dislike the tropical rainforest for the same reason: unless one is a native, it invites disorientation. I like the desert because it is an open map, with the sun serving as a dependable marker of east and west, and with sharply etched landforms—visible from miles away—that unmistakably tell the visitor where he is.

But my dread of rainforest and love of desert hint at something deeper than just orientation. Underneath such likes and dislikes are questions of one's fundamental attitude toward life and death. In the rainforest, all I can see and smell—perversely I admit—is decay. In the desert, by contrast, I see not lifelessness but purity. I sometimes say teasingly to environmentalists that, unlike them, I am a genuine lover of nature. But by nature I mean the planet Earth, not just its veneer of life, and I mean the whole universe, which is overwhelmingly inorganic.

This leads to my most serious reply to the question "Why are you a geographer?" I take up geography because I have always wondered, perhaps to a neurotic degree, about the meaning of existence: I want to know what we are doing here, what we want out of life. Big questions of this kind, which occur to most children as they approach puberty, have never left me. But rather than seek an answer in the great abstractions of philosophy and religion, I sought to begin my quest at the down-to-earth level of how people make a living in different places and environments. This, to me, was and is the substantive core of human geography. But I could never be satisfied with just learning about the economics and politics of survival. The word *survival* itself, which appears rather often in ecological literature, seems to me unduly restrictive and harsh. It evokes images of nature "red in tooth and claw," of people constantly fighting and struggling, rising up the ladder of well-being only to sink again. Its message to me is that people can do little more than cope. Maybe that pretty much summarizes the human story, as it does the animal story. But I am not altogether convinced. My almost pathological need to find meaning presses me to ask, again and again, What else is there? What goes beyond—even far beyond—coping and survival, the vocabulary of nature and of ecological studies?

## Finding One's Way

I went to Oxford as an undergraduate believing that it offered the best program in human geography. Not so. Geography at Oxford after the Second World War was in the doldrums. Its human geography lacked all inspiration.[5] The only part of the curriculum that gave me any intellectual sustenance was geomorphology—the study of landforms. So I specialized in geomorphology. At Oxford, I came across a paper by a California geographer, John Kesseli, called "The Concept of the Graded River."[6] I liked it: it satisfied my desire for clear reasoning, for the grounding of arguments on evidence that can be seen and touched. In 1951, I went to University of California at Berkeley to study geomorphology with Kesseli and human geography with Carl Sauer. My graduate-student years there were exceptionally happy and productive. Although I spent most of my time working on a geomorphological dissertation, my intellectual engagement with the intangibles of human existence never weakened. I read much else besides technical papers and books on geography. A cousin of mine, who was then a professor of mathematics at the University of Washington, visited me in Berkeley and found a copy of Robin Collingwood's *Idea of History* and several works on existentialism on my desk. He smiled in a worldly-wise way and said, "You will outgrow it."

Well, I never did. I couldn't jettison the metaphysical questionings of childhood. I continued to tussle with, What does it mean to live on this watery planet called Earth? Are we determined by our environment, including its physical elements, or are we free? If we possess even a small degree of freedom, what have we done with it? All human beings surely want the good life. But what is it? Can one properly and fruitfully speak of the good life in the singular—that is, as a goal toward which humankind as a whole might aspire—or is the good life inevitably relative, a vast array of barely compatible goals dictated by the accidents of taste?

These questions, even when they are grounded in the tangible facts of ge-

---

5. The most frequently offered excuse I heard at the time was that Oxford geographers wore themselves out compiling Admiralty Handbooks as a part of their war effort.

6. John E. Kesseli, "The Concept of the Graded River," *Journal of Geology* 49 (1941): 561–88.

ography, remain hard to grapple with. How will I ever find the time to study them? By happy chance, my first tenure-track position was at the University of New Mexico. In the early 1960s, New Mexico was a small university located in a modest-size town that lay in the midst of broad expanses of shrub and desert. The university's geography department was so small that, by hiring me, it doubled its size. We two geographers had to carry a horrendous teaching load, but felt little pressure to publish.[7] This lack of pressure to publish meant that I had the time and, more important, peace of mind to shift my commitment back to human geography. But I needed more than time and peace of mind. I also needed encouragement and stimulation. These I received from my one geography colleague—a learned man who was also a kindred spirit—and from J. B. Jackson, who lived in Santa Fe and was the founding editor of a magazine called *Landscape*. Under his editorship, this magazine was as much philosophy as geography; many of its articles moved far beyond descriptions of particular landscapes to seminal reflections on culture and human nature. I lived, then, in a New Mexico oasis, a setting that in its isolation is close to the ideal of a lighthouse on an inaccessible coast recommended by Einstein for the serious postdoctoral student. The desert itself provided me with discipline and inspiration. Its harshness scotched my natural love of ease; its pure lines cleansed my mind of rank growths.

### Digging below the Surface

In 1969, thirty-eight years old, I felt that, at last, I knew enough of life and world to engage seriously with systematic humanistic geography. I was then a full professor at the University of Minnesota. With that status achieved, I convinced myself that I had no wish for greater worldly success. Ambition, which remained a powerful drive, could henceforth concentrate without distraction on intellectual tasks, including ones that might seem foolish. *Ambition* is not, however, quite the right word, for it suggests too much the limelight. I was never anxious for the limelight. As a child, I had no interest in becoming president or famous scholar. As an adult, the idea that I should struggle to increase

---

7. For several years, my only colleague was B. L. Gordon, a cultural geographer who earned his Ph.D. from Carl Sauer in 1954.

the length of my obituary by another half inch is just too absurd to entertain. On the other hand, ever since I entered Oxford, I worked hard; moreover, all my life I have lived in and loved the campus atmosphere of work. My sense of well-being fed and continues to feed on an awareness of students swarming just beyond my field of vision and on the hum of activity in classrooms and libraries that portends high purpose and worthy goals. Official days of rest—weekends and national holidays—were and still are unwelcome. I need my daily dose of mental stimulation. Each day I require the reassurance of having taken a step forward in my own enlightenment; and I have simply assumed that this is true of most people in a university community.

By the time I went to Minnesota, I was a driven man, driven ultimately by the big and woozy questions that, as I said earlier, started to haunt me as a child. They took on an extra edge in young adulthood. My great good fortune was that the Oxford of my undergraduate years made them seem respectable, even fashionable. In the aftermath of a horrible war, the university welcomed students and teachers of a philosophical, theological, or literary bent, willing to address the timeless puzzles of life. If I showed exceptional diligence as an undergraduate, it was because this deeply rooted personal need found satisfaction and stimulation in the books, lectures, and, above all, endless late-night talks that were a part of the university's ambience.[8]

There was another reason. At Oxford, my eyes were opened to the richness and sheer fascination of human reality. Like most people, I felt the world's magic as a child but had lost it in middle school's unholy mix of regimented work and relentless competition. At Oxford, I regained it. I was drawn to wonder not only at the glories and miseries of human existence, observable on

---

8. Today when one thinks of Christianity, one thinks of Bible-thumping fundamentalism. No college intellectual would touch it. How strange to think that when I was an undergrad at Oxford, Christianity was at the height of intellectual fashion. J. R. R. Tolkien and C. S. Lewis held forth as local talents. T. S. Eliot hovered at the edges of our life as the outstanding intellect—we saw him as not just a poet but also a profound Christian thinker. His play *The Cocktail Party* (1950), with its subtext of suffering and redemption, provoked endless discussion. We undergrads also read the Continental Christian philosophers—Gabriel Marcel, Simone Weil, Emil Brunner, and Paul Tillich. Non-Christian thinkers and intellectuals were equally concerned with the meaning of life and other perennials of philosophy: the two outstanding figures were Albert Camus and Jean-Paul Sartre.

the streets as well as in colleges, but also at the common routines of cooking and washing, planting and harvesting, buying and selling, traffic flowing and ebbing in city streets, houses being raised and torn down, institutions inhaling and exhaling men, women, and children—these and other pulsations of life that make up the content of human geography. They are thoroughly worthy of study. It is sad for me to say that geographers have all too often made them seem rather pedestrian. The challenge to the humanist geographer is to see afresh these ordinary objects and motions of life, take delight in them for what they are, as they *appear*, then dive under for hidden relations and meanings, and, finally, present both—surfaces and depths, the seen and the unseen—in a language that is subtle yet vigorous.

With the onset of middle age, my drive to learn could easily have lost momentum, turned into studious habit or chores dutifully performed, if I thought that I was merely consuming the world's stock of knowledge and wisdom—that is, observing reality through what others had said and written. I needed to feel that I had something to add, not so much in factual details as in frameworks and perspectives, which, by permitting facts and ideas to bed one another in new and fruitful ways, could subtly reshape our understanding of the world. An ambitious goal, indeed. I worked hard toward it in the various places I taught between 1956 and 1968—Indiana, New Mexico, and Toronto —but it was only at Minnesota that I came anywhere close to touching the hem of success.

Yet, in 1983, I left Minnesota for Wisconsin. Why? I have wondered. I was happy at Minnesota and expected to stay there till my three score years and ten ran their course. Perhaps it was the awareness that a late-middle-age crisis lurked down the road, or that I needed a change that would give me a final surge of energy. Wisconsin's faculty was as distinguished as Minnesota's, but younger. I could draw on their youth and vitality. I would be the oldest member there, itself a pleasing thought to someone of Asian descent. So, after fourteen good years at Minnesota, I moved to Wisconsin, where I spent another fourteen good years. The late-middle-age crisis I vaguely anticipated did break while I was trying to settle down in my new academic home. Fortunately, it was brief. I managed to overcome depression—a sense that I had studied life but had not lived it—with the help of understanding colleagues and friends.

## Successes and Discouragements

Let me now turn to my small successes. They are so in my own eyes rather than the world's, although *Topophilia* might be counted an exception. This book, which is a systematic study of how people come to be attached to place, was published in 1974, at a time when the environmental movement needed a work in the tradition of humane letters to complement the flood of publications that poured out of the factories of applied science. *Topophilia*'s popularity may be gauged by the fact that at one college bookstore, it was put on a shelf labeled "Astrology and Occult." To my surprise, late-blooming Flower Children found it sympathetic, as did its bible *The Whole Earth Catalog*. *Topophilia*'s romance with Counter Culture was, however, brief and would have been briefer had not Counter Culture dovetailed into the environmental movement. The book eventually achieved the respectability of a required text for college courses in landscape architecture. More important than marketplace success in motivating me to persevere, despite many dead ends and failures, was the occasional psychological reward—success in the sense of an unexpected shift in my pattern of thought. Allow me to illustrate what I mean with two books that I published in the 1980s.

The first is *Segmented Worlds and Self: Group Life and Individual Consciousness*. I was led to write this book in the following manner. Human emotional and mental makeup has consistently been my point of departure. Humanistic geography is the geography of people, and, of course, people have feelings, language, and ideas. So *topophilia* is the love of place, as *topophobia* is the fear of place, a theme I took up in a book called *Landscapes of Fear*. Love and fear are basic human emotions transformed to varying degree by imagination and culture. Having touched base with these emotions, I then sought to explore a unique quality of human consciousness—self-consciousness. How is it related to the development of individuality and the need for privacy? How would an individual's growing sense of self affect social cohesion and group life? It is a topic of predictable appeal to the human geographer, for a heightening of self-consciousness and individuality clearly evolved with the progressive partitioning of space. Partitioned space promotes privacy by allowing individuals to be alone, engage in separate activities, have separate thoughts, or explore each other's world in sustained conversation. My plan was to refurbish

and extend my knowledge of the European house from the Middle Ages to the
late nineteenth century, concentrate on the number and arrangement of
rooms, the purposes they served, and correlate these changes in spatial organization with the history of consciousness.

It seemed a worthwhile project. Yet after a year or two of reading and writing, I grew discouraged. I could see a decade of hard work ahead, in the
course of which I would be wiser in details but not necessarily in understanding, for my arc of vision would have stayed essentially the same. I minded the
meagerness of the recompense because I am so far from being the "utmost
scholar," caricatured by George Steiner as someone who can happily spend a
lifetime studying Korean chamber pots of the ninth century and little else.[9]
To confess that I lack patience for minutiae and the long haul is to invite dismissal as a dilettante. Dilettante in the sense of someone who takes delight in
the world, yes, I admit to fitting the description. But I like to claim, or rather
confess to, something else, namely, a yearning for the sudden illumination—
an excuse to shout "Eureka!"—that is perhaps more common among physical
scientists, artists, and essayists than among humanist scholars. Be that as it
may, just as I was about to give up my project on specialized rooms and individuality, it suddenly occurred to me that the story could be given a thicker
texture and far greater resonance by having it placed between two other narratives—one on food and eating, the other on the theater.

The medieval manor house was little more than its centerpiece—the hall,
an unsegmented area in which all sorts of activities took place. Food eaten
there, too, was either a whole animal, a hefty shank, or a stew of many ingredients thrown together with little consideration for compatibility. Life in the
Middle Ages was public and gregarious. Food was eaten heartily; table manners were minimal. People had few places to withdraw to and did not seem to
mind. There was no doubt a strong sense of self, but rather little of self-
consciousness. Obviously I cannot trace the complexly interwoven stories of
house, food, and table manners here. But the gist of what I am getting at is
clear enough. By late nineteenth century, the great European house had probably reached a maximum degree of partitioning and specialization. A room
existed for every purpose—including that of being alone with one's books and

9. George Steiner, "The Cleric of Treason," *The New Yorker*, December 8, 1980, 184.

thoughts. Dining reached new heights of refinement, not so much in flavor as in the quality and multiplicity of utensils. Meat in vulgar bulk, except for the roast in England, was banished from the table. Foods were served separately, not indiscriminately mixed, as in the Middle Ages and even in the seventeenth century. To proper Victorians, dining was a ritual at which drinking a wine inappropriate to the meat or confusing the fish knife with the butter knife was an embarrassing breach of etiquette. Whereas in earlier times guests shared a bench at the dining table, now each guest was ensconced in his own chair and had before him his own private world of sparkling glasses and silverware. He was expected to eat as though nothing so gross had crossed his mind and that the real purpose for sitting down to table was to engage in polite conversation with his charming neighbors.

More exciting and potentially more revealing than the story of food and eating is the story of the theater. Social scientists seek models of society and yet have curiously neglected the theater as model. The theater is a model both in its sociospatial organization and in the plays enacted on its stage. Again, I am able to offer here only a few pointers. As physical space, the medieval theater was, like the church building, a cosmos, embracing heaven, earth, and underworld. In the market square where plays were periodically performed, actors and spectators freely intermingled; no proscenium arch and curtain, no lighted stage and darkened hall—all much later inventions—separated them. The sort of promiscuous mixing I noted in the medieval hall and in medieval cooking was also a feature of the medieval theater. As for the play's theme, what could it be other than the salvation of man? Plays were morality plays. Even those of a much later time—Shakespeare's, for example—might be considered morality plays, with the scope and trappings of the medieval worldview still lingering about them, if only because of their cosmic resonance and religious underpinning, their performance in an all-embracing space called the Globe. How strikingly different the theater became in the eighteenth and nineteenth centuries, when, rather than the cosmos, there was landscape, a much more subjective concept, and, in the end, even landscape was too open and public, and the human drama of cues and miscues, failed communication and loneliness, moved to interior space—the living room. Late-nineteenth-century plays that depict individual separateness within rooted, communal life find a parallel in the theater's own physical arrangement—its compartmentalized spaces. Actors and spectators do not mix. The sense of one world—the Globe—is missing. On the one side is the illuminated stage, on the other, the darkened hall in which spectators sit in their separate chairs, as if alone.

## Later Viewpoints

My second example of a psychological reward—an unexpected insight that made for a period of personal satisfaction—if little else—is the book *Dominance and Affection: The Making of Pets*. This may seem an unlikely work to come out of a geographer, yet I consider it mainstream geography, but mainstream with a twist. And what was the twist? To answer the question, I must first say something about the mainstream. There are, in fact, several mainstreams, several historically rooted, generally accepted approaches to geography, one of which is studying how humans have transformed the earth. This particular approach received a major impetus in 1955, when three distinguished scholars, Carl Sauer, Marston Bates, and Lewis Mumford organized an international symposium to demonstrate its fruitfulness, and a further impetus when the results were published in a much acclaimed volume called *Man's Role in Changing the Face of the Earth*. I was a student at Berkeley in the early '50s, which were also Carl Sauer's last years before retirement. I could sense the excitement—the importance of what was going on. Since the '60s, this flurry of excitement on university campuses became the well-organized, sometimes well-financed, often passionate, nationwide, and, eventually, worldwide environmental movement. Publications quickly reached floodtide. Overwhelmingly they emphasized, as they still do, how human economic activities, driven by necessity and even more by greed, have radically altered and all too often despoiled the earth.

*Dominance and Affection* follows the mainstream insofar as it too can be taken as a study of how humans have transformed the earth. But, as the title indicates, the book's point of departure is psychological rather than economic. My concern is more with human nature than with nature "out there." In the book, changes on the earth's surface matter only to the degree that they throw light on the sort of animal we are. One thing is clear. We are the sort of animal that has amassed enormous power, which we have used for a large number of ends, among them being—at a fundamental level—to temper nature's harshness and to make it produce food and other desirables regularly and abundantly. In advanced societies or civilizations, this power has often been abused. Not only nature but weaker peoples suffer in consequence. Economic exploitation is the name of the game.

Now, the twist I introduce shifts attention away from economic to aesthetic exploitation, which is the mistreatment of nature, including human beings, for purposes of pleasure and art. I ask the reader to picture not cattle yoked to the plow or trees cut down to make homes, but rather the laptop poodle, the potted garden, and the pampered human underlings of a potentate. In the latter set of examples, power is applied to create comeliness and beauty, and it is applied with a certain affection for the manipulated objects. Hence, the book's title, which is dominance *and* affection, and the book's subtitle, which is the making of pets. Power so used tends to be regarded as benign. It may also be considered benign because it does not bring about massive changes on the earth's surface or cause serious problems of pollution. I cast a shadow on this attractive picture by saying that power can be even more uninhibited, more arbitrary and cruel, when it is playful. *Play* is such a sunny word that we forget its dark side. Bad as it is to be used, it can be worse to be played with. Economic exploitation has a limited end, the efficient attainment of which calls for an awareness of the nature of the thing to be manipulated and a deference to tried-and-true rules of procedure. Play, by contrast, is open-ended and freely experimental. The manner of play is not guided by anything other than the manipulator's fantasy and will.

In my book, the first pet I introduce is water. It may seem poetic license to call water a pet, for a pet has to be animate, and water is not animate, not an organism. Yet, in the imagination, water is almost universally considered "alive." This moving and living force is harnessed for many economic ends. But it is also a plaything. We train it, we force it to act against its nature for our amusement, outstandingly, as fountains that leap and dance. Everyone likes a fountain. It imparts an air of spontaneous jollity to the garden, even though there is nothing spontaneous about it. Rather than signify natural exuberance, great fountains are a clear example of submission to power. For them to exist, it may be necessary to bring in water from distant sources through canals, tunnels, and aqueducts. A complex organization must be set up in which managers, engineers, skilled workers and a large labor force all effectively

cooperate. Moreover, fountains depend on the development of a sophisticated hydraulic science. Without it, a choreography of sprouting water is impossible. From the sixteenth to the eighteenth century, fountains were among
the most showy pets of European princes.

Following this excursus into the playful use of water, I move on to the
"petification" of plants and animals, and, in the process, tread—perhaps
rather heavy-footedly—on the idea that the great or aesthetic garden can be
considered as belonging to the sphere of nature and innocence. Some backyard horticultural garden might be so categorized, but not the princely garden, which is as artificial and artifactual as the palatial house to which it is
attached. Indeed, my favorite image of the human domination of nature in
preindustrial times is not monumental architecture, but rather the diminutive
garden, known as *penjing* in Chinese and *bonsai* in Japanese. The bonsai is
wilderness reduced to potted landscape, domesticated in the literal sense, and
by exquisite means of torture—torture in the literal sense of twisting and
bending—sustained with loving attention to detail over an extended period
of time.

### Oppositions and Meanings

I have used the two books *Segmented Worlds and Self* and *Dominance and Affection* to show how a slightly different angle of vision can impart a fresh glow
to even the most familiar themes. Being able to come up with a different angle
is the sweetener that has motivated me to press on in a career of four decades.
But these sweeteners, though necessary, are not sufficient. The deeper motivation is the quest for meaning that I have repeatedly alluded to—above all,
the meaning of life, and I am the sort of person who cannot abandon that
quest without also saying quits to life.

What have I discovered about the world that is, equally, self-discovery? As I
look over my books, I am struck by their binary titles and themes: topophilia
and topophobia, space and place, community and self, dominance and affection, cosmos and hearth. I used to think that I had treated the two poles of
each binary evenhandedly, but this I eventually realized was naïve. In the
book, *Cosmos and Hearth*, I openly acknowledged my bias in the subtitle,
which is *A Cosmopolite's Viewpoint. Hearth* I love. Who doesn't? Hearth

stands for home, community, familiar customs—stability. *Cosmos,* by contrast, stands for world, society, human achievements, and aspirations touched by the strange and the unpredictable. My predilection is for the latter and, in particular, for the latter as goal: one starts from hearth and moves toward cosmos. This trajectory seems to me entirely natural, but it may also be that I am disposed toward it as a consequence of certain experiences in childhood. Bear with me therefore if I add here an autobiographical note.

Born in China, I left it with my family at age ten for Australia. I have been rootless—on the whole happily rootless—since. My childhood in China corresponded to the period of war with Japan. We were constantly on the road, escaping from the invading army. We ended up in the wartime capital of Chongqing. The economy had broken down. We barely had enough to eat. The elementary school I attended was a single, ill-equipped room. Yet, astonishingly, we were given a thoughtfully packaged, cosmopolitan education. We read elevating stories from the Chinese, European, and American pasts, stories about great scientists and inventors such as Isaac Newton, Louis Pasteur, and Benjamin Franklin that were meant to stimulate our intellectual ambition, and moral tales—ones of filial piety, naturally, but also Oscar Wilde's "The Happy Prince"—that were intended to help us grow into compassionate adults.

Although I loved school, I was wary of the daily trek getting there, for in our path was a village. From time to time, we had to stop at the village to let a funeral procession go by. How clearly I remember even now the procession's centerpiece, the corpse. It was wrapped in a bamboo sheet, on top of which was tied a rooster, which served as an advance warning system to the carriers and mourners, for it would crow if the corpse stirred. I both dreaded and hated the funeral procession, which I came to identify with traditional culture and hearth, drenched in fear and superstition, compared with which school was liberation—a sunny world where we learned about a man who sought to bring down electricity with a kite, and where we were inspired by a code of behavior—as in "The Happy Prince"—that in its idealism went far beyond filial piety and kinship obligation.

Thus, early in life, in the midst of war and poverty, I had a taste of the True, the Beautiful, and the Good, culled from different civilizations. It excited in me a yearning for the great triad, especially for the Good, that has lasted into old age. I began the search for what good means to different peoples in the facts of human geography. Good, I was to find out, means at bottom nurturance and stability. Nature provides nurturance, though rather meagerly; and it provides stability, though not one that can be counted on. Culture is how humans, by imagination and skill, escape into more predictable, responsive, and flattering worlds of their own making. These worlds are immensely varied. On the material plane, they range from grass hut to skyscraper, village to metropolis; on the mental plane, from magical beliefs to great systems of religion and metaphysics. They are all, in different ways, aspirations to the good.

Three intractable problems arise out of this human venture, darkening it. One is violence and destructiveness. On a material plane, nothing is ever built without prior destruction—a fact too soon forgotten since the new is tangible presence, whereas the destroyed is at best memory. The second problem is this. The ability to destroy and build presupposes power, and power is morally suspect and can all too easily become monstrous. Historically, construction on any scale was possible only because the elite exercised control over the sinews of laborers and draft animals. Civilization could not come into existence without rather rigid chains of command and obedience—without, that is, social inequality, a condition that had cast a long shadow over civilized life even as it strove to diminish its worst effects. The third problem derives from the feeling that even if we put aside violence and the abuses of power, and just look at our accomplishments, our works, what do we find? Wonders galore, but many of these can seem curiously unreal—aeries that have risen too far above the ground of common sense and ordinary experience. They amaze and entertain, they are sources of great prestige, but they do not satisfy deep human longings for—for what? The short answer is, for the real and the good. But what is real? More answerably, what do people mean when they use the word? And what is its relation to the good?

## The Balance Sheet

I am now at a stage in life when some sort of accounting seems called for. Have I done the best I can with my talent and the opportunities that have come my way? What have I learned? Am I wiser or happier for it? Predictably, my answers to these questions are ambivalent. Take talent. I have read somewhere that we only use about 10 percent of what we are given. In biblical language, we bury our talent; we are unproductive servants. But I cannot believe this to be true of me. I have used whatever I was born with to the full. "What you see is all there is," I say to friends who encourage me by hinting that I, like most people, have seeds of creativity as yet undeveloped. It must be extremely frustrating to have talents that fail to come to fruition for lack of opportunity. But that has not been my fate. Mine is the humbling one of knowing that I have only a small talent to begin with, and that its full flowering under even excellent conditions has produced only a very modest bouquet.

Still, however meager the result, how can I not be grateful for the opportunities that the profession of geography has provided? Geography has allowed me to roam from the physical to the human—from climate and landforms to morals and ethics—and still remain within its capacious borders. The downside is isolation—isolation from fellow geographers who may roam the same grounds but come up with quite different questions and answers; and isolation from scholars in philosophical disciplines who, though they may share my questions, find no reason to heed the cogitations of an outsider.[10] But how can it be otherwise? Even at a cocktail party, it is not easy to break into the conversation of people who already know one another well. How immeasurably more difficult, then, it is to break into the conversation of a long line of philosophers in apostolic succession that began with Plato.

Isolation is a subterranean motif of my life as an academic. It did not start that way, for I was once a student of landforms—a specialist on the pediments and arroyos of the American Southwest. I have wondered what life would be like if I had remained in that field, cultivating it side by side with fellow geomorphologists. How warmly communal it would have been, even in the midst of arguments and debates! If, as such a specialist, I gave the Haskins Lecture, I would be able to acknowledge my indebtedness to esteemed predecessors and fellow workers. I would have the satisfaction of seeing myself adding my tiny sheaf to the abundant harvest of geomorphological knowledge. But I

10. An exception is the Kantian scholar—and my friend—Donald W. Crawford, who invited me to write a paper for a volume called *Landscape, Natural Beauty, and the Arts,* edited by Salim Kemal and Ivan Gaskell (Cambridge: Cambridge Univ. Press, 1993). My paper bears the title "Desert and Ice: Ambivalent Aesthetics," 139–57

chose otherwise. For all the excitement and satisfaction of that choice, it has distinct disadvantages, not the least of which is the appearance of egotism. For what is a maverick scholar but one who cannot readily name his forebears? Of course he has them, and they are a multitude, but they tend to be individuals from different fields who do not know one another rather than conversing members of a single discipline and established tradition.

As I enter my crepuscular years, I wonder about Socrates's famous dictum, "the unexamined life is not worth living." A scholar certainly examines. But what he or she examines is *other* people's lives—the world out there, not his or her own life. I can devote an entire career studying desert landforms or traffic flows in congested cities without reflecting on who I am and what I have made of my existence. Indeed, paying attention to the world may be a way of escaping from the intractable dilemmas of selfhood. While this is plausible, it is also plausible—and a central tenet of postmodernist thought—that any serious and prolonged intellectual engagement with the world transmutes into a marriage of self and the other, such that, as with old married couples, the two may even begin to look alike. My own type of work, ostensibly about "people and environment," draws so much on the sort of person I am that I have wondered whether I have not written an unconscionably long autobiography. By tiny unmarked steps, examination turns into self-examination. Is it worth doing? Will it lead one to the good life? Or will it, as Saul Bellow believes, make one wish one were dead?[11] I oscillate between the two possibilities. In the end, I come down on the side of Socrates, if only because the *un*examined life is as prone to despair as the examined one; and if despair—occasional despair—is human, I would prefer to confront it with my eyes open, even convert it into spectacle, than submit to it blindly as though it were implacable fate.

---

11. Mel Gussow, "For Saul Bellow at 81, Seeing with Fresh Eyes," *New York Times*, May 26, 1997, B7.

# Autobiographical Essay

## Gilbert F. White

Courtesy of Gilbert F. White

*Gilbert White (b. 1911) received all three of his degrees from the University of Chicago (Ph.D., 1942) and served in the Executive Office of the President in Washington, D.C., before helping refugees in Vichy France, with the American Friends Service Committee. He was eventually interned in Germany from 1942 until liberated in 1945. From 1946 to 1955, he was president of Haverford College but returned to teaching and research in geography at Chicago. He later directed the Institute of Behavioral Science at the University of Colorado. Internationally renowned for his lifelong work on water, he has received many honors, including the Anderson Medal for Applied Geography from the Association of American Geographers, the Daly Medal from the American Geographical Society, the Hubbard Medal from the National Geographical Society, and the Vautrin Lud International Prize. He is a member of the National Academy of Sciences and the American Academy of Arts and Letters, and is a foreign member of the Russian Academy of Sciences. He has also chaired or served as a member on innumerable local, national, and international committees, including the National Science Foundation Committee on Environmental Science as well as various United Nations and other international advisory committees dealing with damming rivers such as the Volta, the Nile, and the Lower Mekong, and with problems of the Aral Sea Basin. In addition, he has contributed to the appraisal of many environmental problems, including drought, flood, earthquakes, global warming, coastal erosion, and nuclear waste disposal. He is the Gustavson Distinguished Professor of Geography emeritus at the University of Colorado.*

341

A s I have made my way as a geographer in a rapidly changing world, there seems to have been six chronological periods during which my concern with roughly four fields of major scientific and policy interest have unfolded. Looking back, I think I can perceive a few of the factors that influenced those interests.

## Early Influences

Growing up as a boy in the Hyde Park neighborhood of Chicago, I had several family influences. My mother had visited that area at the time of the great Chicago World's Fair in 1893. She had come with her mother from Atchison, Kansas, and had stayed at a Baptist boarding house next to a house in which a young professor named Harper and his wife had come the previous year to found a new institution. When she was married a few years later to my father, who was transferring his job on the Burlington Railroad to Chicago, she said that because neither of them had attended college, they should locate near that new institution so that their children, as yet unborn, could attend it. It was called the University of Chicago. And, indeed, three of us did take degrees there. Carrying mail for the Faculty Exchange and going to school with the children of professors, I early on had a healthy respect for the institution.[1]

I spent my summers working on a ranch in which my father was a partner in the Tongue River Valley, below the river's flow out of the Big Horn Mountains in Wyoming. Helping clean the irrigation ditches, stack the hay crop, and tend sheep camp in the national forest sensitized me to some of the environmental problems there. When and how was the forest land overgrazed? When was there inadequate drainage, leading to soil degradation? What systems of land use made for prosperity or poverty in the local community? We debated these problems and many related ones around the mess table and when we went across the Montana line to harvest wild hay in the Crow Indian reservation. I was continually alerted to issues of use and deterioration of natural resources. And, incidentally, because one of the chores involved in tend-

1. Martin Reuss, *Water Resources People and Issues* (Fort Belvoir, Va.: U.S. Army Corps of Engineers, 1993).

ing sheep camp was to count black sheep periodically as a means of telling whether or not the band was intact, I was regularly exposed to the problems of inference from statistical samples. In the early days, one of our neighbors across the street in Chicago was Stephen Mather, and I recall neighborhood gossip about his tribulations in establishing the National Park Service. It was he who first showed my sister and me what Yellowstone Park looked like to a senior administrator.

Mixed in with life in the university and ranching communities were two strong currents of belief about individual responsibility. There was the belief of Quaker forebears that each person should try to express in his or her daily life their belief in a divinely guided way. Beside it, among several of my kin, was an agnostic view that each person should look for what was effective in building a healthy community and follow that. As a young man, I initially took the approach of preparing for military service through the ROTC. After that trial, I decided to become a conscientious objector to any violence and to look for ways of expressing my belief in constructive, peaceful action. Stimulated by a long talk with a visiting Quaker philosopher, Rufus Jones, I set out to make my professional training and action consistent with pursuit of a constructive, nonviolent life.

All three of these forces were at work in my choosing to study geography. I had great respect for Chicago professors such as John Morrison, Charles Colby, Wellington Jones, and Griffith Taylor. I was especially influenced by Harlan Barrows, who had entitled his presidential address in 1923 to the American Association of Geographers "Geography as Human Ecology" and who encouraged me to take courses in plant ecology and urban ecology. I went on a summer field course on land-use types in the United Kingdom with Henry Leppard and collected data leading to a master's thesis on Humberside. Ideas of regional definition and social-environmental relations were exciting. Other graduate students—such as Donald Hudson, Willis Miller, and Malcolm Proudfoot—were stimulating, and Edith Parker kept reminding us that one test of our fruitfulness was in what could be taught to students in lower grades.

## Periods of Development

### The New Deal

It was against that background that in 1934, with a Ph.D. dissertation to write, I accepted an offer from Harlan Barrows to assist him for a few months with the Mississippi Valley Committee in Washington, D.C.. Those months stretched into eight years before I returned with a completed dissertation and ready to take my final examination. The Washington years were a rapidly moving project of sharing in innovative approaches to resource management in the United States. After the Mississippi Valley Committee report, I served as a staff member on the National Resources Committee as it canvassed water problems in all drainage basins in the country. Then I was involved in the reports of the National Resources Planning Board on regional planning, a summary of current thinking as to low dams, and the special report of the president's Great Plains Drought Committee on the impacts and possible responses to the drought of 1936. During the two years of 1940–42, I served in the Bureau of the Budget (BOB) in the Executive Office of the President on Pennsylvania Avenue, where I shared responsibility, under the assistant director, for preparing the papers going to President Roosevelt every morning he was in town—papers on proposed reports to the Congress, on response to legislation passed by Congress, and on draft executive orders. My responsibility was for basic work on land and water matters, summarizing the evidence and the views of cabinet members and other responsible administrators, and drafting statements or messages for the president to consider. He typically would respond with approval, disapproval, or directions to revise, to consult others, or to try again. Occasionally, I would be instructed to consult further with interested cabinet officers to make sure they agreed.

The scientific papers I wrote during those years were not published in geographical journals, but rather in publications intended for economists, engineers, hydrologists, or planners, and dealt with concepts that had spatial implications, such as water supply adequacy, distribution of project costs and benefits, and floodplain zoning.

## Vichy France, and Germany

One Sunday in late 1941, walking to the office after Quaker meeting, I found a row of cabinet limousines outside the building and learned that the Japanese had bombed Pearl Harbor. Preparations were being made for the president to go to Congress the next day with a declaration of war. In the subsequent weeks, I helped with various emergencies, such as assisting Milton Eisenhower, coordinator of land use in the Department of Agriculture, to recruit sensitive people to manage the relocation of Japanese civilians. I realized that it would be very difficult to avoid direct involvement in the war effort. There was a congressional mandate against any able-bodied conscientious objector leaving the country, but notwithstanding my declaration as an objector, my draft board had ruled that because of my position in the executive office I was not subject to the draft. When I talked with the director of the budget about seeking permission to go to Vichy France, to do Quaker relief work, he said that if asked, he would refuse permission, but that if the draft board did not ask, he would remain silent. The board heard my request, reminded me that I could remain exempt, and gave me permission to leave the country on "work of national importance." With the help of Anne Underwood, whom I hoped to marry, I finished the dissertation at nights, took the final examination, and in June set out for Lisbon and Marseille. The boat was filled with Nazi sympathizers who were being deported after being stripped of any paper record.

The scene in Vichy France, south of the limit of Nazi occupation, was of a stagnant economy, with the Spanish and Jewish refugees in various degrees of deprivation. My role was to assist a dedicated network of workers administer a wide assortment of efforts to help those in need, without distinction of race or politics or religion. The activities included providing canteens for French children, hostels for Jewish children, and a school for Spanish children; distributing food, medical supplies, and clothing in concentration camps; settling refugees in deserted villages; setting up a small factory to produce artificial arms and legs to enable Spanish amputees to qualify for work permits to get out of camps; and procuring visas to enable Jewish refugees to leave the country. The visa opportunity terminated when Eisenhower landed in North Africa in late 1942, and the German army took control of the entire country.

The Germans ordered all remaining American diplomats and relief workers to be interned in Lourdes, but two of us decided to remain in a place to which we moved the supplies from the Marseille docks. At the same time, all relief activities were placed in French hands. The German general staff moved to the same town, but, to the surprise of the French officials, never bothered us, and we learned that the commander had been fed by Quakers when he was a boy in Germany after the First World War. In early 1943, acting on misinformation from the French secret police, we joined the American group in Lourdes. We were transported the next day to a place of detention in Germany. There, a group of 134 Americans spent the following thirteen months.

The sojourn in Baden-Baden was chiefly a time of waiting until repatriation. In the "university" we created to keep ourselves busy, I found a few activities of geographic interest: a class in geography for school children in the group; a cooperative seminar on the wartime German economy, led by a newspaperman we later learned was an Office of Strategic Services (OSS) agent; German conversation with a Gestapo guard; a class in Russian by a diplomat's wife; supervised walks in the Black Forest; and a seminar I gave on Haushofer's geographic theories.

## The American Friends Service Committee

Returning to Philadelphia in the spring of 1944, I was caught up in the management of various Quaker overseas relief activities and promptly married Anne, who had patiently awaited my uncertain return. One of the urgent operations was in getting medical supplies to survivors of the Bengal famine, where approximately two million people had lost their lives as a result of wartime interruption of agricultural and shipping activities. Another was the continuing movement of medical supplies and personnel to Chinese civilians behind the Japanese lines.

## Haverford College

By the end of 1945, the war was over, and I was a member of the team from American voluntary agencies that inspected Germany under U.S. occupation and recommended the organization of relief for civilians in cities that had suffered from bombing and transport losses. It now was time, in 1946, for me to take up the tentative offer from the University of Chicago to return as an assistant professor of geography. As I was about to accept, Haverford College offered me its presidency. I decided to take the Haverford post because of the opportunities to go on with the concern for education emphasizing Quaker principles at the undergraduate level and training for relief and reconstruction at the graduate level. When I went to Chicago to discuss this decision with Robert Hutchins, he leaned back, flicked a cigarette ash, and remarked that Haverford did not have a really good undergraduate program, but was probably the best among small colleges. At Haverford, Anne and I found the whole social setting congenial, and while my time was spent largely in administration and fund-raising, I did offer one class each year in natural-resources conservation and on the side was able to serve on various government commissions.

Although Anne and I greatly enjoyed the Haverford community and the close contact with individual students, we felt an uneasiness that came forcefully to the front one evening when Margaret Mead, the University of Pennsylvania anthropologist, was our guest at dinner. The group had been discussing a specific research problem, and I had said that I would hope to study one aspect of it. Margaret responded, "Gilbert, you will never work on that." "Why not?" I asked. "Because," she said in effect, "you are on a one-way street called 'educational administration,' and you never will get back to doing real research. You will have opportunities to go on down the street with other types of administration and advice but not for original study."

Anne and I talked over that comment most of the night. We concluded that Margaret was correct, and that if we were ever to make a U-turn, we would have to do it soon. The problem would be to find an institution that would take a chance in offering a teaching position to a worn administrator. Indeed, I shortly thereafter received offers of an Ivy League presidency and the vice presidency of a large foundation. I held out and finally was delighted when Chicago renewed its old offer, this time as chair of the geography department. At the end of 1955, I moved back to Chicago in a professorial role and began searching for support for research investigation that I had been dreaming about at Haverford.

## University of Chicago

My first research emphasis was directed at floodplains and related problems. This work advanced rapidly through the great good fortune of the admission, over the next few years, of imaginative and highly competent students who proceeded to tackle questions in their own ways. The result was a long series of doctoral dissertations that broke new ground, as will be noted later. My one significant innovation in the academic program was to help the department put together a required course for all beginning graduate students in which every faculty member participated, beginning with fieldwork. The result of this new course was that by the end of the first quarter each student had an assessment of every faculty member, and the faculty had a joint appraisal of the strengths and weaknesses of every new student.[2]

A special experience in the Chicago period was that of serving as a visiting professor at the University of Oxford in 1962–63. Wes Calef had arranged for me to exchange places with Robert Beckinsale at Oxford. When I arrived and was given a schedule of lectures in the School of Geography, the chair, Billy Gilbert, asked me if I would care to teach. I replied that I thought that was what I would be doing, but he explained that real teaching was at the tutorial level. I enthusiastically tried my hand at that.

## University of Colorado

One long-term consequence of going back to Chicago was the opportunity to use my summers for study instead of fund-raising. My work at Haverford had entailed a reduction of student enrollment to 450, while enlarging the faculty and nearly doubling the total endowment. Now I was attracted by the thought of giving our three children a summer experience similar to my boyhood sojourns on the Tongue River ranch. From one of the graduate students (Jackie Beyer), I learned that a Colorado student (Melvin Marcus) and his wife had lived the preceding summer on the home place of a ranch outside of Boulder, Colorado, while the rancher and his family took their cattle to their

2. G. F. White, "Introductory Graduate Work for Geographers," *The Professional Geographer* 10, no. 2 (1958): 6–8.

national forest permit grazing above Nederland. I wrote to the rancher, whose wife replied that they would be glad to rent us the home building for the summer if we would agree to take care of the stallion and any sick animals, take phone calls, and pay $150. We accepted sight unseen and drove out the following June to find the Wittemyer family ready to show us the facilities and then head for the high country. But we were to see much more of that gracious family as they involved us in branding calves, working in a two-day cattle drive to the national forest, eradicating some rangeland weeds, and haying. We promptly fell in love with Sunshine Canyon, and I asked Leonard if he would sell us a lot from his holding of several thousand acres. The previous year he had found he could sell one building lot for as much cash as he made in a year's operation, so he would consider only that type of transaction. With his son's help, we looked farther up the canyon and found several hundred acres we could buy, which became our summer place after we built a cabin, and its delightful location on the upper rim of Twomile Creek convinced us that we would like to retire there. Thus, when the possibility of moving from Chicago to a post at the University of Colorado was raised in 1969, we enthusiastically decided to move, with a view to settling in Boulder after retirement. The original offer from the University of Colorado was for a professorship in geography and in the Institute of Behavioral Science. However, in the last days of the state legislature session, the acting president of the university phoned to say that although appropriation cuts forced them to delete the simple professorship, they could offer me the recently vacated post of director of the institute. We accepted the revised appointment and cast a vote for Boulder as a lifetime residence.

My work at Boulder placed many of the geography graduate students in a somewhat different position than those at Chicago. The Institute of Behavioral Science emphasized interdisciplinary research, which meant that many of the geographers working with me found themselves in varying associations with anthropologists, economists, political scientists, psychologists, and sociologists. Thus, those who took part in the assessment of research on natural hazards were obliged to deal regularly with students from other fields. After I retired as a teacher in 1978, I had occasion to work with a few advanced geographers but did not carry principal responsibility for their research.

Looking back over the active years at Haverford, Chicago, and Boulder, I reckon that I have taught approximately forty classes for undergraduates and thirty-seven classes for graduate students. Thinking of graduate students for whom I was the major advisor, I estimate that there were approximately twenty-one who completed master's degrees and twenty-four who completed doctor's degrees.

## Scientific and Policy Interests

From the early days in Washington, my research interests have focused on four major problems that are intertwined: the human management of floods and other natural hazards; integrated water management; the environmental effects of water management; and the postaudit of the consequences of environmental projects. Exciting aspects of the initial organization of agencies concerned with geographic problems under the United Nations were the early concerns for water management and for scientific collaboration in dealing with natural hazards. In both directions, there were new opportunities for strengthening research as well as the application of its results in those fields.

In all of this effort, my underlying concern was, "What shall it profit a professor if it fabricates a nifty discipline about the world while that world and the human spirit are degraded?" [3]

### Floods and Other Natural Hazards

My concern with the problems of floodplains and of other areas subject to extreme events in nature began with the Mississippi Valley Committee. At that time, the devastation of the Mississippi floods of 1927 was still fresh in mind, and the Corps of Engineers had launched a wide set of basin studies in which levee and dam structures to deal with floods played a major role. The Tennessee Valley Authority (TVA) was getting under way, and there was anticipation of national flood-control legislation. The president and a few members of the Congress favored extending the TVA pattern to other basins, but Congress soundly defeated this idea. There were reviews of the problems of every basin

---

3. G. F. White, "Geography and Public Policy," *The Professional Geographer* 24 (1972): 302–9.

in the nation, and many involved flood hazard. "Little waters" and "upstream engineering" were popular descriptions of collaborative scientific enterprises.

I began to puzzle over the effects of the various public measures and examined the various sets of criteria used to judge water projects. The results of this work were not always popular. For example, when I reported at a national planning conference in 1936 that only three states then had any regulations of encroachments on natural stream channels and went on informally to suggest that 100 percent federal financing of flood-control dams in California be contingent on enactment of state regulation of further occupation of floodplains, one of the Congressmen asked for an investigation of my preaching "un-American ideas." Nothing came of the investigation, and I continued to look for a systematic way of analyzing the full physical and social implications of land use in areas subject to floods. The result was my 1942 dissertation, published as one of the Chicago geography series in 1945.[4] My approach there was to classify the range of possible human uses of land vulnerable to flooding, the types of actions that could be taken to achieve them, and the full social costs and benefits that might be associated with each.

At Chicago during the late 1950s, a joint study was launched on what had happened to land use and vulnerability to flooding in seventeen sample areas during the twenty years following the enactment of the national flood-control legislation in 1936.[5] They contributed to the establishment by the Corps of Engineers of a national floodplain information service.[6]

Federal policies began to change very slowly. The further studies of Burton on agricultural uses, of Kates on manufacturing uses, of Sheaffer on flood proofing, and of others contributed to the report of the BOB task force that in 1966 led to a presidential executive order on actions of federal agencies and,

---

4. R. Platt, T. O'Riordan, and G. F. White, "Classics in Human Geography Revisited," *Progress in Human Geography* 21, no. 2 (1997): 243–50.

5. G. F. White, W. C. Calef, J. W. Hudson, H. M. Mayer, J. R. Sheaffer, and D. J. Volk, *Changes in Urban Occupance of Flood Plains in the United States* (Chicago: Univ. of Chicago Department of Geography, 1958).

6. J. W. Moore and D. P. Moore, *The Army Corps of Engineers and the Evolutions of Federal Flood Plain Management Policy* (Boulder, Colo.: Institute of Behavioral Science, 1989). See also J. M. Wright, *Floodplain Management: Its Origins and Evolution* (Madison, Wisc.: Association of State Flood Plain Managers, 1997), 11–17.

with a parallel Housing and Urban Development report, suggested the creation of a federal flood insurance program.[7] That program was enacted in 1968. Researchers collaborated very closely with the TVA, which, under the guidance of James E. Goddard, was pioneering in integrated management of floodplains.[8]

Parallel to the floodplain studies and going back to my experience with the Great Plains Drought Committee was my involvement in international scientific studies of drought and desertification. In tandem with water planning, these studies stimulated a broader approach to the study of natural hazards. For example, this approach encouraged examination of what there was in common and what was different in the ways in which people perceived and dealt with an inundated floodplain or a field parched by sustained drought. An example was Saarinen's study of differing perceptions of drought hazard.[9]

As one of its early ventures into collaborative science, the United Nations Education, Scientific, and Cultural Organization (UNESCO) launched an examination of desertification of arid lands. The steering committee, of which I was a member from 1953 to 1956, held a series of discussions in various arid countries to ask what was known and what were desirable research agendas to pursue. The results were published proceedings, a general summary, and the initiation or strengthening of a network of desert research stations located at universities or attached to government agencies. An early summary volume grew out of a conference that was organized by the American Association for the Advancement of Science (AAAS),[10] and there were periodic attempts to review research progress, as in the AAAS meeting in Tucson in 1970 or in the a later session in 1985.[11]

7. U.S. Bureau of the Budget Task Force, *A Unified National Program for Managing Flood Losses*, 89th Congress, 2d sess., 1966, House Document No. 465.

8. Wright, *Floodplain Management*.

9. T. F. Saarinen, *Perception of the Drought Hazard on the Great Plains* (Chicago: Univ. of Chicago Department of Geography, 1966).

10. G. F. White, ed., *The Future of Arid Lands* (Washington, D.C.: American Association for the Advancement of Science, 1956). G. F. White, *Science and the Future of Arid Lands* (Paris: UNESCO, 1960).

11. Harold E. Dregne, ed., *Arid Lands in Transition* (Washington, D.C.: AAAS, 1970). E. E. Whitehead, C. F. Hutchinson, B. N. Timmermann, and R. G. Varady, eds., *Arid Lands: Today and Tomorrow* (Boulder, Colo.: Westview, 1988).

An aspect of my work in natural hazards that led to my involvement in related fields was risk perception. An example was the problem of how people judge the risk of exposure to high-level nuclear waste. I served for seven years in the 1980s and 1990s as chair of the state of Nevada advisory committee on studies related to the proposed Yucca Mountain disposal site and joined in several published judgments about the policy involved. The problem was in part one of how serious the hazard was and in part how people judged the risk to themselves and how they valued the judgments of agencies and investigators.[12]

While working on theory and policy of floodplain management on a national scale, I tried both to test and to apply the findings near to home. In Chicago, this involved chairing a committee of the Northeast Illinois Regional Planning Commission that prepared a report on recommended floodplain management, which arranged, with financial help from the Cook County Forest Preserve District, for the preparation by the U.S. Geological Survey of a map of flood-hazard lands in the Desplaines River basin that was the first of its kind.

In Boulder, my concern for local problems involved the 1958 report on changes in floodplain occupance (written jointly for the city and county of Boulder in 1969, this report on floodplain regulation contributed to the enactment of zoning ordinances by both governments) and a series of studies by graduate students leading to the initiation of community flood forecasting and disaster response organizations following the Big Thompson flood of 1969. Over the next twenty-five years, I participated in the joint Corps of Engineers and city review of flood problems and in community appraisals of possible public action on management of the Flatirons property along South Boulder Creek. We also established and funded a web site to be used at some future time to survey promptly the effects of a major flood on Boulder Creek, and published a report summarizing the past history, the flood results, and the choices possibly open to the citizenry. I continue to hope that the united city groups in Boulder will see a positive way to prevent flood catastrophes in future.

12. James Flynn, James Chalmers, Doug Easterling, Roger Kasperson, Howard Kunreuther, C. K. Mertz, Alvin Mushkatel, David Pijawka, Paul Slovic, and Lydia Dotto, *One Hundred Centuries of Solitude* (Boulder, Colo.: Westview, 1995).

## Integrated Water Management

When the new United Nations in 1949 organized at Lake Success the first international review of experience and problems in management and conservation of natural resources, there was thoughtful assessment of river-basin planning. I led the discussion in a session dealing with accomplishments and difficulties in such basins as the Tennessee, the Rhine, the Rhône, the Tigris-Euphrates, the Ganges, and the Yellow. It was exciting to feel the enthusiasm of scientists and engineers from around the world who were beginning to learn a little from the experience of others.

Out of those discussions and subsequent ones came the ideas that the United Nations should sponsor a critical review of integrated river-basin experience and should share the lessons learned to that time. In 1957, I was asked by the Department of Economic and Social Affairs to chair an international panel of experts to prepare a consensus report. The panel included French, British, and Russian engineers, a Pakistani planner, a Colombian attorney, and a Dutch social scientist. It issued its findings and recommendations in 1958.[13] The major recommendations were for improving basic hydrologic services; sharpening analytical tools on water use; expanding scientific and technical investigations; aiding national initiatives; and encouraging measures to reconcile conflicting interests.

Much critical application followed, and in 1970 a revised report was issued that gave special attention to the lack of careful evaluation of earlier efforts and to the need to accompany engineering works with other types of measures to affect resource use.[14] The enlarged perspective would require greater attention, for example, to national economic plans. At the same time, the involvement of international scientific agencies was, in fact, being expanded. In 1968, I tried to sum up my own views on the essentials of truly integrated management.[15]

13. Panel of Experts, *Integrated River Basin Development* (New York: United Nations, Department of Economic and Social Affairs, 1970).

14. United Nations, Preface to *Integrated River Basin Development* (New York: Department of Economic and Social Affairs, 1970).

15. G. F. White, *Strategies of American Water Management* (Ann Arbor: Univ. of Michigan Press, 1969).

During the same period and later, I was involved in a long series of U.S. government efforts to improve the quality of its water management. My participation began with service on the committee on natural resources of the Hoover Commission on government organization in 1948, continued with service as vice chairman of President Truman's Water Resources Policy Commission in 1950, and later involved serving as a consultant to the Senate Select Committee on Water Resources and to the National Water Commission chaired by Charles Luce in 1971–73. Both the Senate Select Committee and the National Water Commission were under the staff direction of Theodore Schad and had major impacts, such as the establishment of the federal support for state water-resource research programs and the creation of the Federal Water Resources Council with its subsidiary river-basin commission.[16]

My most recent effort to influence the course of international action in the water-management field was in the promotion of the creation of the World Water Council, which began its activities in 1998[17] and initiated publication of *Water Policy*.[18]

In addition, over the years following the BOB task force report of the 1960s came a few opportunities to help organize international efforts to translate research insights into action in major river basins. The first was in the Lower Mekong basin of Cambodia, Laos, Thailand, and Vietnam in 1961, when I chaired an advisory board with members from the Netherlands, the United Kingdom, and the United States that advised on social and economic aspects of water planning. This collaborative effort led to a reorientation of the cooperative efforts and contributed to the formulation of further studies in 1970, directing attention to work on tributaries rather than large, mainstream projects.[19] We had hoped for a few days that President Johnson would

16. Theodore Schad, *Water Resources People and Issues* (Fort Belvoir, Va.: U.S. Army Corps of Engineers, 1998).

17. G. F. White and Y. A. Mageed, "Critical Analysis of Existing Institutional Arrangements," *International Journal of Water Resources Development* 11, no. 3 (1995): 103–11.

18. G. F. White, "Reflections on the 50-Year International Search for Integrated Water Management," *Water Policy* 1, no. 1 (1998): 21–27.

19. H. Hori, *The Mekong: Its Development and the Environment* (Tokyo: United Nations Univ. Press, 1999).

suggest that international cooperation on the Mekong might be a substitute for looming war, but this idea soon evaporated.[20]

As a member of the United Nations Environment Program's committee on water, I had a hand in planning and carrying out studies of water planning in the Lake Chad and Zambezi Basins, but I was probably most useful in advocating geographic concepts in the Aral Sea Basin of Central Asia. The committee, including scientists and engineers from the five newly independent countries of Central Asia, with other participants from France, Germany, Russia, and the United States, prepared a report diagnosing the land and water problems and needs of the basin. That report was accepted at a meeting of interested international agencies and nations in Paris in 1993 and was the basis for pledges in the neighborhood of thirty million dollars in financial support for remedial work. I also helped initiate an international review of Yellow River Basin problems in 1989. More recently, I have chaired a committee, established as their first cooperate undertaking, of the science academies in Israel, Jordan, the West Bank and the Gaza Strip, and the United States to appraise the quality and quantity of water resources of that Middle Eastern area.[21]

Long after the 1970 revision of the BOB task force report, I also took part in a series of activities aimed at translating research findings into global international water policy: the Mar de Plata conference on water in 1970 and the Dublin conference in 1992. The latter drafted the major water recommendations adopted later that year in Rio.

Another analysis of resource management with national policy implications was the study of national energy policy of the Ford Foundation, which was undertaken with the advice of a board I chaired during 1971–74.[22] An excellent staff under the direction of David Freeman canvassed the whole range of policy alternatives and offered an innovative "zero energy growth" scenario.

20. G. F. White, "Vietnam: The Fourth Course," in *The Vietnam Reader*, edited by M. Raskin and B. Fall, 351–57 (New York: Random House, 1965).

21. National Research Council Committee on Sustainable Water Supplies for the Mid-East, *Water for the Future: The West Bank and Gaza Strip, Israel, and Jordan*, report (Washington, D.C.: National Research Council, 1999).

22. Energy Policy Project, *A Time to Choose: America's Energy Future* (Cambridge: Ballinger, 1974).

We strongly challenged the notion that electric power production needed to continue to grow at the prevailing rate.

A relatively ignored aspect of water-resource management was the strategy of dealing with nonpoint water pollution. The problem of constructive responses to pollution beginning in the 1930s had focused on means of preventing, reducing, and treating pollution originating in cities or factories. In 1997, the National Geographic Society and the Conservation Fund established the special Forum on Nonpoint Source Pollution, and I was the only academic on the thirty-two member committee. The forum published a report recommending policy changes,[23] and in twenty-five different U.S. communities it also initiated demonstrations of concrete action to remedy the problem. One of the demonstrations was at the office site of the Land and Water Fund in Boulder, where the building and parking lot was to be treated by design structures and landscaping so as to reduce runoff and pollution resulting from small, frequent storms. I chaired the committee overseeing the design, construction, and monitoring of the project, jointly with the city of Boulder in 1996–98, and I enjoyed seeing a down-to-earth effort to express a national vision in plants and rocks on a nearby office building parking lot.

Although all of the water-related research in which I was involved was interesting and usually pleasant, none yielded greater personal satisfaction than the studies initiated in 1996 of domestic water use in East Africa. I had been offered a grant by the Rockefeller Foundation to undertake fieldwork of my choice, and at the same time the foundation with the Carnegie Corporation asked me if I would spend enough time in the newly independent countries of East Africa to help the appropriate educational institutions put together a joint budget proposal that the foundations could fund for a new University of East Africa. They had no experience with such a venture, and I had worked on increasing cooperation among Bryn Mawr, Haverford, and Swarthmore colleges in joint programs. I did help fashion a fundable budget request to the foundations, but while at Makerere College in Kampala I encountered a medical researcher, David Bradley, who was also funded by Rockefeller and had converging research interests.

23. Terrene Institute, *Taking a New Tack on Nonpoint Water Pollution* (Washington, D.C.: n.p., 1997).

David and I had been puzzling about how significant decisions concerning water were made. The most elementary and the most widespread decision about water in the world was, and still is, that made every day by the women who decide to go to some source and carry a certain quantity home. We wanted to find out how the drawers of water decided where to go and how much to bring back. East Africa would be an attractive area because of the tremendous diversity it offered in landscape—from dry grasslands to tropical rain forest, from wealthy suburbs to urban slums to remote farms—as well as in social organization.

A decisive factor in whether or not to go to East Africa was that the task should be congenial for my wife Anne. She was experienced in field interviewing techniques, and with our youngest child going off to Quaker boarding school at West Branch, Iowa, she finally was at liberty to work with me in the field. David and I and Anne agreed on a plan in which, using selected students from Dar es Salaam, Nairobi, and Kampala, we investigated more than seven hundred households in thirty-four different sites. We tramped along terraces and forest trails and on city streets. Out of this investigation came the first precise observations of how much water is used for domestic purposes; what it costs the family in time, calories, and money; what its health implications may be; and how the decision is made to draw so much water from what source. The results were published,[24] and after we assembled the preliminary findings, we shared them with international development agencies that were beginning to put money into domestic water projects. We also analyzed the results for human health of providing potable water in Egypt and encouraged a fresh approach to the necessity of altering behavior as well as supply.

It was heartening that in 1997 a combination of British, Danish, Dutch, and Swedish development agencies paid funds to repeat the whole study at exactly the same sites. A team from London, Dar es Salaam, and Nairobi reviewed the preliminary findings at the Stockholm Water Symposium in 1998. Anne would have been pleased to see that a joint effort had again contributed to a more accurate and sensitive understanding of how people can be helped to help themselves meet a basic need.

24. G. F. White, D. J. Bradley, and A. U. White, *Drawers of Water* (Chicago: Univ. of Chicago Press, 1972).

## Environmental Effects of Water Management

Along with my continuing concerns for natural hazards and water management, there grew a concern for the interrelationship between human management of natural resources and the enduring health of those ecosystems. It came into sharp focus in the case of dams as one water-management device. In 1965, I was invited by the administrator of the United Nations Development Program (UNDP) to head a task force to examine a number of major water-storage and -diversion projects under construction in Africa from the standpoint of their possible effects on the environment. The projects were in the Nile, Volta, Senegal, and Zambezi drainages.[25] The review required an interdisciplinary and international committee of scientists who examined the work in progress and recommended measures that UNDP or other agencies could initiate to deal with environmental problems that appeared to be neglected. Thus, in the case of the Aswan high dam, the Soviet engineer in charge of constructing the dam explained that his job was to build effectively and not to look at its effects. We recommended that a new research institute be established to study the likely problems of aquatic life in the lake being created.

In 1970, the Scientific Committee on Problems of the Environment (SCOPE) of the International Council of Scientific Unions organized a special review of the environmental man-made lakes.[26] I helped that effort and joined with William Ackermann and Barton Worthington in organizing an international symposium on the problems of those lakes.[27] I was president of SCOPE for six years. In 1977, I edited the results of an international symposium held afloat on the Volga River, on the environmental effects of complex river development.[28]

25. G. F. White, "Interdisciplinary Studies of Large Reservoirs in Africa," in *Proceedings of Conference on Interdisciplinary Analysis of Water Resource Systems*, edited by J. E. Flack, 63–85, 102–4 (New York: American Society of Civil Engineers, 1975).

26. SCOPE Working Group, *Man-made Lakes as Modified Ecosystems*, SCOPE Report no. 17 (Washington, D.C.: SCOPE, 1973).

27. W. Ackermann, G. F. White, and E. B. Worthington, eds., *Man-made Lakes: Their Problems and Environmental Effects*, Geophysical Monograph no. 17 (Washington, D.C.: AGU, 1973).

28. Gilbert F. White, ed., *Environmental Effects of Complex River Development in a Variety of Areas* (Boulder, Colo. Westview, 1977).

A major aspect of water management is its reciprocal relationship to the health of ecosystems. It had been a concern in the New Deal planning, but, as I have noted in a recent review,[29] a number of critical distinctions in research and policy were neglected. Thus, the concept of small drainage areas as subjects for unified study was negated by the assignment of responsibility for upstream studies to the Soil Conservation Service and Forest Service, whereas downstream areas were examined by the Corps of Engineers. Unfortunately, the problem of water pollution was addressed largely in terms of point sources in urban areas and industrial installations, but nonpoint sources were largely neglected. Similarly, whereas the Russian analysis of soils in terms of genetic causes was rapidly adopted—chernozems, for example—the Russian/German system of landscape or *landschaft* classification was largely ignored.

In 1986, in a related expression of this concern, the United Nations Environment Program (UNEP) formed the International Lake Environment Committee, hosted by the Japanese government and based on Lake Biwa in Japan. The committee, of which I was a member, set out to promote scientific exchange and collaboration among all agencies concerned with the health of lakes, both natural and artificial. It has sponsored a variety of cooperative review and training programs, and oversaw the preparation and publication of the first comprehensive inventory of the state of the world's lakes, in several volumes.

While working with SCOPE, I was one of those scientists who became concerned about the possible consequences of human alterations in the global geochemical cycles. There were discussions of ways of assessing changes in carbon, sulphur, and nitrogen flows, and one of the results was the report *The Greenhouse Effect, Climatic Change, and Ecosystems*, written by two geographers, among others.[30]

In 1979, I had joined with Mostafa Tolba, director of UNEP, in a statement calling attention to the looming problem of human interventions in global life-support systems, and now in the 1980s the results of cooperative research began to call for major international response.[31]

29. White, "Reflections."

30. B. Bolin, E. R. Doos, J. Jager, and R. A. Warrick, eds., *The Greenhouse Effect, Climatic Change, and Ecosystems,* SCOPE Report no. 29 (Chichester: J. Wiley, 1986).

31. R. W. Kates and I. Burton, *Geography, Resources, and Environment,* (Chicago: Univ. of Chicago Press, 1986), 1: 414–17.

## Postaudits

From an early concern with the actual effects on the ground of federal flood-control policy, I have made a recurrent effort to encourage concrete audits of what in fact happened as a result of specific measures and policies. There have been very few careful appraisals of results of public resource-management programs, but without such postaudits it is difficult to design sound new measures for desirable change in either planning or practice.

Studies of change in floodplain occupance contributed to a new federal program of floodplain information and to activities of the Association of State Flood Plain Managers. Studies of the environmental consequences of large water projects added support to federal requirements of environmental impact statements. The review of environmental effects of nuclear war was the basis for a United Nations Assembly resolution and continues to be an urgent need. Only in 1998 did the Federal Insurance Administration give systematic attention to a possible postaudit of the effects of its operations over the previous thirty years—a type of study that had been suggested in the task force report in 1966. I still regard the postaudit as an essential block in a foundation for sound water management.[32]

## Professional Geographers

In the years after receiving a Ph.D., I was active during two short periods in helping advance the status of geographers as a professional group. The first was during 1961–62, when, as president and president-elect of the Association of American Geographers (AAG), I worked on a four-point program, of which three were completed.

We first established a regular scientific association with salaried personnel. Until then the association was managed by contributed labor from interested members, using a paid secretary to keep membership and mailing records. It seemed important to release a professional geographer for full-time promo-

---

32. Task Force on Federal Flood Control Policy, *A Unified National Program for Managing Flood Control Losses,* 89th Congress, 2d sess., House Document no. 465 (Washington, D.C.: U.S. Government Printing Office, 1966).

tion of association concerns, which would require decisions and money. Second, we initiated a program to support improvement and dissemination of teaching of the latest thinking and practice of geography at the college level. Third, we instituted systematic efforts to raise the quality of geography teaching in high schools. This clearly was a concern of the National Council for Geographic Education, but that group had not then established the tax-exempt status necessary to qualify for financial help from the National Science Foundation, which was supporting such efforts in other scientific fields. Fourth, we needed a thoughtful appraisal of the circumstances and effects of applying geographic research results to public policy and program.

As we undertook the first three parts of the program, we corrected a glaring deficiency in the national geographic enterprise: the long-standing lack of any formal communication between the AAG and the National Geographic Society (NGS). For reasons I never fully understood, the two groups had gone for many years without collaboration. Through a publishing friend, I got in touch with Melville Grosvenor, then president of the NGS, and said I would like to talk with him about possible cooperation. He arranged for me to meet with a group of his staff and, after introducing me, turned to me and in his usual direct, cheerful manner said, "What do you want?" I explained that I wanted to see the two organizations work together and that we needed assistance in establishing a national AAG office. After appropriate discussion, Mel stated that NGS would provide free office space for the new base in the Sixteenth Street building of the NGS and would contribute ten thousand dollars a year toward its expenses. We were in business.

The College Program was established under a new committee, and the High School Geography Project, outlined by Bill Pattison, began operations under the leadership of Nicholas Helburn.[33] The latter program probably improved the quality of work by participating senior geographers at the university level more than the quality of high school teaching, but it was the beginning.

The applied geography initiative under Ed Ackerman never got underway; Ed was too busy applying his research to current water-policy issues. Relations with NGS flourished.

---

33. Nicholas Helburn, "The High School Geography Project: A Retrospective View," *Social Studies* 89, no. 5 (1998): 212–18.

My other direct involvement in the status of geographers came after 1973, when I received word that I had been elected a member of the National Academy of Sciences (NAS). At that time, there was no active geographer in the academy, although Isaiah Bowman had been very influential in the years 1930–50. My nomination had come from scientists in other disciplines. I looked into the election process, consulted with colleagues in related areas, and was helpful in the election of Brian Berry (1975), John Borchert (1976), and Robert Kates (1978). By 1998, there were ten geographers in total.

I was also involved in a special nonacademic concern related to geography and natural resources: Resources for the Future (RFF). This nongovernment agency was organized in 1952 in connection with the national midcentury conference held in Washington, D.C., after publication of the Paley Commission on Materials Policy, and was supported substantially from its establishment by the Ford Foundation. It was from RFF that I obtained the grant to initiate the studies of changes in floodplain occupance in 1956. It had drawn together a remarkable team of researchers, including John Krutilla.

I was a member of its board from 1967 to 1979 and was president at the crucial time when the Ford Foundation announced that it was terminating substantial support but would continue on a reduced scale if RFF would become a part of the Brookings Institution in Washington. This would have meant a major shift in its orientation and independence. The absorption by Brookings seemed almost inevitable. Trying to canvas alternatives, I turned to other sources and found a few willing to take a principal share of the burden of supporting RFF while it moved toward a broader base.

## Retrospect

Looking back over six decades of activity in the geographic field, I find satisfaction in considering its implications in several directions. I made some contributions to structures of thought in a few fields: human adjustments to natural hazards, alternatives in water management, and environmental effects of natural-resource management. Related to these contributions were several efforts to alter relevant public policy: for example, developing floodplain-management policy, creating strategies for water management, dealing with the effects of large dams, improving domestic water use in developing countries, curbing nuclear war, and appraising nuclear waste disposal. I undertook all of this activity in a spirit of trying to be helpful to needy fellow humans by cooperative, nonviolent methods. The greatest personal satisfaction has come from observing the performance of former students who have made more significant contributions than I.

I undertook much of the influential research as a member of various teams of scientists drawn from geography and a range of other disciplines—including both human and physical geography, as well as ecology, economics, engineering, hydrology, political science, and sociology. A persistent underlying challenge was the prospect that the quality of the lives of ordinary people might be improved as a result of the work we did. There is an old Quaker saying that what counts is not how a Friend talks but how a Friend walks. An observer should be able to judge people's values in terms of how they act rather than in terms of what they say. In that spirit, a basic judgment of geography is what it contributes cooperatively in fashioning a sustainable earth. With that aim, the geographic discipline has solid performance and expanding opportunities.

## Additional References

Burton, I., R. W. Kates, and G. F. White. *The Environment as Hazard*. 2d ed. New York: Guilford, 1993.

White, G. F. "Organizing Scientific Investigations to Deal with Environmental Impacts." In *The Careless Technology,* edited by M. T. Farvar and J. P. Milton, 914–26. Garden City: Natural History Press, 1972.

———. "Watersheds and Streams of Thought." In *Reviews in Ecology: Desert Conservation and Development,* edited by H. H. Barakat and A. K. Hogagy, 89–98. Cairo: UNESCO, 1997.

# Index

Tocumwal, 125
Tokugawa Japan, 289
Tolba, Mostafa, 360
Tolkien, J. R. R., 329
Tomlinson, Roger F., 318
Topalov, Christian, 174
Topologically Integrated Geographic Encoding
    and Referencing (TIGER) files, 315
Torres, Flor, 177
torus sampling, 284
Toynbee, Arnold, 59, 64, 79, 197, 204
transformations, 293
transition probabilities, 92
Transvaal, 78
travel time isochrones, 311
Trewartha, Glenn T., 37, 39, 41, 44, 72, 276
Troll, Carl, 9, 24, 64
tropical rainforest, 326
turbulence theory, 89
Turner, Frederick Jackson, 25, 200
Turner, Ralph, 204

Ullman, Edward L., 26, 86, 104, 166, 216,
    220, 305
Ulrich, Roger, 256
United Nations, 86
United Nations Development Program
    (UNDP), 359
United Nations Education, Scientific, and
    Cultural Organization (UNESCO), 352
United Nations Environment Program
    (UNEP), 356, 360
University College, London, 22, 323
University of Auckland, 81
University of Bonn, 52, 65, 72
University of Bristol, 167
University of British Columbia, 124, 269, 293
University of Buenos Aires, 149
University of California, Berkeley, 99, 256,
    323
University of California, Santa Barbara, 124,
    140, 293
University of California system, 141
University of Cambridge, 149, 154, 164
University of Canterbury, 124, 135, 137
University of Chicago, 39, 52, 76, 79, 88, 190,
    341, 347

University of Colorado, 313, 341, 349
University of Durham, 81, 84
University of East Africa, 357
University of Hawaii, 269–70
University of Idaho, 193, 274
University of Illinois, 34, 99
University of Iowa, 124, 137
University of Lund, 211, 219
University of Maryland, 81, 88
University of Michigan, 218, 237, 253, 269,
    275, 293, 308, 355
University of Minnesota, 27, 43, 45, 323, 328
University of New England, 124, 131
University of New Mexico, 323, 328
University of Oregon, 269, 280
University of Oxford, 149, 179, 323
University of Pennsylvania, 99, 105, 237, 248,
    256, 307, 347
University of Pittsburgh, 99, 256, 269, 285
University of St. Andrews, 201
University of Sydney, 140
University of Texas, 2, 52, 80
University of Toronto, 81, 89, 323
University of Uppsala, 237, 239
University of Utah, 189, 199, 202
University of Washington, 81, 87, 99, 104, 123,
    189, 194, 211, 219, 269, 293, 302
University of Waterloo, 323
University of Wisconsin, 27, 37, 52, 68, 75, 79,
    87, 323, 330
University of Zurich, 52, 293, 311
Uppsala, Sweden, 147
Urban and Regional Information Systems
    Association (URISA), 119, 307
U. S. Aid Mission, 285
U. S. Geological Survey (USGS), 34, 353
utilization function, 93
Uyemura, Fukushichi, 283

vacancy chains, 93, 95
variance spectrum statistics, 89
Vatican Library, 317
Vautrin Lud International Prize, 149, 341
vector algebra, 129
Verdery, Katherine, 169
Vichy France, 345
victory gardens, 56